LONG-MEMORY TIME SERIES

BICENTENNIAL 1807 WILEY 2007 BICENTENNIAL

The Wiley Bicentennial–Knowledge for Generations

Each generation has its unique needs and aspirations. When Charles Wiley first opened his small printing shop in lower Manhattan in 1807, it was a generation of boundless potential searching for an identity. And we were there, helping to define a new American literary tradition. Over half a century later, in the midst of the Second Industrial Revolution, it was a generation focused on building the future. Once again, we were there, supplying the critical scientific, technical, and engineering knowledge that helped frame the world. Throughout the 20th Century, and into the new millennium, nations began to reach out beyond their own borders and a new international community was born. Wiley was there, expanding its operations around the world to enable a global exchange of ideas, opinions, and know-how.

For 200 years, Wiley has been an integral part of each generation's journey, enabling the flow of information and understanding necessary to meet their needs and fulfill their aspirations. Today, bold new technologies are changing the way we live and learn. Wiley will be there, providing you the must-have knowledge you need to imagine new worlds, new possibilities, and new opportunities.

Generations come and go, but you can always count on Wiley to provide you the knowledge you need, when and where you need it!

William J. Pesce
President and Chief Executive Officer

Peter Booth Wiley
Chairman of the Board

LONG-MEMORY TIME SERIES

Theory and Methods

Wilfredo Palma
Pontificia Universidad Católica de Chile

BICENTENNIAL 1807 WILEY 2007 BICENTENNIAL

WILEY-INTERSCIENCE
A JOHN WILEY & SONS, INC., PUBLICATION

Published by John Wiley & Sons, Inc., Hoboken, New Jersey.
Published simultaneously in Canada.

Wiley Bicentennial Logo: Richard J. Pacifico

Library of Congress Cataloging-in-Publication Data is available.

ISBN 978-0-470-11402-5

CONTENTS

PREFACE

During the last decades long-memory processes have evolved into a vital and important part of the time series analysis. Long-range-dependent processes are characterized by slowly decaying autocorrelations or by a spectral density exhibiting a pole at the origin. These features change dramatically the statistical behavior of estimates and predictions. As a consequence, many of the theoretical results and methodologies used for analyzing short-memory time series, for instance, ARMA processes, are no longer appropriate for long-memory models.

This book aims to provide an overview of the theory and methods developed to deal with long-range-dependent data as well as describe some applications of these methodologies to real-life time series. It is intended to be a text for a graduate course and to be helpful to researchers and practitioners. However, it does not attempt to cover all of the relevant topics in this field.

Some basic knowledge of calculus and linear algebra including derivatives, integrals and matrices is required for understanding most results in this book. Apart from this, the text intends to be self-contained in terms of other more advanced concepts. In fact, Chapter 1 of this book offers a brief discussion of fundamental mathematical and probabilistic concepts such as Hilbert spaces, orthogonal projections, stationarity, and ergodicity, among others. Definitions and basic properties are presented and further readings are suggested in a bibliographic notes section. This chapter ends with a number of proposed exercises. Furthermore, Chapter 2 describes some fundamental concepts on state space systems and Kalman filter equations. As discussed in this chapter, state space systems offer an alternative representation of time series models which may be very useful for calculating estimates and predictors, especially in the presence of data gaps. In particular, we discuss applications of state space techniques to parameter estimation in Chapter 4, to missing values in Chapter 11, and to seasonal models in Chapter 12.

Even though it seems to be a general agreement that in order to have long memory a time series must exhibit slowly decaying autocorrelations, the formal definition of a long-range-dependent process is not necessarily unique. This issue is discussed in Chapter 3 where several mathematical definitions of long memory are reviewed. Chapter 4 is devoted to the analysis of a number of widely used parameter estimation methods for strongly-dependent time series models. The methodologies are succinctly presented and some asymptotic results are discussed. Since a critical problem with many of these estimation methods is their computational implementation, some specific aspects such as algorithm efficiency and numerical complexity are also analyzed. A number of simulations and practical applications complete this chapter. The statistical analysis of the large-sample properties of the parameter estimates of long-memory time series models described in Chapter 4 is different and usually more complex than for short-memory processes. To illustrate this difference, Chapter 5 addresses some of the technical aspects of the proof of the consistency, central limit theorem, and efficiency of the maximum-likelihood estimators in the context of long-range-dependent processes.

Chapter 6 and Chapter 7 deal with heteroskedastic time series. These processes, frequently employed to model economic and financial data, assume that the conditional variance of an observation given its past may vary with time. While Chapter 6 describes several widely used heteroskesdatic models, Chapter 7 characterizes these processes in terms of their memory and the memory of some of their nonlinear transformations. On the other hand, Chapter 8 discusses Bayesian methods for dealing with strongly dependent data. Special attention is dedicated to iterative procedures such the Metropolis-Hastings algorithm and the Gibbs sampler. Prediction of long-memory time series models is reviewed in Chapter 9. This chapter summarizes several results on the prediction of stationary linear processes and discusses some specific methods for heteroskedastic time series. Linear regression models with strongly dependent disturbances are addressed in Chapter 10. In particular, some large sample statistical properties of least squares and best linear unbiased estimators are analyzed, including consistency, asymptotic normality, and efficiency. Furthermore, these results are applied to the estimation of polynomial and harmonic regressions.

Most of the methods reviewed up to Chapter 10 are only applicable to complete time series. However, in many practical applications there are missing observations. This problem is analyzed in Chapter 11 which describes some state space techniques for dealing with data gaps. Finally, Chapter 12 examines some methodologies for the treatment of long-memory processes which display, in addition, cyclical or seasonal behavior. Apart from discussing some theoretical and methodological aspects of the maximum-likelihood and quasi-maximum-likelihood estimation, this chapter illustrates the finite sample performance of these techniques by Monte Carlo simulations and a real-life data application. It is worth noting that similarly to Chapter 1, every chapter of this book ends with a bibliographic notes section and a list of proposed problems.

I wish to express my deep gratitude to Jay Kadane for many insightful discussions on time series statistical modeling, for encouraging me to write this book, and for valuable comments on a previous version of the manuscript. I am also indebted to many coauthors and colleagues, some of the results described in this text reflect part of that fruitful collaboration. I would like to thank Steve Quigley, Jacqueline Palmieri, Christine Punzo, and the editorial staff at John Wiley & Sons for their continuous support and for making the publication of this book possible. Special thanks to Anthony Brockwell for several constructive suggestions on a preliminary version of this work. I am also grateful of the support from the Department of Statistics and the Faculty of Mathematics at the Pontificia Universidad Católica de Chile. Several chapters of this book evolved from lecture notes for graduate courses on time series analysis. I would like to thank many students for useful remarks on the text and for trying out the proposed exercises. Financial support from Fondecyt Grant 1040934 is gratefully acknowledged.

W. PALMA

Santiago, Chile

January, 2007

ACRONYMS

ACF	autocorrelation function
AIC	Akaike's information criterion
ANOVA	analysis of variance
AR	autoregressive
ARCH	autoregressive conditionally heteroskedastic
ARFIMA	autoregressive fractionally integrated moving-average
ARMA	autoregressive moving-average
BLUE	best linear unbiased estimator
DFA	detrended fluctuation analysis
DWT	discrete wavelet transform
EGARCH	exponential generalized autoregressive conditionally heteroskedastic
fBm	fractional Brownian motion
FFT	fast Fourier transform
fGn	fractional Gaussian noise

FI	fractionally integrated
FIGARCH	fractionally integrated generalized autoregressive conditionally heteroskedastic
FIEGARCH	fractionally integrated exponential generalized autoregressive conditionally heteroskedastic
FN	fractional noise
GARCH	generalized autoregressive conditionally heteroskedastic
GARMA	Gegenbauer autoregressive moving-average
HTTP	hyper text transfer protocol
IG	inverse gamma distribution
IM	intermediate memory
LM	long memory
LMGARCH	long-memory generalized autoregressive conditionally heteroskedastic
LMSV	long-memory stochastic volatility
LSE	least squares estimator
MA	moving average
MCMC	Markov chain Monte Carlo algorithm
ML	maximum likelihood
MLE	maximum-likelihood estimator
MSE	mean-squared error
MSPE	mean-squared prediction error
PACF	partial autocorrelation function
QMLE	quasi-maximum-likelihood estimator
R/S	rescaled range statistics
SARFIMA	seasonal autoregressive fractionally integrated moving-average
SD	standard deviation
SM	short memory
SV	stochastic volatility

CHAPTER 1

STATIONARY PROCESSES

This chapter discusses some essential mathematical and probabilistic concepts for the analysis of stationary stochastic processes. Section 1.1 begins with a description of the concepts of *norm* and *completeness* and then proceeds to the definition of *inner product* and *Hilbert spaces*. There is a strong motivation for introducing these spaces. Prediction, interpolation, and smoothing are key aspects of time series analysis, and they are usually obtained in terms of orthogonal projections onto some vector subspaces. A Hilbert space does not only guarantee the existence of the best linear predictor, interpolator, or smoother but also provides an explicit tool for obtaining them through the *projection theorem*. Several other fundamental concepts about linear processes such as *stationarity*, *singularity*, *regularity*, *causality*, *invertibility*, and *ergodicity* are also discussed in this section. In addition, Section 1.1 succinctly introduces the notions of *martigale*, *fractional Brownian motion* and *wavelets* which will be used later in this book.

Some of the theorems presented in this chapter are stated without proof since they are well-known results. However, Section 1.2 provides some references to the reader interested in their proofs and other related subjects. This chapter concludes with a number of proposed problems.

1.1 FUNDAMENTAL CONCEPTS

Let $\mathcal{M}$ be a vector space. An *inner product* is a function $\langle\cdot,\cdot\rangle : \mathcal{M} \to \mathbb{R}$ such that for all $x, y, z \in \mathcal{M}$ and $\alpha, \beta \in \mathbb{R}$ it satisfies (i) $\langle \alpha x + \beta y, z\rangle = \alpha\langle x, z\rangle + \beta\langle y, z\rangle$, (ii) $\langle x, y\rangle = \langle y, x\rangle$, and (iii) $\langle x, x\rangle \geq 0$ and $\langle x, x\rangle = 0$ if and only if $x = 0$. The vector space $\mathcal{M}$ endowed with an inner product $\langle\cdot,\cdot\rangle$ is said to be an *inner product space*.

A *norm* on a vector space $\mathcal{M}$ is a function $\|\cdot\| : \mathcal{M} \to [0, \infty)$ satisfying for all $x, y \in \mathcal{M}$ and $\alpha \in \mathbb{R}$ (i) $\|x\| = 0$ if and only if $x = 0$, (ii) $\|\alpha x\| = |\alpha|\|x\|$, and (iii) $\|x + y\| \leq \|x\| + \|y\|$. The inner product $\langle\cdot,\cdot\rangle$ induces the norm $\|x\| = \sqrt{\langle x, x\rangle}$ in $\mathcal{M}$. The sequence $\{x_t\}$ in $\mathcal{M}$ is said to be a *Cauchy sequence* if and only if for all $\varepsilon > 0$ there is an integer n such that $\|x_t - x_s\| < \varepsilon$ for all $t, s > n$. The vector space $\mathcal{M}$ is *complete* if and only if every Cauchy sequence has a limit in $\mathcal{M}$. With these definitions we are ready to introduce the Hilbert spaces.

A *Hilbert* space is a complete inner product vector space. These spaces are particularly important for time series analysis mainly because of the following *projection theorem*.

Theorem 1.1. *Let $\mathcal{M}$ be a closed subspace of the Hilbert space $\mathcal{H}$ and let $x \in \mathcal{H}$. Then, (a) there is a unique point $y \in \mathcal{M}$ such that $\|x - y\| = \inf_{z\in\mathcal{M}} \|x - z\|$ and (b) $y \in \mathcal{M}$ and $\|x - y\| = \inf_{z\in\mathcal{M}} \|x - z\|$ if and only if $y \in \mathcal{M}$ and $\langle x - y, z\rangle = 0$ for all $z \in \mathcal{M}$.*

Proof. (a) By definition, there is a sequence $\{y_t\}$ in $\mathcal{M}$ such that $\|x - y_t\| \to \alpha$ where $\alpha = \inf_{z\in\mathcal{M}} \|x - z\|$. This is a Cauchy sequence since $(y_t + y_s)/2 \in \mathcal{M}$ and by the parallelogram law (see Problem 1.5) we have

$$\|(y_t - x) - (y_s - x)\|^2 + 4\|(y_t + y_s)/2 - x\|^2 = 2\|y_t - x\|^2 + 2\|y_s - x\|^2.$$

Thus, $\|y_t - y_s\|^2 = 2\|y_t - x\|^2 + 2\|y_s - x\|^2 - 4\alpha^2$, and then $\|y_t - y_s\| \to 0$ as $t, s \to \infty$. Since $\mathcal{H}$ is complete, there is $y \in \mathcal{H}$ such that $\|y - y_t\| \to 0$ as $t \to \infty$. Furthermore, since $\mathcal{M}$ is closed, we conclude that $y \in \mathcal{M}$. Now, by the continuity of the norm we have that $\|x - y\| = \lim_{t\to\infty} \|x - y_t\| = \alpha$. To show that this point is unique assume that there is another vector $z \in \mathcal{M}$ such that $\|x - z\| = \alpha$. By the parallelogram law we have

$$\|y - z\|^2 = 2\|y - x\|^2 + 2\|z - x\|^2 - 4\|(y + z)/2 - x\|^2 = 0,$$

so that $y = z$. (b) Suppose that $y \in \mathcal{M}$ and $\|x - y\| = \inf_{z\in\mathcal{M}} \|x - z\|$. If $z \in \mathcal{M}$, then $y + z \in \mathcal{M}$ and

$$\|x - y\|^2 \leq \|x - (y + z)\|^2 = \|(x - y) - z\|^2 = \|x - y\|^2 + \|z\|^2 - 2\langle x - y, z\rangle.$$

Therefore, $2\langle x - y, z\rangle \leq \|z\|^2$ for any $z \in \mathcal{M}$. For fixed z, $\lambda z \in \mathcal{M}$ for any $\lambda \geq 0$. Assuming that $\langle x - y, z\rangle \geq 0$ (otherwise take $-z$), we have $0 \leq 2\lambda\langle x - y, z\rangle \leq \lambda^2\|z\|^2$, that is, $0 \leq 2\langle x - y, z\rangle \leq \lambda\|z\|^2$. Consequently, as $\lambda \to 0$ we have

$\langle x - y, z \rangle = 0$. Conversely, consider a point $y \in \mathcal{M}$ satisfying $\langle x - y, z \rangle = 0$ for all $z \in \mathcal{M}$. Then, for any $z \in \mathcal{M}$ we have

$$\|x - z\|^2 = \|x - y\|^2 + \|y - z\|^2 \geq \|x - y\|^2.$$

Therefore, $\|x - y\| = \inf_{z \in \mathcal{M}} \|x - z\|$. □

Given a subset M of a Hilbert space $\mathcal{H}$, the *span* of M, denoted by sp$M$, is the subspace generated by all finite linear combinations of elements of M and $\overline{\text{sp}}\, M$ denotes its closure in $\mathcal{H}$, that is, $\overline{\text{sp}}\, M$ contains all the limits of sequences in sp$M$. In what follows, we illustrate the concept of Hilbert space with several well-known examples.

■ **EXAMPLE 1.1**

Let n be a positive integer and consider the space $\mathbb{C}^n$ endowed with the Euclidean inner product

$$\langle x, y \rangle = x\overline{y} = \sum_{j=1}^{n} x_j \overline{y}_j,$$

where $x = (x_1, \ldots, x_n)'$, $y = (y_1, \ldots, y_n)'$, and $\overline{y}$ is the complex-conjugate of y. Then, $\mathbb{C}^n$ is a Hilbert space with norm $\|x\| = (\sum_{j=1}^{n} |x_j|^2)^{1/2}$.

■ **EXAMPLE 1.2**

Let $\{y_t : t \in \mathbb{Z}\}$ be a real, zero-mean stochastic process defined on a probability space $(\Omega, \mathcal{F}, \mathsf{P})$. Then, $\mathcal{L}_2(\Omega, \mathcal{F}, \mathsf{P})$ denotes the Hilbert space with inner product $\langle x, y \rangle = E(xy)$ and norm $\|y\| = \sqrt{E(y^2)}$, where $E(\cdot)$ stands for the expectation operator. This space will be simply called $\mathcal{L}_2$ throughout this book.

■ **EXAMPLE 1.3**

Let F be a distribution function and let $f(\lambda)$ and $g(\lambda)$ be two complex-valued functions with domain $[-\pi, \pi]$ such that

$$\int_{-\pi}^{\pi} |f(\lambda)|^2 \, dF(\lambda) < \infty, \qquad \int_{-\pi}^{\pi} |g(\lambda)|^2 \, dF(\lambda) < \infty.$$

Then,

$$\langle f, g \rangle_F = \int_{-\pi}^{\pi} f(\lambda) \, \overline{g(\lambda)} \, dF(\lambda)$$

is an inner product and the generated Hilbert space is denoted by $\mathcal{L}_2(F)$. Observe that by Hölder's inequality this inner product is well defined since

$$|\langle f, g \rangle|^2 \leq \int_{-\pi}^{\pi} |f(\lambda)|^2 \, dF(\lambda) \int_{-\pi}^{\pi} |g(\lambda)|^2 \, dF(\lambda) < \infty.$$

EXAMPLE 1.4

As a particular case of Example 1.3, if $F(\lambda)$ corresponds to the Lebesgue measure over $[-\pi, \pi]$ given by $dF(\lambda) = d\lambda/2\pi$, then the resulting Hilbert space is denoted by $\mathcal{L}_2(d\lambda)$.

1.1.1 Stationarity

A stochastic process $\{y_t : t \in \mathbb{Z}\}$ is said to be *second-order* or *weakly stationary* if there exists a symmetric *autocovariance* function $\gamma(\cdot)$ with $\gamma(0) < \infty$ such that $\langle y_t, y_{t+h} \rangle = \gamma(h)$ for all $t, h \in \mathbb{Z}$.

Now, let $Y_h(\omega) = \{y_{t_1+h}(\omega), \ldots, y_{t_n+h}(\omega)\}$ be a trajectory of the process $\{y_t\}$ with $t_1 + h, \ldots, t_n + h \in \mathbb{Z}$. The process is said to be *strictly stationary* if and only if the distribution of Y_h is the same regardless of h.

Observe that a process can be strictly stationary but not necessarily weakly stationary and vice versa. For instance, the process $\{y_t : t \in \mathbb{Z}\}$ where y_t are independent and identically distributed Cauchy random variables is strictly stationary but not weakly stationary since the first and the second moments do not exist. A more sophisticated example of this is the *fractionally integrated generalized autoregressive conditionally heteroskedastic* (FIGARCH) model introduced in Chapter 6. Conversely, let $\{\varepsilon_t\}$ be a sequence of independent and identically distributed normal random variables with zero mean and unit variance, and let $\{\eta_t\}$ be a sequence of independent and identically distributed exponential random variables with rate 1. Then the process generated by $y_t = \varepsilon_t[t/2] + (\eta_t - 1)[(t+1)/2]$, where $[\cdot]$ denotes the integer part function, is weakly stationary but not strictly stationary. Nevertheless, these two concepts are equivalent for Gaussian processes.

A *strict white noise* process is a sequence of independent and identically distributed random variables while a *weak white noise* process is a sequence of uncorrelated random variables with zero mean and constant finite variance, that is, with an autocovariance function satisfying $\gamma(0) < \infty$ and $\gamma(h) = 0$ for all $h \neq 0$.

The autocovariance function of second-order stationary process, $\gamma(\cdot)$, can be written as

$$\gamma(h) = \int_{-\pi}^{\pi} e^{ih\lambda}\, dF(\lambda),$$

where the function F is right-continuous, non-decreasing, bounded over $[-\pi, \pi]$, and satisfies the condition $F(-\pi) = 0$. F is called the *spectral distribution* of $\gamma(\cdot)$. Furthermore, if

$$F(\lambda) = \int_{-\pi}^{\lambda} f(\omega)\, d\omega,$$

then $f(\cdot)$ is called the *spectral density* of $\gamma(\cdot)$.

Remark 1.1. For simplicity, hereafter *stationary process* will always mean *second-order stationary process* or *weakly stationary process*, unless specified otherwise.

Similarly, a *second-order white noise* or *weak white noise* will be simply called *white noise*.

1.1.2 Singularity and Regularity

Let $\mathcal{F}_t = \overline{\text{sp}}\{y_s : s < t\}$ be the *past* of the process at time t and $\mathcal{F}_{-\infty} = \bigcap_{t=-\infty}^{\infty} \mathcal{F}_t$. The process is said to be *deterministic*, that is, *perfectly predictable*, if and only if $y_t \in \mathcal{F}_{-\infty}$ for all $t \in \mathbb{Z}$. In other words, a process is deterministic if and only if $\mathcal{F}_{-\infty} = \mathcal{F}_{\infty}$. These processes are also called *singular*. On the other hand, a process is said to be *purely nondeterministic* or *regular* if and only if $\mathcal{F}_{-\infty} = \{0\}$.

1.1.3 Wold Decomposition Theorem

The following result, known as the *Wold decomposition theorem*, is a fundamental tool for analyzing stationary processes; see Wold (1938).

Theorem 1.2. *Any stationary process is the sum of a regular process and a singular process; these two processes are orthogonal and the decomposition is unique.*

According to the Wold representation theorem, a stationary purely nondeterministic process may be expressed as

$$y_t = \sum_{j=0}^{\infty} \psi_j \varepsilon_{t-j} = \psi(B)\varepsilon_t, \tag{1.1}$$

where $\psi_0 = 1$, $\sum_{j=0}^{\infty} \psi_j^2 < \infty$, $\{\varepsilon_t\}$ is a white noise sequence with variance σ^2. The *Wold expansion* (1.1) is unique and $\varepsilon_t \in \mathcal{F}_{t+1}$ for all $t \in \mathbb{Z}$.

Recall now that a complex-valued stochastic process $\{\varepsilon(\lambda) \colon \lambda \in [-\pi, \pi]\}$ is said to be a *right-continuous orthogonal-increment process* if it satisfies the following conditions: (i) $\|\varepsilon(\lambda)\| < \infty$, (ii) $\langle 1, \varepsilon(\lambda)\rangle = 0$, (iii) $\langle \varepsilon(\lambda_2) - \varepsilon(\lambda_1), \varepsilon(\omega_2) - \varepsilon(\omega_1)\rangle = 0$ for all disjoint intervals $(\lambda_1, \lambda_2]$ and $(\omega_1, \omega_2]$, and (iv) $\|\varepsilon(\lambda+\delta) - \varepsilon(\lambda)\| \to 0$, as $\delta \to 0_+$, where $\|\cdot\|$ is the $\mathcal{L}_2$ norm. The following result, usually called the *spectral representation theorem*, establishes that a stationary process can be written as an integral of a deterministic function with respect to a right-continuous orthogonal-increment sequence; see, for example, Rozanov (1967, Theorem 4.2), Hannan (1970, Section II, Theorem 2), and Brockwell and Davis (1991, Section 4.7 and Theorem 4.8.2).

Theorem 1.3. *Let $\{y_t\}$ be a zero-mean stationary process with spectral distribution F. Then, there exists a right-continuous orthogonal-increment process $\{\varepsilon(\lambda)\}$ such that $F(\lambda) = \|\varepsilon(\lambda) - \varepsilon(-\pi)\|^2$ for $\lambda \in [-\pi, \pi]$ and the process $\{y_t\}$ can be written as*

$$y_t = \int_{-\pi}^{\pi} e^{it\lambda}\, d\varepsilon(\lambda),$$

with probability one.

In time series analysis we often need to apply a *linear filter* $\{\psi_j : j \in \mathbb{Z}\}$ to a stationary process $\{x_t\}$, that is, $y_t = \sum_{j=-\infty}^{\infty} \psi_j x_{t-j} = \psi(B)x_t$. The following result provides general conditions on the linear filter $\{\psi_j : j \in \mathbb{Z}\}$ so that the filtered process $\{y_t\}$ is also stationary. Furthermore, it establishes conditions to ensure that this filtering procedure is reversible; see, for instance, Rozanov (1967, Section III.5), Hannan (1970, Section II.4), and Brockwell and Davis (1991, Theorem 4.10.1).

Theorem 1.4. *Let $\{x_t\}$ be a zero-mean stationary process with spectral representation*

$$x_t = \int_{-\pi}^{\pi} e^{i\lambda t}\, d\varepsilon(\lambda)$$

and spectral distribution F. Let $\{\psi_j : j \in \mathbb{Z}\}$ be a sequence such that for $\lambda \in (-\pi, \pi]$

$$\sum_{j=-n}^{n} \psi_j e^{i\lambda j} \to \psi(e^{i\lambda}) = \sum_{j=-\infty}^{\infty} \psi_j e^{i\lambda j},$$

as $n \to \infty$ in the Hilbert space $\mathcal{L}(F)$ defined in Example 1.3. Then,

(a) The filtered process $\{y_t\}$ defined by

$$y_t = \sum_{j=-\infty}^{\infty} \psi_j x_{t-j}$$

is zero-mean stationary.

(b) The spectral distribution of $\{y_t\}$ is given by

$$F_y(\lambda) = \int_{-\pi}^{\lambda} |\psi(e^{-i\omega})|^2\, dF(\omega).$$

(c) The process $\{y_t\}$ can be expressed as

$$y_t = \int_{-\pi}^{\pi} e^{i\lambda t} \psi(e^{-i\lambda})\, d\varepsilon(\lambda).$$

(d) Assume that the sequence $\{\psi_j\}$ is such that $\psi(e^{-i\lambda}) \neq 0$ for $\lambda \in \Lambda$, where Λ^c has zero F-measure. Then, the process $\{x_t\}$ can be written as

$$x_t = \int_{-\pi}^{\pi} e^{i\lambda t} \pi(e^{-i\lambda})\, d\varepsilon_y(\lambda),$$

where $\pi(e^{i\lambda}) = 1/\psi(e^{i\lambda})$ and $d\varepsilon_y(\lambda) = \psi(e^{-i\lambda})\, d\varepsilon(\lambda)$.

The next result, which is a consequence of Theorem 1.4, will be extensively employed in Chapter 3 to prove the stationarity, causality and invertibility of a class of long-memory models.

Theorem 1.5. *Let $\{x_t\}$ be a zero-mean stationary process and let $\{\varphi_j : j \in \mathbb{Z}\}$ be an absolutely summable sequence. Then,*

(*a*) *The process $\{y_t\}$ defined by*

$$y_t = \sum_{j=-\infty}^{\infty} \varphi_j x_{t-j} = \varphi(B) x_t$$

is zero-mean stationary.

(*b*) *Suppose that the stationary process $\{x_t\}$ is given by*

$$x_t = \sum_{j=-\infty}^{\infty} \eta_j \varepsilon_{t-j} = \eta(B) \varepsilon_t.$$

Then, we can write

$$y_t = \varphi(B)\eta(B)\varepsilon_t = \eta(B)\varphi(B)\varepsilon_t = \psi(B)\varepsilon_t,$$

where these equations are in the mean square sense and $\psi(B) = \eta(B)\varphi(B)$.

1.1.4 Causality

The process defined by (1.1) is said to be *causal* since the representation of y_t is based only on the present and past of the input noise, $\{\varepsilon_t, \varepsilon_{t-1}, \dots\}$. Causality is a key property for predicting future values of the process.

■ **EXAMPLE 1.5**

Stationarity and causality are two related concepts but not necessarily equivalent. To illustrate this point let $\{\varepsilon_t\}$ be a white noise sequence with $\mathrm{Var}(\varepsilon_t) < \infty$ and consider, for instance, the process $y_t = \varepsilon_t + \varepsilon_{t+1}$. This process is stationary but not causal, since y_t depends on future values of the sequence $\{\varepsilon_t\}$. On the other hand, the process $y_t = \varepsilon_t + \varepsilon_{t-1}$ is causal and stationary.

1.1.5 Invertibility

A linear regular process (1.1) is said to be *invertible* if there exists a sequence of coefficients $\{\pi_j\}$ such that

$$\varepsilon_t = -\sum_{j=0}^{\infty} \pi_j y_{t-j}, \tag{1.2}$$

where this expansion converges in $\mathcal{L}_2$. From (1.2), and assuming $\pi_0 = -1$, the process $\{y_t\}$ may be expressed as

$$y_t = \varepsilon_t + \sum_{j=1}^{\infty} \pi_j y_{t-j}. \tag{1.3}$$

1.1.6 Best Linear Predictor

Let $\{y_t\}$ be an invertible process as in (1.3). Following Theorem 1.1, the best linear prediction of the observation y_t based on its history $\overline{\text{sp}}\{y_t, y_{t-1}, y_{t-2}, \dots\}$ is given by

$$\widehat{y}_t = \sum_{j=1}^{\infty} \pi_j y_{t-j}. \tag{1.4}$$

Consequently, $\varepsilon_t = y_t - \widehat{y}_t$ is an innovation sequence, that is, an orthogonal process with zero mean and constant variance representing the part of $\{y_t\}$ that cannot be linearly predicted from the past.

On the other hand, suppose that we only have a finite trajectory of the stationary process, $\{y_1, y_2, \dots, y_n\}$. According to Theorem 1.1, the best linear predictor of y_{n+1} based on its finite past $\overline{\text{sp}}\{y_1, y_2, \dots, y_n\}$, $\widehat{y}_{n+1}$, satisfies the equations

$$\langle y_{n+1} - \widehat{y}_{n+1}, y_j \rangle = 0,$$

for $j = 1, \dots, n$. By writing the best linear predictor as

$$\widehat{y}_{n+1} = \phi_{n1} y_n + \phi_{n2} y_{n-1} + \cdots + \phi_{nn} y_1,$$

we conclude that the coefficients ϕ_{nj} satisfy the equations

$$\sum_{i=1}^{n} \phi_{ni} \langle y_{n+1-i}, y_j \rangle = \langle y_{n+1}, y_j \rangle,$$

for $j = 1, \dots, n$. If $\langle y_i, y_j \rangle = \gamma(i - j)$, where $\gamma(\cdot)$ is an autocovariance function, then we have that

$$\sum_{i=1}^{n} \phi_{ni} \gamma(n + 1 - i - j) = \gamma(n + 1 - j).$$

Further details on this subject are given in Chapter 4 where we discuss an efficient technique for obtaining the best linear forecasts, the so-called Durbin-Levinson algorithm, and in Chapter 9 where we address the prediction of long-memory processes.

1.1.7 Szegö-Kolmogorov Formula

The spectral measure of the purely nondeterministic process (1.1) is absolutely continuous with respect to the Lebesgue measure on $[-\pi, \pi]$ and has spectral density

$$f(\lambda) = \frac{\sigma^2}{2\pi} |\psi(e^{-i\lambda})|^2.$$

Hence, its autocovariance function $\gamma(\cdot)$ may be written as

$$\gamma(h) = \int_{-\pi}^{\pi} f(\lambda) e^{-i\lambda h} \, d\lambda. \tag{1.5}$$

An explicit relationship between the spectral density and the variance of the noise sequence $\{\varepsilon_t\}$ is given by the Szegö-Kolmogorov formula

$$\sigma^2 = \exp\left\{\frac{1}{2\pi}\int_{-\pi}^{\pi} \log[2\pi f(\lambda)]\, d\lambda\right\}. \tag{1.6}$$

From this expression, observe that for a purely nondeterministic process, $\sigma^2 > 0$ or equivalently, $\int_{-\pi}^{\pi} \log f(\lambda)\, d\lambda > -\infty$.

1.1.8 Ergodicity

Let $\{y_t : t \in \mathbb{Z}\}$ be a strictly stationary process and consider the σ-algebra $\mathcal{F}$ generated by the sets $S = \{\omega\colon Y_0(\omega) = [y_{t_1}(\omega), \ldots, y_{t_n}(\omega)] \in \mathcal{B}\}$ where $\mathcal{B}$ is a Borel set in $\mathbb{R}^n$. Let T be a *shift* operator $T\colon \Omega \to \Omega$ such that $TS = \{\omega\colon Y_1(\omega) = [y_{t_1+1}(\omega), \ldots, y_{t_n+1}(\omega)] \in \mathcal{B}\}$. This operator is measure preserving since the process is strictly stationary. Let $T^{-1}S$ be the preimage of S, that is, $T^{-1}S = \{\omega\colon T(\omega) \in S\}$. A measurable set is said to be *invariant* if and only if $T^{-1}S = S$. Similarly, a random variable X is said to be *invariant* if and only if $X(\omega) = X(T^{-1}\omega)$ for all $\omega \in \Omega$.

Let $\mathcal{S}$ be the set containing all the *invariant* measurable events. The process $\{y_t : t \in \mathbb{Z}\}$ is said to be *ergodic* if and only if for all the events $S \in \mathcal{S}$, either $P(S) = 0$ or $P(S) = 1$. Loosely speaking, this means that the only invariant sets are the entire space or zero measure events.

The concept of *ergodic process* arises from the study of the long run behavior of the average of a stationary sequence of random variables. Consider, for example, the strictly stationary process $\{y_t\}$ and the average $\bar{y}_n = \sum_{t=1}^{n} y_t/n$. The following result establishes that the sequence $\{\bar{y}_n\}$ converges under some mild conditions; see Hannan (1970, p. 201).

Theorem 1.6. *If $\{y_t\}$ is a strictly stationary process with $E(|y_t|) < \infty$, then there is an* invariant *random variable $\bar{y}$ such that $E(|\bar{y}|) < \infty$,*

$$\lim_{n\to\infty} \bar{y}_n = \bar{y}, \tag{1.7}$$

almost surely, and

$$E(\bar{y}) = E(y_t). \tag{1.8}$$

A natural question that arises at this point is under what circumstances the limit $\bar{y}$ is a constant. For instance, as discussed in Example 1.6 below, if $\{y_t\}$ is a sequence of independent random variables with $E\,|y_t| < \infty$, then the law of the large numbers guarantees that the limit $\bar{y}$ is equal to $E(y_t)$. Nevertheless, if a process is such that all invariant random variables are constant with probability 1, by the above theorem we can conclude that the limit $\bar{y}$ is a constant with probability 1. Stationary sequences satisfying this requirement are called *ergodic processes*.

EXAMPLE 1.6

A simple example of a strictly stationary ergodic process is a sequence of independent and identically distributed random variables $\{\varepsilon_1, \varepsilon_2, \dots\}$ with $E(|\varepsilon_t|) < \infty$. In this case, if $\mu = E(\varepsilon_t)$ and

$$\bar{\varepsilon}_n = \frac{1}{n} \sum_{t=1}^{n} \varepsilon_t,$$

is the sample mean of size n, then by the strong law of large numbers we have that $\lim_{n\to\infty} \bar{\varepsilon}_n = \mu$; see, for example, Stout (1974, Theorem 3.2.2).

The following result is important for the theoretical treatment of ergodic systems since it establishes that strict stationarity and ergodicity is preserved by measurable transformations; see (Stout, 1974, Theorem 3.5.8) or (Taniguchi and Kakizawa, 2000, Theorem 1.3.3).

Theorem 1.7. *If $\{y_t : t \in \mathbb{Z}\}$ is a strictly stationary ergodic process and $\phi : \mathbb{R}^\infty \to \mathbb{R}$ is a measurable transformation such that $x_t = \phi(y_t, y_{t-1}, \dots)$, then $\{x_t\}$ is a strictly stationary ergodic process.*

An important consequence of Theorem 1.7 is that a linear filter of an independent and identically distributed sequence is strictly stationary and ergodic as stated in the following result.

Theorem 1.8. *Let $\{\varepsilon_t\}$ be a sequence of independent and identically distributed random variables with zero mean and finite variance. If $\sum_{j=0}^{\infty} \alpha_j^2 < \infty$, then*

$$y_t = \sum_{j=0}^{\infty} \alpha_j \varepsilon_{t-j}, \tag{1.9}$$

is a strictly stationary ergodic process.

Proof. This is an immediate consequence of Theorem 1.7 since

$$\phi(\varepsilon_t, \varepsilon_{t-1}, \dots) = \sum_{j=0}^{\infty} \alpha_j \varepsilon_{t-j},$$

with $\sum_{j=0}^{\infty} \alpha_j^2 < \infty$ is a measurable transformation and the sequence $\{\varepsilon_t\}$ is a strictly stationary ergodic process. □

EXAMPLE 1.7

For a trivial example of nonergodic process consider $y_t = \varepsilon_t + \theta\varepsilon_{t-1} + \eta$ where $\{\varepsilon_t\}$ is a sequence of independent and identically distributed random variables with zero mean and unit variance, and η is a random variable independent of

$\{\varepsilon_t\}$ with $E(|\eta|) < \infty$. The process $\{y_t\}$ is strictly stationary but not ergodic since $\bar{y}_n \to \eta$ as $n \to \infty$ and η is an invariant random variable which is not necessarily a constant.

Ergodicity is an essential tool for carrying out statistical inferences about a stochastic process. Usually, only one trajectory of the process is available for analysis. Consequently, there is only one sample of the process at any given time t. Thanks to the ergodicity of the process we can make up for this lack of information by using the trajectory to estimate the moments of the distribution of a strictly stationary process $\{y_t : t \in \mathbb{Z}\}$. Specifically, assume that $E(|y_t^i y_s^j|) < \infty$ for positive integers i, j and define $\mu_{ij} = E(y_t^i y_s^j)$. Observe that this moment does not depend on t or s since the process is strictly stationary. Therefore, we may write $\mu_{ij} = E(y_0^i y_h^j)$, where $h = s - t$. The moment μ_{ij} can be estimated by the time-average

$$\widehat{\mu}_{ij}(n) = \frac{1}{n} \sum_{t=1}^{n-h} y_t^i y_{t+h}^j. \tag{1.10}$$

The following result, known as the *pointwise ergodic theorem for stationary sequences*, states that this estimate is strongly consistent for every fixed h; see (Stout, 1974, Theorem 3.5.7) or (Hannan, 1970, Theorem 2).

Theorem 1.9. *If $\{y_t : t \in \mathbb{Z}\}$ is a strictly stationary and ergodic process such that $E(|y_t|^{i+j}) < \infty$, then $\widehat{\mu}_{ij}(n) \to \mu_{ij}$ almost surely as $n \to \infty$.*

Proof. Let us define the random variable $x_t = y_t^i y_{t+h}^j \equiv \phi(y_t, y_{t+h})$. Observe that by Hölder's inequality

$$E(|x_t|) = E\left(|y_t^i y_{t+h}^j|\right) \leq \left(E|y_t|^{i+j}\right)^{i/(i+j)} \left(E|y_{t+h}|^{i+j}\right)^{j/(i+j)} < \infty.$$

Since $\{y_t\}$ is strictly stationary ergodic and $\phi(\cdot)$ is a measurable transformation, an application of Theorem 1.7 shows that $\{x_t\}$ is strictly stationary and ergodic. Consequently, from Theorem 1.6 and the ergodicity of x_t we conclude that $\bar{x}_n \to \mu_{ij}$ as $n \to \infty$. □

The following two definitions are important for the analysis of non-Gaussian time series, that is, when the innovation sequence of the Wold expansion is not assumed to be normal, independent, and identically distributed. The concept of martingale will be particularly important for the treatment of heteroskedastic time series studied in Chapter 6. On the other hand, the concept of cumulant will be relevant for the analysis of the asymptotic properties of, for instance, the least squares estimates of the coefficients of a linear regression model with long-range-dependent errors, as discussed in Chapter 10.

1.1.9 Martingales

Let $(\Omega, \mathcal{F}, P)$ be a probability space and let $\{y_1, y_2, \ldots\}$ be a sequence of integrable random variables on $(\Omega, \mathcal{F}, P)$. Let $\{\mathcal{F}_t\}$ be an increasing sequence of σ-algebras of $\mathcal{F}$, that is, $\mathcal{F}_1 \subset \mathcal{F}_2 \subset \cdots$ where y_t is $\mathcal{F}_t$-measurable.

A sequence $\{y_t\}$ is a *martingale* relative to $\mathcal{F}_t$ if and only if

$$E(y_t|\mathcal{F}_{t-1}) = y_{t-1},$$

almost everywhere for all $t = 1, 2, \ldots$

On the other hand, a sequence $\{y_t\}$ is a *martingale difference* if and only if

$$E(y_t|\mathcal{F}_{t-1}) = 0,$$

almost everywhere for all $t = 1, 2, \ldots$

EXAMPLE 1.8

Let $\{\varepsilon_t\}$ be a sequence of independent and identically distributed random variables with zero mean and define the process $y_t = \sum_{j=1}^{t} \varepsilon_j$. Observe that $\{y_t\}$ is a martingale sequence since

$$E[y_t|\mathcal{F}_{t-1}] = E[\varepsilon_t + y_{t-1}|\mathcal{F}_{t-1}] = E[\varepsilon_t|\mathcal{F}_{t-1}] + E[y_{t-1}|\mathcal{F}_{t-1}] = y_{t-1}.$$

1.1.10 Cumulants

Let $\{y_1, y_2, \ldots, y_k\}$ be a set of random variables and define the function

$$g(t_1, t_2, \ldots, t_k) = \log \left[E \left(e^{i \sum_{j=1}^{k} t_j y_j} \right) \right].$$

The *cumulant* $\text{cum}(y_1, y_1, \ldots, y_k)$ is the coefficient of $(it_1, it_1 \cdots, it_k)$ in the Taylor expansion of $g(t_1, t_2, \ldots, t_k)$. Cumulants will be used throughout this book for the theoretical analysis of estimators for both Gaussian and non-Gaussian processes; see, for example, Chapters 4, 5, and 10.

The next subsection introduces the concepts of Brownian motion and fractional Brownian motion which will be useful for defining the fractional Gaussian noise in Chapter 3 and for analyzing the asymptotic properties of some estimation methods in Chapter 4.

1.1.11 Fractional Brownian Motion

A Gaussian stochastic process $B(t)$ with continuous sample paths is called *Brownian motion* if it satisfies (1) $B(0) = 0$ almost surely, (2) $B(t)$ has independent increments, (3) $E[B(t)] = E[B(s)]$, and (4) $\text{Var}[B(t) - B(s)] = \sigma^2|t - s|$.

On the other hand, a standard *fractional Brownian motion* $B_d(t)$ may be defined as

$$B_d(t) = c(d) \int_{-\infty}^{\infty} (t-s)_+^d - (-s)_+^d \, dB(s), \tag{1.11}$$

where

$$s_+ = \begin{cases} s & \text{if} \quad s \geq 0 \\ 0 & \text{if} \quad s < 0, \end{cases}$$

and

$$c(d) = \frac{\sqrt{\Gamma(2d+2)\cos \pi d}}{\Gamma(d+1)}.$$

For a precise definition of the stochastic integral in (1.11), see, for example, Taqqu (2003). The term *standard* in this definition corresponds to the fact that the process $B_d(t)$ has unitary variance at time $t = 1$, that is, $\mathrm{Var}[B_d(1)] = 1$. In addition, the standard fractional Brownian motion (1.11) satisfies for $d \in (-\frac{1}{2}, \frac{1}{2})$

$$E[B_d(t)] = 0, \tag{1.12}$$

$$\mathrm{Cov}[B_d(t), B_d(s)] = \tfrac{1}{2}\left(|t|^{2d+1} + |s|^{2d+1} - |t-s|^{2d+1}\right). \tag{1.13}$$

A real-valued stochastic process $\{y(t) : t \in \mathbb{R}\}$ is *self-similar* with index $H > 0$ if for any constant $c > 0$ and for any time $t \in \mathbb{R}$ it satisfies

$$y(ct) \sim c^H y(t),$$

where the symbol $\sim$ means that both terms have the same distribution. The self-similarity index H is called the *scaling exponent* of the process or the *Hurst exponent*, in honor of the British hydrologist H. E. Hurst who developed methods for modeling the long-range dependence of the Nile River flows.

Observe that from (1.12) and (1.13) we conclude that

$$B_d(t) \sim N(0, |t|^{2d+1}).$$

Hence, for any positive constant c we have

$$B_d(ct) \sim N(0, c^{2d+1}|t|^{2d+1}),$$

and then

$$B_d(ct) \sim c^{d+\frac{1}{2}} B_d(t).$$

Thus, the fractional Brownian motion $B_d(t)$ is a self-similar process with self-similarity index $H = d + \frac{1}{2}$.

In the next subsection we address briefly the concept of *wavelets* which will be useful for defining a wavelet-based estimation method in Chapter 4. In the bibliographic section we provide some references for works covering this subject in detail.

1.1.12 Wavelets

A *wavelet* is a real-valued integrable function $\psi(t)$ satisfying

$$\int \psi(t)\, dt = 0. \tag{1.14}$$

A wavelet has n vanishing moments if

$$\int t^p \psi(t)\, dt = 0,$$

for $p = 0, 1, \ldots, n-1$.

Consider the following family of *dilations* and *translations* of the wavelet function ψ defined by

$$\psi_{jk}(t) = 2^{-j/2}\psi(2^{-j}t - k),$$

for $j, k \in \mathbb{Z}$. In this context, the terms j and 2^j are usually called the *octave* and the *scale*, respectively. It can be shown that (see Problem 1.17)

$$\int \psi_{jk}^2(t)\, dt = \int \psi^2(t)\, dt.$$

The *discrete wavelet transform* (DWT) of a process $\{y(t)\}$ is then defined by

$$d_{jk} = \int y(t)\psi_{jk}(t)\, dt,$$

for $j, k \in \mathbb{Z}$.

Provided that the family $\{\psi_{jk}(t)\}$ forms an orthogonal basis, that is,

$$\int \psi_{ij}(t)\psi_{k\ell}(t)\, dt = 0,$$

for all i, j, k, ℓ, excepting $i = j = k = \ell$, we obtain the following representation of the process $\{y(t)\}$:

$$y(t) = \sum_{j=-\infty}^{\infty} \sum_{k=-\infty}^{\infty} d_{jk}\psi_{jk}(t).$$

■ **EXAMPLE 1.9**

The *Haar wavelet system*

$$\psi(t) = \begin{cases} 1 & \text{if } t \in [0, \frac{1}{2}) \\ -1 & \text{if } t \in [\frac{1}{2}, 1) \\ 0 & \text{otherwise,} \end{cases}$$

is a simple example of a function satisfying (1.14). Observe that

$$\int t^p \psi(t)\, dt = \frac{1}{p+1}\left\{\frac{1}{2^p} - 1\right\}.$$

Therefore, this wavelet has a vanishing moment only for $p = 0$.

EXAMPLE 1.10

The so-called *Daubechies wavelets* are a family of wavelets that extends the previous example achieving a greater number of vanishing moments. This family forms an orthogonal wavelet basis and it is built in terms of the *multiresolution analysis*. Only a hint of this procedure is given here and further references are provided in the bibliographic section.

Starting from a *scaling function* ϕ that satisfies

$$\phi\left(\tfrac{t}{2}\right) = \sqrt{2}\sum_j u_j \phi(t-j),$$

we obtain the *mother wavelet* ψ by defining

$$\psi(t) = \sqrt{2}\sum_j v_j \phi(2t-j).$$

Thus, the Haar wavelet described in the previous example is obtained by setting $\phi(t) = \mathbf{1}_{[0,1]}(t)$, $u_0 = u_1 = v_0 = -v_1 = 1/\sqrt{2}$ and $u_j = v_j = 0$ for $j \neq 0, 1$, where $\mathbf{1}_A$ denotes the indicator function of the set A, that is,

$$\mathbf{1}_A(t) = \begin{cases} 1 & \text{if} \quad t \in A, \\ 0 & \text{if} \quad t \in A^c. \end{cases}$$

1.2 BIBLIOGRAPHIC NOTES

There is an extensive literature on Hilbert spaces. The first two chapters of Conway (1990) offer an excellent introduction to this subject. Chapter 9 of Pourahmadi (2001) discusses stationary stochastic processes in Hilbert spaces. A very readable proof of the Szegö-Kolmogorov formula is given in Hannan (1970). The concept of ergodicity is also described in Chapter IV of that book. A good overview of stochastic processes is given in the first chapter of Taniguchi and Kakizawa (2000). Carothers (2000) is a helpful reference on real analysis. The book by Stout (1974) is a very good source for convergence results, especially its Chapter 3 about limiting theorems for martingales and stationary random variables. Wavelets are vastly discussed in the literature. A nice recent revision, especially related to long-range dependence, is the article by Abry, Flandrin, Taqqu, and Veitch (2003) while Chapter 9 of Percival and Walden (2006) offers a comprehensive revision of wavelet methods for long-memory processes. Additionally, Chapter 2 of Flandrin (1999) and Chapter 13 of Press, Teukolsky, Vetterling, and Flannery (1992) provide overviews about this topic, including several details about the Daubechies wavelets. Finally, there are several other versions of Theorem 1.5; see, for example, Rudin (1976, Theorem 3.50) and Kokoszka and Taqqu (1995, Theorem 2.3).

Problems

1.1 Consider the Hilbert space $\mathcal{L}_2$ and a subspace $\mathcal{M} \subseteq \mathcal{L}_2$. Show that the orthogonal projection of $y \in \mathcal{L}_2$ onto $\mathcal{M}$ is given by the conditional expectation

$$\widehat{y} = E(y|\mathcal{M}).$$

1.2 Consider the Hilbert space $\mathcal{H} = \overline{\text{sp}}\{e_t : t = 0, 1, 2, \dots\}$ where $\{e_t\}$ is an orthonormal basis, that is, $\langle e_t, e_s \rangle = 0$ for all $t \neq s$ and $\langle e_t, e_t \rangle = 1$ for all t.

a) Let $x \in \mathcal{H}$, verify that this element may be written as

$$x = \sum_{t=0}^{\infty} \langle x, e_t \rangle e_t.$$

b) Show that $\|x\|^2 = \sum_{t=0}^{\infty} \langle x, e_t \rangle^2$.

c) Let $\mathcal{M} = \overline{\text{sp}}\{e_1, \dots, e_N\}$ and let $\widehat{x}$ be the orthogonal projection of x on $\mathcal{M}$. Show that $\widehat{x} = \sum_{t=0}^{N} \langle x, e_t \rangle e_t$.

d) Verify that $\|x - \widehat{x}\|^2 = \sum_{t=N+1}^{\infty} \langle x, e_t \rangle^2$.

1.3 Let $\{y_t : t \in \mathbb{N}\}$ be a sequence in a Hilbert space $\mathcal{H}$ such that $\sum_{t=1}^{\infty} \|y_t\| < \infty$. Show that $\sum_{t=1}^{\infty} y_t$ converges in $\mathcal{H}$.

1.4 Let $\mathcal{H}$ be a Hilbert space and suppose that $x, y \in \mathcal{H}$ are orthogonal vectors such that $\|x\| = \|y\| = 1$. Show that $\|\alpha x + (1 - \alpha) y\| < 1$ for all $\alpha \in (0, 1)$. From this, what can you say about the set $\{y \in \mathcal{H} : \|y\| \leq 1\}$?

1.5 (Parallelogram law) Show that if $\mathcal{H}$ is an inner product space then

$$\|x + y\|^2 + \|x - y\|^2 = 2\|x\|^2 + 2\|y\|^2,$$

for all $x, y \in \mathcal{H}$.

1.6 Consider the following inner product of two functions f and g in $\mathcal{L}_2[-\pi, \pi]$:

$$\langle f, g \rangle = \frac{1}{2\pi} \int_{-\pi}^{\pi} f(\lambda) \overline{g}(\lambda)\, d\lambda.$$

a) Let $e_j(\lambda) = e^{i\lambda j}$ for $j \in \mathbb{Z}$. Show that the functions $\{e_j(\lambda)\}$ are orthonormal, that is,

$$\langle e_t, e_s \rangle = 0,$$

for all $s, t \in \mathbb{Z}$, $t \neq s$, and

$$\langle e_t, e_t \rangle = 1,$$

for all $t \in \mathbb{Z}$.

1.7 Suppose that a function $f \in \mathcal{L}_2(d\lambda)$ is defined by

$$f(\lambda) = \sum_{j=-m}^{m} \alpha_j e_j(\lambda),$$

where e_j are the functions defined in Problem 1.6.

a) Verify that the coefficients α_j are given by

$$\alpha_j = \frac{1}{2\pi} \int_{-\pi}^{\pi} f(\lambda) e^{-i\lambda j} d\lambda.$$

The coefficients α_j are called the *Fourier coefficients* of the function $f(\cdot)$.

b) Prove that

$$\|f(\lambda)\| = \sqrt{\sum_{j=-m}^{m} \alpha_j^2}.$$

1.8 A strictly stationary process is said to be a *mixing process* if for all $A, B \in \mathcal{A}$

$$\lim_{n \to \infty} P(A \cap T^{-n} B) = P(A)\, P(B).$$

Show that a mixing process is ergodic.

1.9 An endomorphism or measure preserving map T satisfies (i) T is surjective, (ii) T is measurable, and (iii) $\mu(T^{-1}S) = \mu(S)$ for all $S \in \mathcal{F}$. Show that the transformation T defined at the beginning of Subsection 1.1.8 is an endomorphism.

1.10 Show that the autocovariance function of the process defined by the Wold expansion (1.1) is given by

$$\gamma(h) = \sigma^2 \sum_{j=0}^{\infty} \psi_j \psi_{j+|h|}.$$

1.11 Prove that if $\{y_t : t \in \mathbb{Z}\}$ is a stationary process such that $\sum_{k=0}^{\infty} |\gamma(k)| < \infty$, then its spectral density may be written as

$$f(\lambda) = \frac{1}{2\pi} \sum_{h=-\infty}^{\infty} \gamma(h) e^{-i\lambda h},$$

and this function is symmetric and positive.

1.12 Let $\Gamma_n = [\gamma(i-j)]_{i,j=1,\ldots,n}$ be the variance-covariance matrix of a regular linear process with

$$\gamma(h) = \int_{-\pi}^{\pi} f(\lambda) e^{-\imath\lambda h} d\lambda.$$

a) Show that if $f(\lambda) > c$, with $c > 0$, then $\Gamma_n \geq cI_n$, where I_n is the identity matrix of size n.

b) From (a), show that $\|\Gamma_n^{-1}\|$ is uniformly bounded for all n.

Hint: If $z \in \mathbb{C}$, then $\|z\|^2 = z\bar{z}$, where $\bar{z}$ is the conjugate complex of z. If Σ is a positive definite matrix, then there exists a matrix $\Sigma^{1/2}$ such that $\Sigma = \Sigma^{1/2}\Sigma^{1/2}$.

1.13 A *stochastic volatility* process $\{r_t\}$ is defined by

$$\begin{aligned} r_t &= \varepsilon_t \sigma_t, \\ \sigma_t &= \sigma \exp(\nu_t/2), \end{aligned}$$

where $\{\varepsilon_t\}$ is an independent and identically distributed sequence with zero mean and unit variance, and $\{\nu_t\}$ is a regular linear process:

$$\nu_t = \sum_{j=0}^{\infty} \psi_j \eta_{t-j},$$

with $\sum_{j=0}^{\infty} \psi_j^2 < \infty$ and $\{\eta_t\}$ is an independent and identically distributed sequence with zero mean and unit variance, independent of the sequence $\{\varepsilon_t\}$. Show that the process r_t is strictly stationary and ergodic.

1.14 Let $\{B_0(t)\}$ be a fractional Brownian motion with $d = 0$. Verify that

$$\text{Cov}[B_0(t), B_0(s)] = \min\{|t|, |s|\}.$$

1.15 Let $\{B_d(t)\}$ be a fractional Brownian motion with $d \in (-\frac{1}{2}, \frac{1}{2})$. Show that this process has *stationary increments*, that is,

$$B_d(t+h) - B_d(t) \sim B_d(h) - B_d(0),$$

for all $t, h \in \mathbb{R}$.

1.16 Let $\{B_d(t)\}$ be a fractional Brownian motion with $d \in (-\frac{1}{2}, \frac{1}{2})$. Prove that for any $p > 0$,

$$E[B_d(t+h) - B_d(t)]^p = |h|^{p(d+1/2)} E[B(1)^p],$$

and

$$E\left[\frac{B_d(t+h) - B_d(t)}{h}\right]^2 = |h|^{2d-1}.$$

From this result, it can be concluded that the process $B_d(t)$ is nondifferentiable in the $\mathcal{L}_2$ sense.

1.17 Verify that the $\mathcal{L}_2$ norm of ψ and ψ_{ik} are identical for all $j, k \in \mathbb{Z}$, that is,

$$\int \psi_{jk}^2(t)\,dt = \int \psi^2(t)\,dt.$$

1.18 Show that the Haar system generates an orthonormal basis, that is,

$$\int \psi_{ij}(t)\psi_{k\ell}(t)\,dt = \delta_{ij}\delta_{k\ell}.$$

1.19 Consider the so-called Littlewood-Paley decomposition

$$\psi(t) = \begin{cases} 1 & \text{if } |t| \in [1/2, 1), \\ 0 & \text{otherwise.} \end{cases}$$

Prove that the Littlewood-Paley decomposition generates an orthonormal basis.

CHAPTER 2

STATE SPACE SYSTEMS

The linear processes introduced in the previous chapter were described by a Wold decomposition. However, these processes can also be expressed in terms of a *state space linear system*. This chapter is devoted to describe these systems and investigate some of the relationships between Wold expansions and state space representations. As described in this chapter, state space systems are very useful for calculating estimators, predictors, and interpolators. Consequently, Wold expansions and state space representations of linear time series will be extensively used throughout this book.

Section 2.1 introduces the state space linear systems and discusses a number of fundamental concepts such as stability, observability, controllability, and minimality. Additionally, three equivalent representations of a linear process including the Wold decomposition, state space systems, and the Hankel operator are analyzed in Section 2.2. Section 2.3 describes the Kalman filter equations to calculate recursively state estimates, forecasts, and smoothers along with their variances. This section also discusses techniques for handling missing values and predicting future observations. Some extensions of these procedures to incorporate exogenous variables are described in Section 2.4, and further readings on theoretical and practical issues of state space modeling are suggested in Section 2.5. A list of problems is given at the end of the chapter.

2.1 INTRODUCTION

A linear state space system may be described by the discrete-time equations

$$x_{t+1} = Fx_t + H\varepsilon_t, \tag{2.1}$$
$$y_t = Gx_t + \varepsilon_t, \tag{2.2}$$

where $x_t \in \mathcal{H}$ is the *state* vector for all time t with $\mathcal{H}$ a Hilbert space, $y_t \in \mathbb{R}$ is the *observation* sequence, $F : \mathcal{H} \to \mathcal{H}$ is the *state transition operator* or the *state matrix*, $G : \mathcal{H} \to \mathbb{R}$ is the *observation linear operator* or the *observation matrix*, $H : \mathbb{R} \to \mathcal{H}$ is a linear operator, and $\{\varepsilon_t\}$ is the *state white noise* sequence with variance σ^2. Equation (2.1) is called the *state equation* while (2.2) is called the *observation equation*.

2.1.1 Stability

A state space system (2.1)–(2.2) is said to be *strongly stable* if $F^n x$ converges to zero for all $x \in \mathcal{H}$ as n tends to ∞. On the other hand, the system is said to be *weakly stable* if $\langle x_1, F^n x_2\rangle$ converges to zero for all $x_1, x_2 \in \mathcal{H}$ as n tends to ∞. Finally, the system is said to be *exponentially stable* if there exist positive constants c and α such that $\|F^n\| \leq ce^{-\alpha n}$.

Generally speaking, the stability of a state space system means that the state vector does not explode as time increases and that the effect of the initial value of the state vanishes as time progresses. In what follows we will assume that the system is weakly stable unless stated otherwise.

2.1.2 Hankel Operator

Suppose that $\psi_0 = 1$ and $\psi_j = GF^{j-1}H \in \mathbb{R}$ for all $j > 0$ such that $\sum_{j=0}^{\infty} \psi_j^2 < \infty$. Then from (2.1)-(2.2), the process $\{y_t\}$ may be written as the Wold expansion

$$y_t = \sum_{j=0}^{\infty} \psi_j \varepsilon_{t-j}. \tag{2.3}$$

This regular linear process can be characterized by the *Hankel linear operator* or *Hankel matrix* $\mathbf{H} : \ell_2 \to \ell_2$ given by

$$\mathbf{H} = \begin{pmatrix} \psi_1 & \psi_2 & \psi_3 & \cdots \\ \psi_2 & \psi_3 & \psi_4 & \cdots \\ \psi_3 & \psi_4 & \psi_5 & \cdots \\ \vdots & \vdots & \vdots & \end{pmatrix}.$$

Note that this Hankel operator specifies the Wold expansion (2.3) and vice versa. Furthermore, the dimensionality of the state space system (2.1)-(2.2) is closely related to the dimensionality of the matrix $\mathbf{H}$ and to the rationality of the spectral density of the regular process (2.3). Specifically, we have the following result; see Theorem 2.4.1 (ii) of Hannan and Deistler (1988).

Theorem 2.1. *The rank of* $\mathbf{H}$ *is finite if and only if the spectral density of* (2.3) *is rational.*

The class of *autoregressive moving-average* (ARMA) processes have rational spectrum, hence the rank of $\mathbf{H}$ is finite for these models. In turn, as we will see later, this means that any state space system representing an ARMA process is finite dimensional. On the contrary, the class of long-memory processes [e.g., *autoregressive fractionally integrated moving-average* (ARFIMA) models] does not have rational spectrum. Consequently, all state space systems representing such models are infinite dimensional; see Chan and Palma (1998, Corollary 2.1).

Since the state space representation of a linear regular process is not necessarily unique, one may ask which is the minimal dimension of the state vector. In order to answer this question it is necessary to introduce the concepts of *observability* and *controllability*.

2.1.3 Observability

Let $\mathcal{O} = (G', F'G', F'^2G', \dots)'$ be the *observability operator*. The system (2.1)–(2.2) is said to be *observable* if and only if $\mathcal{O}$ is full rank or, equivalently, $\mathcal{O}'\mathcal{O}$ is invertible.

The definition of observability is related to the problem of determining the value of the *unobserved* initial state x_0 from a trajectory of the *observed* process $\{y_0, y_1, \dots\}$ in the absence of state or observational noise. Consider, for example, the deterministic state space system

$$\begin{aligned} x_{t+1} &= Fx_t, \\ y_t &= Gx_t, \end{aligned}$$

and let $Y = (y_0, y_1, \dots)'$ be a trajectory of the process. Since

$$\begin{aligned} y_0 &= Gx_0, \\ y_1 &= GFx_0, \\ y_3 &= GF^2x_0, \\ &\vdots \end{aligned}$$

we may write $Y = \mathcal{O}x_0$. Now, if $\mathcal{O}'\mathcal{O}$ is full rank, we can determine the value of the initial state explicitly as $x_0 = (\mathcal{O}'\mathcal{O})^{-1}\mathcal{O}'Y$.

2.1.4 Controllability

Let $\mathcal{C} = (H, FH, F^2H, \dots)$ be the *controllability operator*. The system (2.1)-(2.2) is said to be *controllable* if and only if $\mathcal{C}$ is full rank or $\mathcal{C}'\mathcal{C}$ is invertible.

The key idea behind the concept of controllability of a system is as follows. Let $\mathcal{E}_{t-1} = (\ldots, \varepsilon_{t-2}, \varepsilon_{t-1})'$ be the history of the noise process at time t and suppose that we want the state to reach a particular value x_t. The question now is whether we can choose an adequate sequence $\mathcal{E}_{t-1}$ to achieve that goal. In order to answer this question, we may write the state at time t as

$$\begin{aligned} x_t &= H\varepsilon_{t-1} + FH\varepsilon_{t-2} + F^2H\varepsilon_{t-3} + \cdots, \\ &= \mathcal{C}\mathcal{E}_{t-1}. \end{aligned}$$

Thus, if the system is controllable, $\mathcal{C}'\mathcal{C}$ is full rank and we may write

$$\mathcal{E}_{t-1} = (\mathcal{C}'\mathcal{C})^{-1}\mathcal{C}'x_t.$$

For a finite-dimensional Hilbert space $\mathcal{H}$, the Cayley-Hamilton theorem establishes that the observability and the controllability operators may be written as

$$\begin{aligned} \mathcal{O} &= (G', F'G', \ldots, F'^{n-1}G')', \\ \mathcal{C} &= (H, FH, \ldots, F^{n-1}H), \end{aligned}$$

where $n = \text{rank}(\mathcal{O}) = \text{rank}(\mathcal{C})$.

2.1.5 Minimality

A state space system is said to be *minimal* if and only if F is of minimal dimension among all representations of the regular linear process (2.3).

If the system is infinite dimensional, the problem of finding minimal representations could be irrelevant. However, for finite-dimensional systems minimality is highly relevant since a state space representation with the smallest dimension may be easier to interpret or easier to handle numerically. The following result specifies when a state space system has the smallest dimension; see Theorem 2.3.3 of Hannan and Deistler (1988).

Theorem 2.2. *A state space system is minimal if and only if it is observable and controllable.*

2.2 REPRESENTATIONS OF LINEAR PROCESSES

A regular linear process may be represented in many different forms, for instance, as a state space system, a Wold decomposition, or by its Hankel matrix. To a large extent, these representations are equivalent and a key issue is how to pass from one representation to another. We have to keep in mind, however, that in most cases these representations are not unique.

2.2.1 State Space Form to Wold Decomposition

As described in Subsection 2.1.2, given a state space system (2.1)-(2.2) with the condition that the sequence $\{GF^jH\}$ is square summable, we may find the Wold representation (2.3) by defining the coefficients $\psi_j = GF^{j-1}H$. Observe that the strong or weak stability of F is not sufficient for assuring the above condition. However, square summability is guaranteed if F is exponentially stable; see Problem 2.5.

2.2.2 Wold Decomposition to State Space Form

A state space representation of the process (2.3) can be specified by the state

$$x_t = [\, y(t|t-1) \quad y(t+1|t-1) \quad y(t+2|t-1) \quad \cdots]', \tag{2.4}$$

where $y(t+j|t-1) = E[y_{t+j}|y_{t-1}, y_{t-2}, \ldots]$ and

$$F = \begin{bmatrix} 0 & 1 & 0 & 0 & 0 & \cdots \\ 0 & 0 & 1 & 0 & 0 & \cdots \\ 0 & 0 & 0 & 1 & 0 & \\ \vdots & \vdots & & & \ddots & \end{bmatrix}, \tag{2.5}$$

$$H = [\, \psi_1 \quad \psi_2 \quad \psi_3 \quad \cdots]', \tag{2.6}$$

$$G = [\, 1 \quad 0 \quad 0 \quad \cdots], \tag{2.7}$$

$$y_t = Gx_t + \varepsilon_t. \tag{2.8}$$

Note that the operator $F : \ell_2 \to \ell_2$ is strongly stable since for all $x \in \ell_2$ we have $\|F^n x\|^2 = \sum_{j=n+1}^{\infty} x_j^2$. Thus, $F^n x$ converges to zero as n tends to ∞.

2.2.3 Hankel Operator to State Space Form

Let A be a linear operator that selects rows of $\mathbf{H}$ such that $\mathbf{H}_0 = A\mathbf{H}$ consists of the basis rows of $\mathbf{H}$. Thus, $\mathbf{H}_0$ is full rank and consequently $\mathbf{H}_0\mathbf{H}_0'$ is invertible. Given the Hankel representation, a state space system can be specified by the state vector

$$x_t = \mathbf{H}_0(\varepsilon_{t-1}, \varepsilon_{t-2}, \ldots)',$$

and the system operators

$$F = A \begin{bmatrix} \psi_2 & \psi_3 & \psi_4 & \cdots \\ \psi_3 & \psi_4 & \psi_5 & \cdots \\ \psi_4 & \psi_5 & \psi_6 & \cdots \\ \vdots & \vdots & \vdots & \end{bmatrix} \mathbf{H}_0'(\mathbf{H}_0\mathbf{H}_0')^{-1},$$

$$H = A(\psi_1, \psi_2, \ldots)',$$

and

$$G = (\psi_1, \psi_2, \ldots)\mathbf{H}_0'(\mathbf{H}_0\mathbf{H}_0')^{-1}.$$

Let $e_j = (0, \ldots, 0, 1, 0, 0, \ldots)$ where the 1 is located at the jth position. Observe that by induction we can prove that $F^{j-1}\mathbf{H}_0 e_1 = \mathbf{H}_0 e_j$. For $j = 1$ is trivial. Suppose that the assertion is valid for j, we will prove that the formula holds for $j + 1$:

$$\begin{aligned} F^j \mathbf{H}_0 e_1 &= F\mathbf{H}_0 e_j = A \begin{bmatrix} \psi_2 & \psi_3 & \psi_4 & \cdots \\ \psi_3 & \psi_4 & \psi_5 & \cdots \\ \psi_4 & \psi_5 & \psi_6 & \cdots \\ \vdots & \vdots & \vdots & \end{bmatrix} e_j \\ &= A \begin{bmatrix} \psi_{j+1} \\ \psi_{j+2} \\ \psi_{j+3} \\ \vdots \end{bmatrix} = A\mathbf{H} e_{j+1} = \mathbf{H}_0 e_{j+1}. \end{aligned}$$

Therefore,

$$\begin{aligned} GF^{j-1}H &= (\psi_1, \psi_2, \ldots)\mathbf{H}_0'(\mathbf{H}_0\mathbf{H}_0')^{-1}F^{j-1}\mathbf{H}_0 e_1 \\ &= (\psi_1, \psi_2, \ldots)\mathbf{H}_0'(\mathbf{H}_0\mathbf{H}_0')^{-1}\mathbf{H}_0 e_j. \end{aligned}$$

On the other hand, since $(\psi_1, \psi_2, \ldots)$ belongs to span of $\mathbf{H}$, it may be written as $(\psi_1, \psi_2, \ldots) = b\mathbf{H}_0$. Thus,

$$GF^{j-1}H = b\mathbf{H}_0 e_j = (\psi_1, \psi_2, \ldots)e_j = \psi_j.$$

2.3 ESTIMATION OF THE STATE

Usually, the state vector in the system (2.1)-(2.2) is not directly observed. Sometimes, it may represent a physical variable, an underlying economic dynamics, an instrumental variable, and the like. On the other hand, we do observe the process $\{y_t\}$. Thus, a relevant problem is estimating the *unobserved* state x_t from the *observed* process $\{y_t\}$.

Depending on what information is available, we may consider the following three situations. When the history of the process $(\ldots, y_{t-2}, y_{t-1})$ is available, we talk about the *prediction* of the state x_t. If the history and the present of the process $(\ldots, y_{t-1}, y_t)$ is available, then we have the problem of *filtering* the state x_t. Finally, if a whole trajectory of the process is available $(\ldots, y_{t-1}, y_t, y_{t+1}, \ldots)$, then the problem is to find a *smoother* of the state x_t. These three estimators of the state have their respective finite past or finite trajectory counterparts which are defined analogously.

In what follows we summarize the Kalman recursive equations to find state predictors, filters, and smoothers. The main purpose of these equations is to simplify the numerical calculation of the estimators. Since in practice we usually have only a finite stretch of data, we focus our attention on projections onto subspaces generated by a finite trajectories of the process $\{y_t : t \in \mathbb{Z}\}$.

2.3.1 State Predictor

Let $\widehat{x}_t$ be the projection of the state x_t onto $\overline{\text{sp}}\{y_s : 1 \leq s \leq t-1\}$ and let $\Omega_t = E[(x_t - \widehat{x}_t)(x_t - \widehat{x}_t)']$ be the state error variance, with $\widehat{x}_1 = 0$ and $\Omega_1 = E[x_1 x_1']$. Then, the state predictor $\widehat{x}_t$ is given by the following recursive equations for $t \geq 1$:

$$\Delta_t = G\Omega_t G' + \sigma^2, \tag{2.9}$$

$$K_t = (F\Omega_t G' + \sigma^2 H)\Delta_t^{-1}, \tag{2.10}$$

$$\Omega_{t+1} = F\Omega_t F' + \sigma^2 HH' - \Delta_t K_t K_t', \tag{2.11}$$

$$\nu_t = y_t - G\widehat{x}_t, \tag{2.12}$$

$$\widehat{x}_{t+1} = F\widehat{x}_t + K_t \nu_t. \tag{2.13}$$

K_t is called the *Kalman gain* and $\{\nu_t\}$ is called the *innovation sequence*, since it is an orthogonal process representing that part of the observation y_t which cannot be predicted from its past.

2.3.2 State Filter

Let $\widehat{x}_{t|t}$ be the projection of the state x_t onto the subspace $\overline{\text{sp}}\{y_s : 1 \leq s \leq t\}$ and let $\Omega_{t|t} = E[(x_t - \widehat{x}_{t|t})(x_t - \widehat{x}_{t|t})']$ be its error variance, with $\widehat{x}_{1|1} = 0$. Then the state filter $\widehat{x}_{t|t}$ is given by the following recursive equations for $t \geq 1$:

$$\begin{aligned}
\Omega_{t|t} &= \Omega_t - \Omega_t G' \Delta_t^{-1} G \Omega_t, \\
\widehat{x}_{t|t} &= \widehat{x}_t + \Omega_t G' \Delta_t^{-1} \nu_t.
\end{aligned}$$

2.3.3 State Smoother

Let $\widehat{x}_{t|s}$ be the projection of the state x_t onto the subspace $\overline{\text{sp}}\{y_j : 1 \leq j \leq s\}$. The state smoother $\widehat{x}_{t|n}$ is given by the following recursive equations for $s \geq t$:

$$\begin{aligned}
\widehat{x}_{t|s} &= \widehat{x}_{t|s-1} + A_{t,s} G' \Delta_s^{-1} \nu_s, \\
A_{t,s+1} &= A_{t,s}(F - K_s \Delta_s^{-1} G)'.
\end{aligned}$$

The state smoother error variance is obtained from the equation

$$\Omega_{t|s} = \Omega_{t|s-1} - A_{t,s} G \Delta_s^{-1} G A_{t,s}',$$

with initial conditions $A_{t,t} = \Omega_{t|t-1} = \Omega_t$ and $\widehat{x}_{t|t-1} = \widehat{x}_t$ from (2.11) and (2.13), respectively.

2.3.4 Missing Observations

When the series has missing values, the Kalman prediction equations (2.11)–(2.13) must be modified as follows. If the observation y_t is missing, then

$$\begin{aligned}\Omega_{t+1} &= F\Omega_t F' + \sigma^2 HH', \\ \nu_t &= 0, \\ \widehat{x}_{t+1} &= F\widehat{x}_t.\end{aligned}$$

This means that the lack of observation y_t affects the estimation of the state at time $t+1$ making the innovation term zero and increasing the state prediction error variance with respect to the known y_t case since the subtracting term $\Delta_t K_t K_t'$ appearing in (2.11) is absent in the modified equations.

Furthermore, when y_t is missing, the modified Kalman filtering equations are

$$\begin{aligned}\Omega_{t|t} &= \Omega_t, \\ \widehat{x}_{t|t} &= \widehat{x}_t,\end{aligned}$$

and the state smoother equations become

$$\begin{aligned}\widehat{x}_{t|s} &= \widehat{x}_{t|s-1}, \\ A_{t,s+1} &= A_{t,s}F', \\ \Omega_{t|s} &= \Omega_{t|s-1}.\end{aligned}$$

More details about this modifications can be found in Subsection 11.2.4.

2.3.5 Steady State System

It is natural to ask whether the Kalman recursions converge to a well-defined limit as time goes to infinity. If the limit exists, then we said that the system has reached its steady state specified by the equations

$$\begin{aligned}\Delta &= G\Omega G' + \sigma^2, \\ K &= (F\Omega G' + \sigma^2 H)\Delta^{-1}, \\ \Omega &= F\Omega F' + \sigma^2 HH' - \Delta KK'.\end{aligned}$$

It would be desirable, at least in theory, that as we accumulate more and more observations $y_1, y_2, \ldots$, the variance of the state estimation error decreases to zero. However, the stability of F is not sufficient to ensure this limit. Other conditions must be satisfied as well. For instance, multiplying equation (2.2) by H and subtracting it from (2.1) we have

$$(I - \Phi B)x_{t+1} = Hy_t, \tag{2.14}$$

where $\Phi = F - HG$. In order to obtain x_{t+1} in terms of y_t, the operator $(1 - \Phi z)$ must to be invertible for $|z| \leq 1$ so that we can write

$$x_{t+1} = (I - \Phi B)^{-1} Hy_t.$$

Observe that if $(1 - zF)$ is invertible for $|z| \leq 1$, then we may write

$$\begin{aligned}\det(I - z\Phi) &= \det(I - zF)\det[I + z(I - zF)^{-1}HG] \\ &= \det(I - zF)[I + zG(I - zF)^{-1}H] \\ &= \det(I - zF)\psi(z),\end{aligned}$$

where $\psi(z) = 1 + \psi_1 z + \psi_2 z^2 + \cdots$ with $\psi_j = GF^{j-1}H$ as in Subsection 2.1.2. Thus, $(I - z\Phi)$ is invertible for $|z| \leq 1$ if $(I - zF)$ and $\psi(z)$ are both invertible for $|z| \leq 1$.

■ EXAMPLE 2.1

Consider the following state space system:

$$x_{t+1} = \phi x_t + \varepsilon_t, \tag{2.15}$$

$$y_t = x_t + \varepsilon_t, \tag{2.16}$$

where ε_t is white noise with unit variance. If $|\phi| < 1$, then the system is exponentially stable and the process y_t is causal and stationary with Wold expansion:

$$y_t = \sum_{j=0}^{\infty} \psi_j \varepsilon_{t-j} = \psi(B)\varepsilon_t,$$

where $\psi(z) = 1 + \psi_1 z + \psi_2 z^2 + \psi_3 z^3 + \cdots = 1 + z + \phi z^2 + \phi^2 z^3 + \cdots$.

On the other hand, for $|\phi| < 1$ we may write

$$x_t = \sum_{j=0}^{\infty} \phi^j \varepsilon_{t-1-j},$$

so that $\mathrm{Var}(x_t) = (1 - \phi^2)^{-1}$.

For this state space system, $\Phi = \phi - 1$ and if $|\phi - 1| < 1$, then the operator $(1 - z\Phi)$ is invertible for $|z| \leq 1$, and we may write the state x_t in terms of the observation sequence $\{y_t\}$ as follows:

$$x_t = \sum_{j=0}^{\infty} (\phi - 1)^j y_{t-1-j}.$$

The Kalman recursion equation for the state prediction error variance Ω_t is

$$\Omega_{t+1} = (\phi - 1)^2 \frac{\Omega_t}{\Omega_t + 1}.$$

Suppose that $\beta = (\phi - 1)^{-2}$ and $\alpha_t = \Omega_t^{-1}$ with $\alpha_1 = 1 - \phi^2$. Then the above equation becomes

$$\alpha_{t+1} = \beta(1 + \alpha_t),$$

with the solution

$$\alpha_t = \beta^{t-1}\alpha_1 + \sum_{j=1}^{t-1} \beta^j.$$

For $\beta > 1$, that is, $0 < \phi < 2$, $\alpha_t \to \infty$ as $t \to \infty$ and therefore $\Omega_t \to 0$ as $t \to \infty$. Hence, $\Delta_t \to 1$ and $K_t \to 1$ as $t \to \infty$. But, recall that in order to have finite state variance we also need the restriction $-1 < \phi < 1$.

On the other hand, for $0 < \beta < 1$, that is, $\phi < 0$ or $\phi > 2$, we have that

$$\Omega_t \to \frac{1-\beta}{\beta} = \phi(\phi - 2),$$

as $t \to \infty$ which means that the state cannot be consistently estimated (e.g., $\|x_t - \widehat{x}_t\| \to 0$ as $t \to \infty$) even if we know the whole trajectory of the observation process $\{y_t\}$. Besides,

$$\Delta_t \quad \to \quad \phi(\phi - 2) + 1, \tag{2.17}$$

$$K_t \quad \to \quad \frac{\phi^2(\phi - 2) + 1}{\phi(\phi - 2) + 1}, \tag{2.18}$$

as $t \to \infty$.

Figure 2.1 displays the evolution of Δ_t, K_t, and Ω_t for $\phi = 0.1$, $\sigma^2 = 1$, and $t = 1, \ldots, 25$. Observe that as indicated by formulas (2.17) and (2.18) for $0 < \phi < 1$, both Δ_t and K_t tend to 1 while $\Omega_t \to 0$ as $t \to \infty$.

On the other hand, Figure 2.2 shows the evolution of Δ_t, K_t, and Ω_t for $\phi = -0.1, \sigma^2 = 1$, and $t = 1, \ldots, 25$. Note that in this case, as indicated by the previous results for $-1 < \phi < 0$, $\Delta_t \to 1.21$, $K_t \to 0.809$, and $\Omega_t \to 0.210$ as $t \to \infty$.

2.3.6 Prediction of Future Observations

The best linear forecast of the observation y_{t+h}, for $h \geq 0$, given its finite past $y_1, \ldots, y_{t-1}$ is readily obtained from the state predictor $\widehat{x}_{t+h}$ as

$$\widehat{y}_{t+h} = G\widehat{x}_{t+h},$$

since the sequence $\varepsilon_{t+1}, \varepsilon_{t+2}, \ldots$ is orthogonal to the space $\overline{\text{sp}}\{y_1, \ldots, y_{t-1}\}$. In turn, the h-step forward state predictor is given by

$$\widehat{x}_{t+h} = F^h \widehat{x}_t.$$

Consequently, we conclude that

$$\widehat{y}_{t+h} = GF^h \widehat{x}_t,$$

with h-step prediction error variance

$$\text{Var}(y_{t+h} - \widehat{y}_{t+h}) = \Delta_{t+h} = G\Omega_t^h G' + \sigma^2,$$

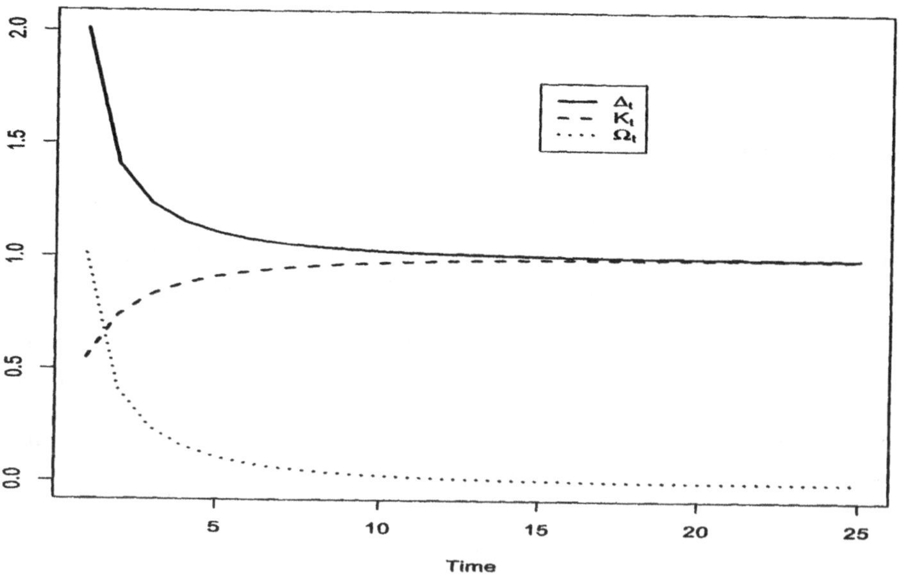

Figure 2.1 State space system example: Evolution of Δ_t (solid line), K_t (broken line), and Ω_t (dotted line) for $\phi = 0.1$, $\sigma^2 = 1$, and $t = 1, \ldots, 25$.

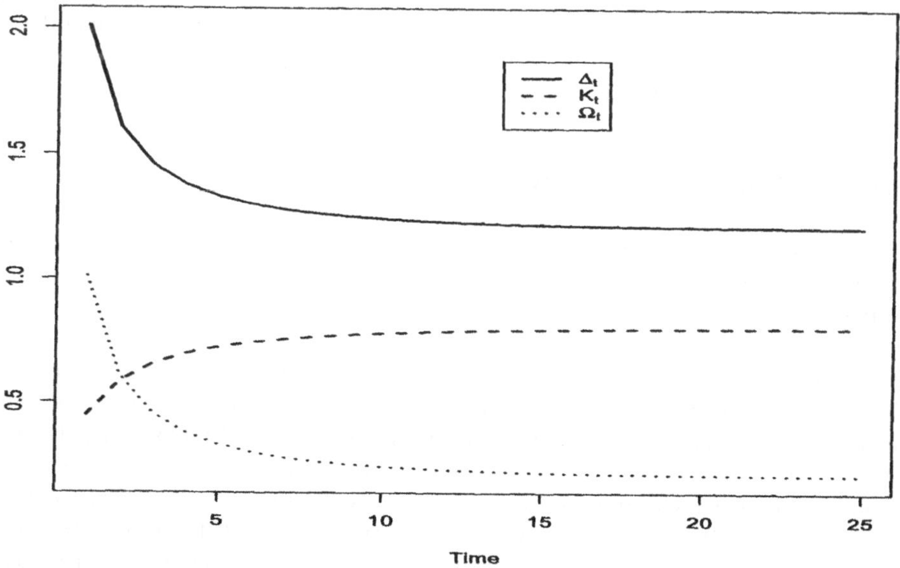

Figure 2.2 State space system example: Evolution of Δ_t (solid line), K_t (broken line), and Ω_t (dotted line) for $\phi = -0.1$, and $t = 1, \ldots, 25$.

where

$$\Omega_t^h = F^h \Omega_t F'^h + \sigma^2 \sum_{j=0}^{h-1} F^j H H' F'^j.$$

2.4 EXTENSIONS

The state space system setup described by (2.1)–(2.2) may be extended to incorporate an exogenous process $\{z_t\}$. This sequence is usually a deterministic component such as a trend or cycle. An extended state space system may be written as

$$\widetilde{x}_{t+1} = F\widetilde{x}_t + Lz_t + H\varepsilon_t, \tag{2.19}$$

$$\widetilde{y}_t = G\widetilde{x}_t + \varepsilon_t. \tag{2.20}$$

The extended model described by equations (2.19)–(2.20) can be modified to fit into the simpler structure (2.1)–(2.2) as follows. Let us define the variable

$$w_t = \sum_{t=1}^{t-1} F^{j-1} L z_{t-j},$$

for $t \geq 1$ and $w_0 = 0$. With this definition we have that

$$w_{t+1} = Fw_t + Lz_t. \tag{2.21}$$

Let $x_t = \widetilde{x}_t - w_t$ be the modified state vector and $y_t = \widetilde{y}_t - Gw_t$ be the modified observation at time t. Then, from (2.19)–(2.20) and (2.21) we conclude that the pair (x_t, y_t) satisfies the system (2.1)–(2.2).

2.5 BIBLIOGRAPHIC NOTES

The book by Hannan and Deistler (1988) gives an excellent theoretical treatment of the linear systems. In particular, the relationships among the different representations of these processes are analyzed in full detail. Most of the theorems presented in this chapter are proved in that book.

The monographs by Anderson and Moore (1979), Harvey (1989), Aoki (1990), and Durbin and Koopman (2001) offer an excellent overview of both methodological and applied aspects of state space modeling. Besides, Chapter 12 of Brockwell and Davis (1991) gives a very good introduction to state space systems in a finite dimensional context, including descriptions of Kalman recursions and treatment of missing values.

The Kalman filter equations were introduced by Kalman (1961) and Kalman and Bucy (1961). Applications of state space systems to the analysis of time series data are reported in fields as diverse as aeronautics [e.g., Kobayashi and Simon (2003)] and oceanography [e.g., Bennett (1992), Chan, Kadane, Miller, and Palma (1996)].

In some texts, the term *reachability* is used instead of *controllability*. Several definitions of stability of state space systems in a very general context of infinite-dimensional systems is given in Curtain and Zwart (1995).

Fitting time series models with missing data has been extensively discussed in the state space context. For ARMA and ARIMA models, see, for instance, Jones (1980), Ansley and Kohn (1983), Kohn and Ansley (1986), and Bell and Hillmer (1991). On the other hand, for ARFIMA models see, for example, Palma and Chan (1997), Palma (2000), and Ray and Tsay (2002).

Problems

2.1 Consider the following state space system:

$$\begin{aligned} x_{t+1} &= \begin{bmatrix} 1 & 0 & 0 \\ 1 & 0 & 0 \\ 0 & 1 & 0 \end{bmatrix} x_t + \begin{bmatrix} 1 \\ 0 \\ 0 \end{bmatrix} \varepsilon_t, \\ y_t &= \begin{bmatrix} 1 & 0 & 0 \end{bmatrix} x_t. \end{aligned}$$

a) Is this system stable?

b) Verify whether this state space system is observable.

c) Verify whether this state space system is controllable.

d) What can you conclude about the system?

2.2 Find a minimal state space representation of the process $y_t = \varepsilon_t + \theta\varepsilon_{t-1}$ where $|\theta| < 1$ and ε_t is white noise.

2.3 Consider the regular linear process $y_t = \phi y_{t-1} + \varepsilon_t$ where $|\phi| < 1$ and ε_t is white noise.

a) Find a minimal state space representation of the process y_t.

b) Verify that the system is stable.

c) Find the Hankel operator representing this process.

d) What is the rank of this Hankel operator?

2.4 Consider the state space system

$$\begin{aligned} x_{t+1} &= \begin{bmatrix} \theta_1 & 0 & 0 \\ 0 & \theta_2 & 0 \\ 0 & 0 & \theta_3 \end{bmatrix} x_t + \begin{bmatrix} 1 \\ 0 \\ 0 \end{bmatrix} \varepsilon_t, \\ y_t &= \begin{bmatrix} 1 & 1 & 1 \end{bmatrix} x_t. \end{aligned}$$

a) For which values of the parameter $\boldsymbol{\theta} = (\theta_1, \theta_2, \theta_3)$ is this system stable?

b) Assume that ε_t is an independent and identically distributed sequence $N(0, 1)$. Simulate several trajectories of the system for a sample size $n = 1000$ and different parameters $\boldsymbol{\theta}$.

c) Implement computationally the Kalman recursions for this state space system.

2.5 Show that if F is exponentially stable, then the sequence $\{GF^jH\}$ is square summable.

2.6 Show the following equations using the definitions in this chapter:

$$\begin{aligned} \det(I - z\Phi) &= \det(I - zF)\det[I + z(I - zF)^{-1}HG], \\ &= \det(I - zF)[I + zG(I - zF)^{-1}H], \\ &= \det(I - zF)\psi(z). \end{aligned}$$

2.7 Consider a finite-dimensional state space system where $x_t \in \mathbb{R}^n$. Write a computer program implementing the Kalman recursion equations (2.9)–(2.13).

2.8 Given a sample $\{y_1, \ldots, y_n\}$, verify that $\Omega_{t|n} \leq \Omega_{t|t} \leq \Omega_t$ for all $t \leq n$, where the matrix inequality $A \leq B$ means that $x'(B - A)x \geq 0$ for all x.

2.9 Consider the following state space system:

$$\begin{aligned} x_{t+1} &= \phi x_t + \varepsilon_t, \\ y_t &= \theta x_t + \varepsilon_t, \end{aligned}$$

where ε_t is white noise with unit variance.

a) For which values of the parameter $\boldsymbol{\theta} = (\phi, \theta)$ is this system stable?
b) For which values of $\boldsymbol{\theta}$ is the system observable or controllable?
c) For which values of $\boldsymbol{\theta}$ are the Kalman recursions stable?
d) Assume that ε_t is an independent and identically distributed sequence $N(0, 1)$. Simulate several trajectories of the system for a sample size $n = 1000$ and different parameters $\boldsymbol{\theta}$.

2.10 Consider the following state space system for $x_t \in \mathbb{R}^n$ with $n \geq 2$:

$$\begin{aligned} x_{t+1} &= Fx_t + H\varepsilon_t, \\ y_t &= Gx_t + \varepsilon_t, \end{aligned}$$

where

$$\begin{aligned} F &= \begin{bmatrix} 0 & 1 & 0 & 0 & \cdots & 0 \\ 0 & 0 & 1 & 0 & \cdots & 0 \\ 0 & 0 & 0 & 1 & \cdots & 0 \\ \vdots & \vdots & & & \ddots & \\ 0 & 0 & 0 & \cdots & 0 & 1 \\ 0 & 0 & 0 & \cdots & 0 & 0 \end{bmatrix}, \\ G &= [\psi_n \quad \psi_{n-1} \quad \psi_{n-2} \quad \cdots], \\ H &= [1 \quad 0 \quad 0 \quad \cdots \quad 0]', \end{aligned}$$

and the coefficients ψ_j are given by the expansion (2.3).

a) Verify that this system is strongly stable.

b) Verify that $\|F\| = \sup_x \|Fx\|/\|x\| = 1$.

c) Is this system exponentially stable?

d) Find the observability matrix for this system $\mathcal{O}$ and verify that it is of full rank if and only if $\psi_n \neq 0$.

2.11 Consider the AR(p) process given by the equation

$$y_t - \phi_1 y_{t-1} \cdots - \phi_p y_{t-p} = \varepsilon_t.$$

a) Show that by defining the state vector $x_t = (y_{t+1-p}, \ldots, y_t)'$, the AR($p$) process may be written in terms of the following state space representation:

$$\begin{aligned} x_{t+1} &= Fx_t + H\varepsilon_{t+1}, \\ y_t &= Gx_t, \end{aligned}$$

where

$$\begin{aligned} F &= \begin{bmatrix} 0 & 1 & 0 & 0 & \cdots & 0 \\ 0 & 0 & 1 & 0 & \cdots & 0 \\ 0 & 0 & 0 & 1 & \cdots & 0 \\ \vdots & \vdots & & & \ddots & \\ 0 & 0 & 0 & \cdots & 0 & 1 \\ \phi_p & \phi_{p-1} & \phi_{p-2} & \cdots & \phi_2 & \phi_1 \end{bmatrix}, \\ G &= [0 \ \ 0 \ \ 0 \ \ \cdots \ \ 0 \ \ 1], \\ H &= [0 \ \ 0 \ \ 0 \ \ \cdots \ \ 0 \ \ 1]'. \end{aligned}$$

b) Write down the observation matrix $\mathcal{O}$ for this state space model. Is this system observable?

c) Find the controllability matrix $\mathcal{C}$ for this state space representation and check whether this system is controllable.

d) Verify whether this system converges to its steady state or not. If yes, write down the steady state system equations.

2.12 Given the state space system

$$\begin{aligned} x_{t+1} &= Fx_t + H\varepsilon_t, \\ y_t &= Gx_t + \varepsilon_t, \end{aligned}$$

where $\{\varepsilon_t\}$ is a white noise sequence, show that the product of the observation matrix and the controllability matrix yields

$$\mathcal{OC} = \begin{pmatrix} GH & GFH & GF^2H & GF^3H & \cdots \\ GFH & GF^2H & GF^3H & GF^4H & \cdots \\ GF^2H & GF^3H & GF^4H & GF^5H & \cdots \\ \vdots & \vdots & \vdots & & \end{pmatrix}.$$

Is this a Hankel operator?

2.13 Suppose that the state transition matrix F satisfies $\|F^j\| \leq ce^{-\alpha j}$ so that the corresponding state space system is strongly stable.

a) Verify that in this case,

$$\sum_{j=0}^{n} F^j z^j,$$

converges for all $|z| \leq 1$.

b) Show that

$$(I - zF)^{-1} = \sum_{j=0}^{\infty} F^j z^j,$$

for all $|z| \leq 1$.

2.14 Consider the following state space model:

$$\begin{aligned} x_{t+1} &= Fx_t + H\varepsilon_t, \\ y_t &= Gx_t + \varepsilon_t, \end{aligned}$$

and let $\psi(z)$ be the operator $\psi(z) = 1 + \psi_1 z + \psi_2 z^2 + \cdots$, where $\psi_j = GF^{j-1}H$ and $|z| \leq 1$.

a) Prove that $\psi(z)$ may be written as

$$\psi(z) = 1 + G(I - zF)^{-1}Hz.$$

b) Assume that $F = 0$. Show that if $|GH| \leq 1$, then $\psi(z)$ is invertible for $|z| \leq 1$.

c) If $F = 0$ and $|GH| \leq 1$, verify that the state x_t may be expressed as

$$x_t = Hy_{t-1} + H\sum_{j=1}^{\infty}(-GH)^j y_{t-1-j}.$$

2.15 Consider the linear state space model where the transition matrix depends on time

$$\begin{aligned} x_{t+1} &= F_t x_t + H\varepsilon_t, \\ y_t &= Gx_t + \varepsilon_t, \end{aligned}$$

for $t \geq 1$, ε_t is a white noise sequence and x_0 is the initial state.

a) Show that the state at time $t+1$ may be written as

$$x_{t+1} = \left(\prod_{j=1}^{t} F_j\right) x_0 + \sum_{j=0}^{t} \varphi_j H\varepsilon_{t-j},$$

and find the coefficients φ_j.

b) Let $z_n = \sum_{j=1}^{n} \log \|F_j\|$ and assume that the limit

$$\gamma = \lim_{n\to\infty} z_n/n$$

exists. Prove that if $\gamma < 0$, then

$$\lim_{t\to\infty} \left(\prod_{j=1}^{t} F_j\right) x_0 = 0, \tag{2.22}$$

in probability.

c) Suppose that $\|F_t\| \leq e^{-\alpha t}$ where α is a positive constant. Show that (2.22) holds in this situation.

d) Assume that $\|F_t\| \leq t^{-\beta}$ where β is a positive constant. Prove that the limit (2.22) holds under these circumstances.

CHAPTER 3

LONG-MEMORY PROCESSES

In the previous chapters we have discussed the theory of stationary linear processes and their Wold and state space representations. In this chapter we focus our attention on a particular class of linear time series called *long-memory* or *long-range-dependent* processes. Long-range dependence may be defined in many ways. Nevertheless, as pointed out by Hall (1997), the original motivation for the concept of long memory is closely related to the estimation of the mean of a stationary process. If the autocovariances of a stationary process are absolutely summable, then the sample mean is root-n consistent, where n is the sample size. This is the case, for instance, for sequences of independent and identically distributed random variables or Markovian processes. Generally speaking, these processes are said to have *short memory*. On the contrary, a process has *long memory* if its autocovariances are not absolutely summable. Formal definitions of long-memory processes are discussed in Section 3.1 and *autoregressive fractionally integrated moving-average* models are introduced in Section 3.2. Relevant issues such as stationarity, causality, and invertibility of these processes are also addressed in this section. Another well-known long-range-dependent process, the *fractional Gaussian noise*, is introduced in Section 3.3. Some technical lemmas are given in Section 3.4 and further readings on this subject are suggested in Section 3.5. This chapter ends with a list of proposed problems.

Long-Memory Time Series. By Wilfredo Palma

3.1 DEFINING LONG MEMORY

Let $\gamma(h) = \langle y_t, y_{t+h} \rangle$ be the autocovariance function at lag h of the stationary process $\{y_t : t \in \mathbb{Z}\}$. A usual definition of long memory is that

$$\sum_{h=-\infty}^{\infty} |\gamma(h)| = \infty. \tag{3.1}$$

However, there are alternative definitions. In particular, long memory can be defined by specifying a hyperbolic decay of the autocovariances

$$\gamma(h) \sim h^{2d-1}\ell_1(h), \tag{3.2}$$

as $h \to \infty$, where d is the so-called *long-memory parameter* and $\ell_1(\cdot)$ is a slowly varying function. Recall that a positive measurable function defined on some neighborhood $[a, \infty)$ of infinity is said to be *slowly varying* in Karamata's sense if and only if for any $c > 0$, $\ell(cx)/\ell(x)$ converges to 1 as x tends to infinity. Examples of slowly varying functions are $\ell(x) = \log(x)$ and $\ell(x) = b$, where b is a positive constant. Hereafter, the notation $x_n \sim y_n$ means that $x_n/y_n \to 1$ as $n \to \infty$, unless specified otherwise.

Another widely used definition of strong dependence on the spectral domain is

$$f(\lambda) \sim |\lambda|^{-2d}\ell_2(1/|\lambda|), \tag{3.3}$$

for λ in a neighborhood of zero and $\ell_2(\cdot)$ is a slowly varying function.

Furthermore, an alternative definition of long-memory behavior is based directly on the Wold decomposition of the process (1.1)

$$\psi_j \sim j^{d-1}\ell_3(j), \tag{3.4}$$

for $j > 0$, where $\ell_3(\cdot)$ is a slowly varying function.

Unfortunately, unless further conditions are imposed, these four definitions are not necessarily equivalent. Some relationships among these definitions of long memory are established in the next result. Two technical lemmas needed to prove this theorem are provided in Section 3.4.

Theorem 3.1. *Let $\{y_t\}$ be a stationary regular process with Wold expansion* (1.1)

$$y_t = \sum_{j=0}^{\infty} \psi_j \varepsilon_{t-j},$$

where $\{\varepsilon_t\}$ is a white noise sequence. Assuming that $0 < d < \frac{1}{2}$ we have

(a) *If the process $\{y_t\}$ satisfies* (3.4), *it also satisfies* (3.2).

(b) *If the process $\{y_t\}$ satisfies* (3.2), *it also satisfies* (3.1).

(c) *If the function $\ell_1(\cdot)$ is quasi-monotone slowly varying, then* (3.2) *implies* (3.3).

Proof. Part (a). From the Wold decomposition (1.1) the autocovariance function of the process $\{y_t\}$ can be written as

$$\gamma(h) = \sigma^2 \sum_{k=0}^{\infty} \psi_k \psi_{k+|h|}.$$

Thus, by condition (3.4) and Lemma 3.2 we conclude (3.2) with

$$\ell_1(h) = \sigma^2 \ell_3(h)^2 B(1-2d, d).$$

Part (b). For any integers $0 < m < n$ we have that

$$\sum_{h=-\infty}^{\infty} |\gamma(h)| \geq \sum_{h=m}^{n} |\gamma(h)|.$$

Now, from (3.2) and Lemma 3.3(b) we have that for large n

$$\sum_{h=m}^{n} |\gamma(h)| \sim \frac{n^{2d}\ell_1(n)}{2d}. \tag{3.5}$$

Since $d > 0$, an application of Lemma 3.3(a) yields $n^{2d}\ell_1(n) \to \infty$ as n tends to infinity, showing that condition (3.2) implies condition (3.1). Part (c). Since ℓ_1 is quasi-monotone slowly varying, an application of Lemma 3.3(c) yields (3.3) with

$$\ell_2(\lambda) = \frac{\ell_1(\lambda)}{2\Gamma(1-2d)\sin(\pi d)} = \frac{\Gamma(1-d)\Gamma(d)}{2\pi\Gamma(1-2d)}\ell_1(\lambda), \tag{3.6}$$

for $\lambda > 0$; see Problem 3.12. □

Observe that there are several alternative definitions of long memory and of slowly varying functions. In particular, formulas (3.2), (3.3), and (3.4) can accommodate negative asymptotic expressions. However, in absence of some type of asymptotic monotonicity, many results about the relationships among the different definitions of strong dependence may not be valid. This is the case, for example, for oscillating asymptotic expressions such as those including sinusoidal functions. In the next subsection we explore a number of additional definitions of a long-memory process.

3.1.1 Alternative Definitions

A set of alternative definitions of strong dependence is obtained by imposing additional restrictions on functions ℓ_1, ℓ_2, and ℓ_3. In this section we discuss a more restrictive definition of slowly varying function given by Zygmund (1959) and the replacement of slowly varying functions by continuous functions. Note that there are many definitions of a slowly varying function. Up to this point, we have used the definition introduced by Karamata (1930). This definition leads to the following representation theorem; see Theorem 1.3.1 of Bingham, Goldie, and Teugels (1987).

Theorem 3.2. *A function $\ell(\cdot)$ is slowly varying if and only if it may be written in the form*

$$\ell(h) = c(h) \exp\left\{ \int_a^h \frac{\varepsilon(x)}{x} \, dx \right\}, \tag{3.7}$$

for $h \geq a$ with some $a > 0$ where $c(\cdot)$ is a measurable function such that $c(h) \to c > 0$ and $\varepsilon(h) \to 0$ as $h \to \infty$.

On the other hand, Zygmund (1959, p. 186) proposed the following alternative definition: A positive function $b(h)$, $h > h_0$, is slowly varying if and only if for any $\delta > 0$, $b(h)h^{\delta}$ is an increasing and $b(h)h^{-\delta}$ is a decreasing function of h for $h \to \infty$. In particular, this implies that

$$b(kh) \sim b(h), \tag{3.8}$$

as $h \to \infty$ for every fixed positive k.

As established by Bojanić and Karamata (1963), the class of slowly varying functions defined by Zygmund is a particular case of Karamata's class when $c(\cdot)$ is taken to be a constant function in (3.7).

Replacing the functions $\ell_1(\cdot)$ in (3.2) and $\ell_2(\cdot)$ in (3.3) by the slowly varying functions in the Zygmund's sense $b_1(\cdot)$ and $b_2(\cdot)$, respectively, the following definitions of long-range dependence are obtained:

$$\gamma(h) \sim h^{2d-1} b_1(h), \tag{3.9}$$

as $h \to \infty$, and

$$f(\lambda) \sim |\lambda|^{-2d} b_2(1/|\lambda|), \tag{3.10}$$

as $\lambda \to 0$.

Observe that the functions $b_1(\cdot)$ and $b_2(\cdot)$ satisfy a relationship analogous to (3.6). Following Section V.2 of Zygmund (1959), these two definitions are equivalent in the sense specified by the next result.

Theorem 3.3. *(a) If $\{y_t : t \in \mathbb{Z}\}$ is a stationary process satisfying* (3.9), *then its spectral density exists and satisfies* (3.10). *(b) If the spectral density of the stationary process $\{y_t : t \in \mathbb{Z}\}$ satisfies* (3.10), *then its autocovariance function satisfies* (3.9).

Another restriction of the definition of long memory is obtained by replacing the slowly varying functions $b_1(\cdot)$ and $b_2(\cdot)$ by the continuous functions $f_1(\cdot)$ and $f_2(\cdot)$ such that

$$\gamma(h) \sim h^{2d-1} f_1(1/h),$$

as $h \to \infty$, and

$$f(\lambda) \sim |\lambda|^{-2d} f_2(|\lambda|), \tag{3.11}$$

as $\lambda \to 0$, where $f_1(\lambda) \to c_1 > 0$ and $f_2(\lambda) \to c_2 > 0$ as $\lambda \to 0$.

The functions $f_1(\cdot)$ and $f_2(\cdot)$ also satisfy the relationship (3.6). An advantage of these definitions is the possibility of using approximations to model strongly dependent processes through the Stone-Weierstrass theorem, as described in the Section 3.2.

3.1.2 Extensions

Up to this point, the behavior of the spectral density of the process (1.1) outside a neighborhood of zero has not been explicitly specified. This is due primarily to the fact that we have been concerned with the square root-n consistency of the sample mean of the process and its relationship with the pole of the spectrum at frequency zero. However, it is possible to consider the presence of poles located at frequencies different from zero. This is the case, for example, of the long-memory seasonal processes discussed extensively in Chapter 12.

For simplicity, in the remaining of this chapter we assume that the spectral density has only one singularity located at the origin. Thus, the spectrum can be assumed to behave well outside a neighborhood of zero. In particular, we shall assume that the spectral density of the process is bounded and continuous for frequencies away from the origin.

3.2 ARFIMA PROCESSES

A well-known class of long-memory models is the autoregressive fractionally integrated moving-average (ARFIMA) processes introduced by Granger and Joyeux (1980) and Hosking (1981). An ARFIMA process $\{y_t\}$ may be defined by

$$\phi(B)y_t = \theta(B)(1-B)^{-d}\varepsilon_t, \tag{3.12}$$

where $\phi(B) = 1 + \phi_1 B + \cdots + \phi_p B^p$ and $\theta(B) = 1 + \theta_1 B + \cdots + \theta_q B^q$ are the autoregressive and moving-average operators, respectively; $\phi(B)$ and $\theta(B)$ have no common roots, $(1-B)^{-d}$ is a fractional differencing operator defined by the binomial expansion

$$(1-B)^{-d} = \sum_{j=0}^{\infty} \eta_j B^j = \eta(B),$$

where

$$\eta_j = \frac{\Gamma(j+d)}{\Gamma(j+1)\Gamma(d)}, \tag{3.13}$$

for $d < \frac{1}{2}$, $d \neq 0, -1, -2, \ldots$, and $\{\varepsilon_t\}$ is a white noise sequence with finite variance.

More generally, if we define

$$\eta(z) = (1-z)^{-d},$$

then this function is analytic in the open unit disk $\{z : |z| < 1\}$ and analytic in the closed unit disk $\{z : |z| \leq 1\}$ for negative d. Under these circumstances, we can write $\eta(\cdot)$ as the Taylor expansion

$$\eta(z) = \sum_{j=0}^{\infty} \eta_j z^j.$$

3.2.1 Stationarity, Causality, and Invertibility

The next theorem examines the existence of a stationary solution of the ARFIMA process defined by equation (3.12), including its uniqueness, causality, and invertibility. Recall that all these concepts were defined in Chapter 1.

Theorem 3.4. *Consider the ARFIMA process defined by* (3.12). *Assume that the polynomials $\phi(\cdot)$ and $\theta(\cdot)$ have no common zeros and that $d \in (-1, \frac{1}{2})$. Then,*

(*a*) *If the zeros of $\phi(\cdot)$ lie outside the unit circle $\{z\colon |z| = 1\}$, then there is a unique stationary solution of* (3.12) *given by*

$$y_t = \sum_{j=-\infty}^{\infty} \psi_j \varepsilon_{t-j},$$

where $\psi(z) = (1 - z)^{-d}\theta(z)/\phi(z)$.

(*b*) *If the zeros of $\phi(\cdot)$ lie outside the closed unit disk $\{z\colon |z| \leq 1\}$, then the solution $\{y_t\}$ is causal.*

(*c*) *If the zeros of $\theta(\cdot)$ lie outside the closed unit disk $\{z\colon |z| \leq 1\}$, then the solution $\{y_t\}$ is invertible.*

Proof. Part (a). Consider first the fractional noise process $x_t = \sum_{j=0}^{\infty} \eta_j \varepsilon_{t-j}$, where the coefficients η_j are given in (3.13). By (3.17), $\sum_{j=0}^{\infty} \eta_j^2 < \infty$ and therefore the trigonometric polynomial $\sum_{j=0}^{n} \eta_j e^{\dot{\imath}\lambda j}$ converges to $\eta(e^{\dot{\imath}\lambda}) = (1 - e^{\dot{\imath}\lambda})^{-d}$ in the Hilbert space $\mathcal{L}_2(d\lambda)$ defined in Example 1.4. Consequently, $\{x_t\}$ is a well-defined stationary process as in (1.1). Since $\phi(z)$ does not have roots on $|z| = 1$, there is $\delta > 1$ such that we can write the absolutely convergent Laurent series

$$\varphi(z) = \sum_{j=-\infty}^{\infty} \varphi_j z^j = \frac{\theta(z)}{\phi(z)},$$

for $\delta^{-1} < |z| < \delta$. Thus, by Theorem 1.5(a),

$$y_t = \varphi(B)x_t = \sum_{j=-\infty}^{\infty} \varphi_j x_{t-j}$$

is a stationary process. Given that $\{\varphi_j\}$ is absolutely summable and $\{x_t\}$ is stationary, by Theorem 1.5(b) we can write $\{y_t\}$ as

$$y_t = \psi(B)\varepsilon_t = \sum_{j=-\infty}^{\infty} \psi_j \varepsilon_{t-j}, \tag{3.14}$$

where $\psi(B) = \varphi(B)\eta(B)$. Now, premultiplying (3.14) by $\phi(B)$ and applying Theorem 1.5 we get

$$\phi(B)y_t = \phi(B)\varphi(B)\eta(B)\varepsilon_t = \theta(B)\eta(B)\varepsilon_t,$$

showing that $\{y_t\}$ is a stationary process that satisfies (3.12). To prove uniqueness, let $\{w_t\}$ be a stationary process satisfying (3.12), that is,

$$\phi(B)w_t = \theta(B)\eta(B)\varepsilon_t. \tag{3.15}$$

Since $\phi(z) \neq 0$ for $|z| = 1$, we can write the absolutely converging Laurent series

$$\frac{1}{\phi(z)} = \sum_{j=-\infty}^{\infty} h_j z^j = h(z),$$

where $\delta^{-1} < |z| < \delta$, for some $\delta > 1$. Therefore, multiplying (3.15) by $h(B)$ and applying Theorem 1.5 we obtain

$$w_t = h(B)\theta(B)\eta(B)\varepsilon_t = \varphi(B)\eta(B)\varepsilon_t = \psi(B)\varepsilon_t,$$

which coincides with $\{y_t\}$ in the mean square sense.

Part (b). Assume that $\phi(z)$ does not have roots in the closed unit disk $\{z\colon |z| \leq 1\}$. Hence, there exists $\varepsilon > 0$ such that

$$\frac{1}{\phi(z)} = \sum_{j=0}^{\infty} g_j z^j = g(z),$$

for $|z| < 1 + \varepsilon$. Thus, for any $\alpha \in (0, \varepsilon)$, the sequence $g_j(1+\alpha)^j$ converges to zero as $j \to \infty$. Therefore, there exists a constant $K > 0$ such that

$$|g_j| \leq K(1+\alpha)^{-j},$$

for $j \geq 0$. Thus, the sequence $\{g_j\}$ is absolutely summable. Now, an application of Theorem 1.5(a) yields

$$y_t = \alpha(B)x_t,$$

where $\{x_t\}$ is the fractional noise process defined in part (a) and $\alpha(B) = g(B)\theta(B) = \sum_{j=0}^{\infty} \alpha_j B^j$. Since $\{\alpha_j : j \geq 0\}$ is absolutely summable and $\{x_t\}$ is stationary, by Theorem 1.5 the sequence $\{y_t\}$ given by

$$y_t = \psi(B)\varepsilon_t,$$

where $\psi(B) = \alpha(B)\eta(B) = \sum_{j=0}^{\infty} \psi_j B^j$, is a well-defined causal process in $\mathcal{L}_2$.

Part (c). Suppose that $\theta(z) \neq 0$ in the closed unit disk $\{z : |z| \leq 1\}$ and let $w_t = \phi(B)y_t$. By Theorem 1.5(a), $\{w_t\}$ is a well-defined stationary process with spectral density

$$f_w(\lambda) = \frac{\sigma^2}{2\pi}|1 - e^{i\lambda}|^{-2d}|\theta(e^{i\lambda})|^2.$$

Consequently, f_w satisfies the conditions of Lemma 3.1. Therefore, the expansion

$$\varepsilon_t = \sum_{j=0}^{\infty} \tilde{\pi} w_{t-j} = \tilde{\pi}(B)w_t,$$

converges in $\mathcal{L}_2$, where $\tilde{\pi}(B) = 1/[\eta(B)\theta(B)]$. Now, an application of Theorem 1.5(b) yields the $\mathcal{L}_2$-convergent series

$$\varepsilon_t = \sum_{j=0}^{\infty} \pi_j y_{t-j} = \pi(B) y_t,$$

where $\pi(B) = \tilde{\pi}(B)\phi(B)$. □

Remark 3.1. Notice that in Theorem 3.4 $d \in (-1, \frac{1}{2})$, extending the usual range $d \in (-\frac{1}{2}, \frac{1}{2})$ of previous results [see, for example, Hosking (1981, Theorem 2) and Brockwell and Davis (1991, Theorem 13.2.2 and Remark 7)]. Bloomfield (1985, Section 4) established the invertibility of a fractional noise for $d \in (-1, \frac{1}{2})$ and the extension of this result to ARFIMA processes is due to Bondon and Palma (2007).

Remark 3.2. An ARFIMA(p, d, q) process is sometimes defined as

$$\phi(B)(1-B)^d y_t = \theta(B)\varepsilon_t. \tag{3.16}$$

However, as noted in Kokoszka and Taqqu (1995), the solution to this equation may not be unique. For instance, let $\{y_t\}$ be a stationary solution of (3.16) with $d > 0$ and let v be a random variable with finite variance. Then, the stationary process $x_t = y_t + v$ is also a solution of (3.16) since in this case the coefficients $\{\pi_j\}$ of the series $(1-z)^d = \sum_{j=0}^{\infty} \pi_j z^j = \pi(z)$ are absolutely summable and $\pi(1) = 0$.

3.2.2 Infinite AR and MA Expansions

According to Theorem 3.4, under the assumption that the roots of the polynomials $\phi(B)$ and $\theta(B)$ are outside the closed unit disk $\{z \colon |z| \le 1\}$ and $d \in (-1, \frac{1}{2})$, the ARFIMA(p, d, q) process is stationary, causal, and invertible. In this case we can write

$$y_t = (1-B)^{-d}\phi(B)^{-1}\theta(B)\varepsilon_t = \psi(B)\varepsilon_t,$$

and

$$\varepsilon_t = (1-B)^d \phi(B)\theta(B)^{-1} y_t = \pi(B) y_t.$$

The MA(∞) coefficients, ψ_j, and AR(∞) coefficients, π_j, satisfy the following asymptotic relationships:

$$\psi_j = \frac{\theta(1)}{\phi(1)} \frac{j^{d-1}}{\Gamma(d)} + \mathcal{O}\left(j^{-1}\right), \tag{3.17}$$

$$\pi_j = \frac{\phi(1)}{\theta(1)} \frac{j^{-d-1}}{\Gamma(-d)} + \mathcal{O}\left(j^{-1}\right), \tag{3.18}$$

as $j \to \infty$; see Kokoszka and Taqqu (1995, Corollary 3.1) and Problem 3.8.

For a fractional noise process with long-memory parameter d these coefficients are given by

$$\psi_j = \prod_{t=1}^{j} \frac{t-1+d}{t} = \frac{\Gamma(j+d)}{\Gamma(d)\Gamma(j+1)},$$

$$\pi_j = \prod_{t=1}^{j} \frac{t-1-d}{t} = \frac{\Gamma(j-d)}{\Gamma(-d)\Gamma(j+1)},$$

for $j \geq 1$ and $\psi_0 = \pi_0 = 1$.

3.2.3 Spectral Density

Under the conditions of Theorem 3.4, the spectral density of the process (3.12) can be written as

$$\begin{aligned} f(\lambda) &= \frac{\sigma^2}{2\pi}|1-e^{-i\lambda}|^{-2d}\frac{|\theta(e^{-i\lambda})|^2}{|\phi(e^{-i\lambda})|^2} \\ &= \frac{\sigma^2}{2\pi}\left[2\sin\frac{\lambda}{2}\right]^{-2d}\frac{|\theta(e^{-i\lambda})|^2}{|\phi(e^{-i\lambda})|^2}, \end{aligned} \tag{3.19}$$

and it is proved in Problem 3.11 that

$$f(\lambda) \sim \frac{\sigma^2}{2\pi}\frac{|\theta(1)|^2}{|\phi(1)|^2}|\lambda|^{-2d}, \tag{3.20}$$

for $|\lambda| \to 0$.

3.2.4 Autocovariance Function

The autocovariance function (ACF) of the ARFIMA$(0, d, 0)$ process is given by

$$\gamma_0(h) = \sigma^2 \frac{\Gamma(1-2d)}{\Gamma(1-d)\Gamma(d)} \frac{\Gamma(h+d)}{\Gamma(1+h-d)}, \tag{3.21}$$

where $\Gamma(\cdot)$ is the gamma function, and the autocorrelation function is

$$\rho_0(h) = \frac{\Gamma(1-d)}{\Gamma(d)} \frac{\Gamma(h+d)}{\Gamma(1+h-d)}.$$

For the general ARFIMA(p, d, q) process, observe that the polynomial $\phi(B)$ in (3.12) may be written as

$$\phi(B) = \prod_{i=1}^{p}(1-\rho_i B).$$

Assuming that all the roots of $\phi(B)$ have multiplicity one, it can be deduced from (1.5) that

$$\gamma(h) = \sigma^2 \sum_{i=-q}^{q} \sum_{j=1}^{p} \psi(i)\xi_j C(d, p+i-h, \rho_j), \tag{3.22}$$

with

$$\psi(i) = \sum_{k=\max(0,i)}^{\min(q,q+i)} \theta_k \theta_{k-i},$$

$$\xi_j = \left[\rho_j \prod_{i=1}^{p} (1-\rho_i \rho_j) \prod_{m \neq j} (\rho_j - \rho_m) \right]^{-1},$$

and

$$C(d, h, \rho) = \frac{\gamma_0(h)}{\sigma^2} [\rho^{2p} \beta(h) + \beta(-h) - 1], \tag{3.23}$$

where $\beta(h) = F(d+h, 1, 1-d+h, \rho)$ and $F(a, b, c, x)$ is the Gaussian hypergeometric function; see Gradshteyn and Ryzhik (2000, Section 9.1),

$$F(a, b, c, x) = 1 + \frac{a \cdot b}{\gamma \cdot 1} x + \frac{a \cdot (a+1) \cdot b \cdot (b+1)}{\gamma \cdot (\gamma+1) \cdot 1 \cdot 2} x^2 + \cdots$$

It can be shown that (see Problem 3.3)

$$\gamma(h) \sim c_\gamma |h|^{2d-1}, \tag{3.24}$$

as $|h| \to \infty$, where

$$c_\gamma = \frac{\sigma^2}{\pi} \frac{|\theta(1)|^2}{|\phi(1)|^2} \Gamma(1-2d) \sin(\pi d).$$

An equivalent expression for c_γ is given in Problem 3.12.

3.2.5 Sample Mean

Let $y_1, y_2, \ldots, y_n$ be a sample from an ARFIMA(p, d, q) process and let $\bar{y}$ be the sample mean. The variance of $\bar{y}$ is given by

$$\text{Var}(\bar{y}) = \frac{1}{n} \left[2 \sum_{j=1}^{n-1} \left(1 - \frac{j}{n}\right) \gamma(j) + \gamma(0) \right].$$

By formula (3.24), $\gamma(j) \sim c_\gamma j^{2d-1}$ for large j. Hence, for large n we have

$$\begin{aligned}\operatorname{Var}(\bar{y}) &\sim 2c_\gamma n^{2d-1}\sum_{j=1}^{n-1}\left(1-\frac{j}{n}\right)\left(\frac{j}{n}\right)^{2d-1}\frac{1}{n} \\ &\sim 2c_\gamma n^{2d-1}\int_0^1 (1-t)t^{2d-1}dt \\ &\sim \frac{c_\gamma}{d(2d+1)}n^{2d-1}. \end{aligned} \tag{3.25}$$

This heuristic result about the asymptotic behavior of the sample mean of an ARFIMA process is formally established in Chapter 10.

3.2.6 Partial Autocorrelations

Explicit expressions for the partial autocorrelation functions (PACF) for the general ARFIMA model are difficult to find. However, the coefficients of the best linear predictor

$$\widehat{y}_{n+1} = \phi_{n1}y_n + \cdots + \phi_{nn}y_1,$$

of a fractional noise process FN(d) are given by

$$\phi_{nj} = -\binom{n}{j}\frac{\Gamma(j-d)\Gamma(n-d-j+1)}{\Gamma(-d)\Gamma(n-d+1)},$$

for $j = 1, \ldots, n$. Thus, the partial autocorrelations are simply

$$\phi_{nn} = \frac{d}{n-d}, \tag{3.26}$$

and then $\phi_{nn} \sim d/n$ for large n.

Despite the lack of explicit formulas for the general ARFIMA case, Inoue (2002) has shown that the asymptotic behavior of the absolute value of the partial autocorrelations is similar to the fractional noise case: If ϕ_{nn} are the partial autocorrelations for a stationary ARFIMA process with $d \in (0, \frac{1}{2})$, then

$$|\phi_{nn}| \sim \frac{d}{n},$$

as $n \to \infty$; see Theorem 9.1.

3.2.7 Illustrations

In the following three simulation examples we illustrate some of the concepts discussed in the previous sections about ARFIMA processes. For simplicity, consider the family of ARFIMA$(1, d, 1)$ models

$$(1+\phi B)y_t = (1+\theta B)(1-B)^{-d}\varepsilon_t,$$

where the white noise sequence satisfies $\{\varepsilon_t\} \sim N(0,1)$. In these examples, the sample autocorrelation function has been calculated by first estimating the autocovariance function by means of the expression

$$\widehat{\gamma}(h) = \frac{1}{n}\sum_{t=1}^{n-h}(y_t - \bar{y})(y_{t+h} - \bar{y}),$$

for $h = 0, \ldots, n-1$, where $\bar{y}$ is the sample mean and then defining the autocorrelation function estimate as:

$$\widehat{\rho}(h) = \frac{\widehat{\gamma}(h)}{\widehat{\gamma}(0)}.$$

The sample partial autocorrelation function, $\widehat{\phi}_{tt}$, was obtained by means of the Durbin-Levinson algorithm described in Subsection 4.1.2, replacing the autocovariance function $\gamma(h)$ by the estimated value $\widehat{\gamma}(h)$.

On the other hand, the spectral density of the process $\{y_t\}$ is estimated by the *periodogram*

$$I(\lambda) = \frac{1}{2\pi n}\left|\sum_{j=1}^{n} y_j e^{i\lambda j}\right|^2. \tag{3.27}$$

The theoretical values of the autocorrelation function were calculated as follows. According to formula (3.22), the autocovariance function for the process $\{y_t\}$ is given by

$$\gamma(h) = \frac{\theta C(d,-h,-\phi) + (1+\theta^2)C(d,1-h,-\phi) + \theta C(d,2-h,-\phi)}{\phi(\phi^2-1)},$$

where the function $C(\cdot,\cdot,\cdot)$ is defined in (3.23). Hence, we have $\rho(h) = \gamma(h)/\gamma(0)$. On the other hand, the theoretical partial autocorrelation function of a fractional noise process is given by (3.26) while the theoretical partial autocorrelation function of the ARFIMA$(1, d, 1)$ model can be computed by means of the Durbin-Levinson algorithm; see Chapter 4 for further details.

■ **EXAMPLE 3.1**

Figure 3.1(a) exhibits 1000 simulated observations from an ARFIMA$(0, d, 0)$ process with $d = 0.4$. Panels (a), (b), and (c) of Figure 3.2 display the sample autocorrelation function, the sample partial autocorrelation function, and the periodogram of this series, respectively. Their theoretical counterparts are plotted in panels (a), (b), and (c) of Figure 3.3, respectively.

Notice from Figure 3.2 that the observed autocorrelations are significant, even after a large number of lags. Besides, the sample partial autocorrelation function decays slowly and the periodogram exhibits a peak around the origin. Observe that these estimates are very close to their theoretical versions shown in Figure 3.3.

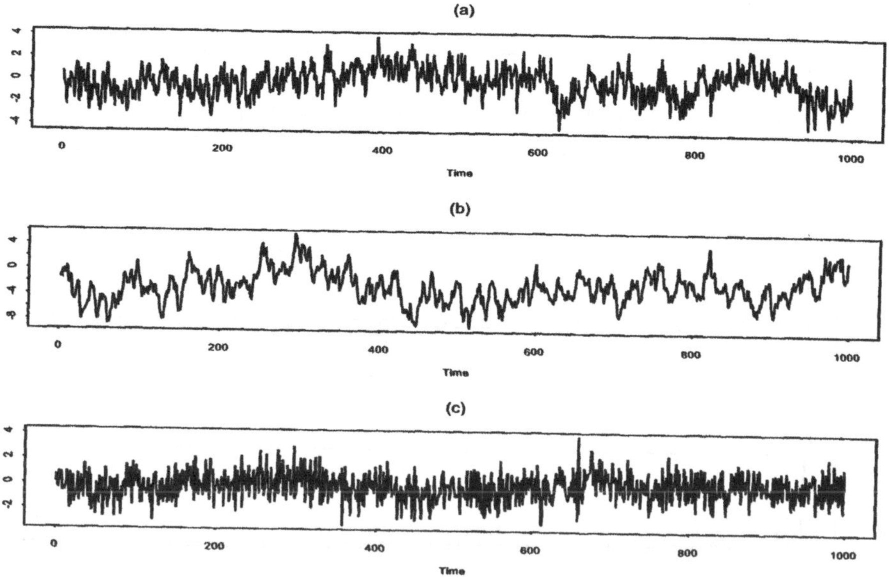

Figure 3.1 ARFIMA processes: 1000 simulated observations from ARFIMA$(1, d, 1)$ models with Gaussian white noise $N(0, 1)$. (a) ARFIMA$(0, d, 0)$ model with $d = 0.4$, (b) ARFIMA$(1, d, 1)$ model with $d = 0.4$, $\phi = -0.7$, $\theta = -0.2$, and (c) ARFIMA$(1, d, 0)$ model with $d = 0.3$, $\phi = 0.5$.

EXAMPLE 3.2

Panel (b) of Figure 3.1 displays 1000 observations from an ARFIMA$(1, d, 1)$ process with $\phi = -0.7$, $d = 0.4$, and $\theta = -0.2$.

The sample autocorrelation function of this simulated series is shown in panel (a) of Figure 3.4 while the sample partial autocorrelation function and the periodogram are plotted in panels (b) and (c) of Figure 3.4, respectively. Notice that the observed autocorrelations are significant for all the lags exhibited. The sample partial autocorrelation function decays very rapidly and the periodogram shows a peak around the origin. The theoretical counterparts including the autocorrelation function, the partial autocorrelation function and the spectral density are displayed in panels (a), (b), and (c) of Figure 3.5. Observe that the estimates of the ACF, PACF and the spectral density shown in Figure 3.4 are close to their theoretical counterparts displayed in Figure 3.5.

EXAMPLE 3.3

Figure 3.1(c) exhibits 1000 observations from an ARFIMA$(1, d, 0)$ process with $\phi = 0.5$ and $d = 0.3$. Panel (a) of Figure 3.6 shows the sample autocorrelation function, Figure 3.6(b) displays the sample partial autocorrelation function, and

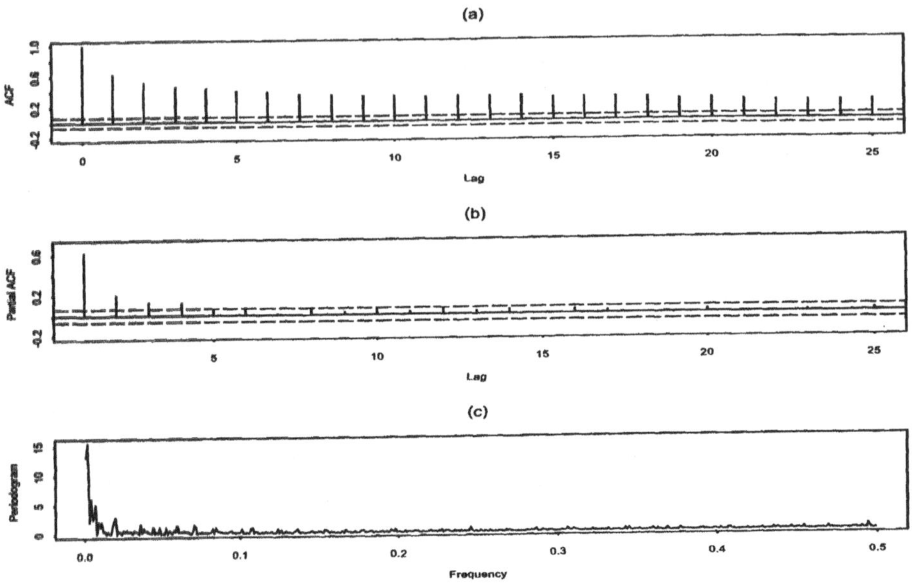

Figure 3.2 Example 3.1, ARFIMA$(0, d, 0)$ model with $d = 0.4$: (a) Sample autocorrelation function, (b) sample partial autocorrelation function, and (c) periodogram.

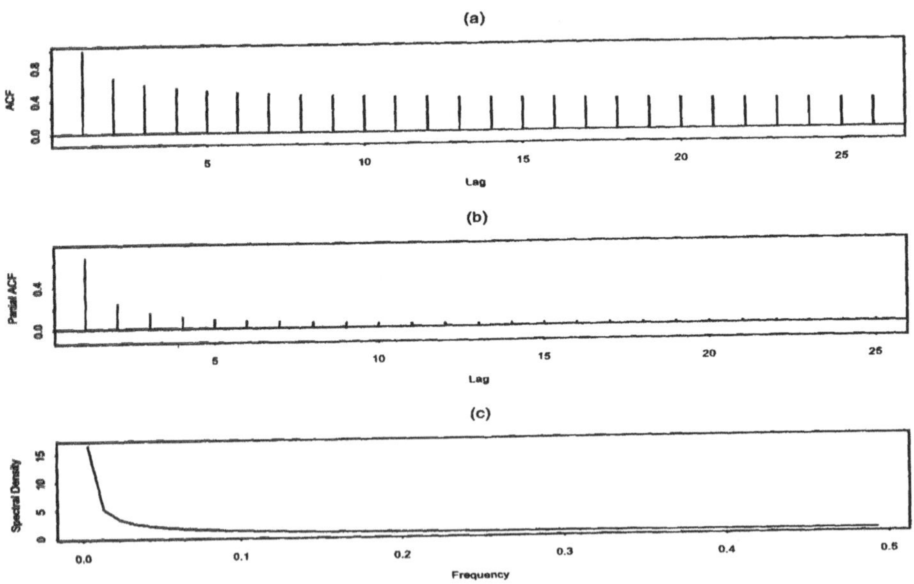

Figure 3.3 Example 3.1, ARFIMA$(0, d, 0)$ model with $d = 0.4$: (a) Theoretical autocorrelation function, (b) theoretical partial autocorrelation function, and (c) spectral density.

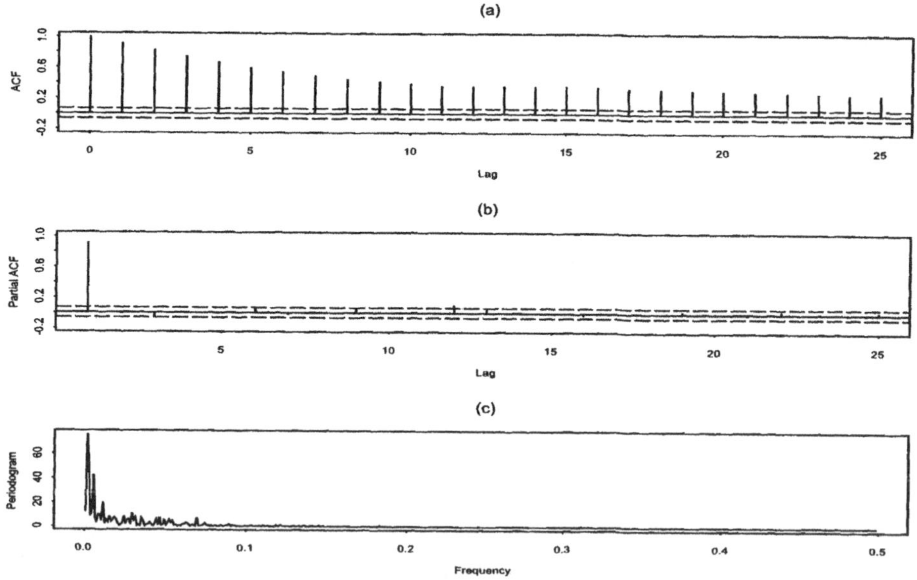

Figure 3.4 Example 3.2, ARFIMA$(1, d, 1)$ model with $\phi = -0.7$, $d = 0.4$, and $\theta = -0.2$: (a) Sample autocorrelation function, (b) sample partial autocorrelation function, and (c) periodogram.

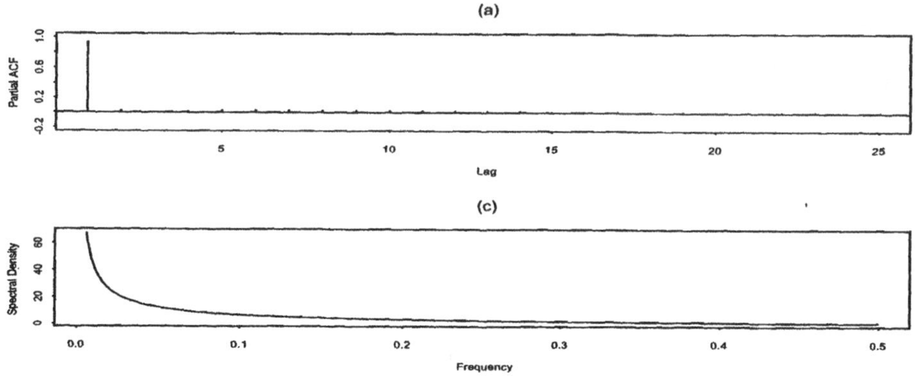

Figure 3.5 Example 3.2, ARFIMA$(1, d, 1)$ model with $\phi = -0.7$, $d = 0.4$, and $\theta = -0.2$: (a) Theoretical autocorrelation function, (b) theoretical partial autocorrelation function, and (c) spectral density.

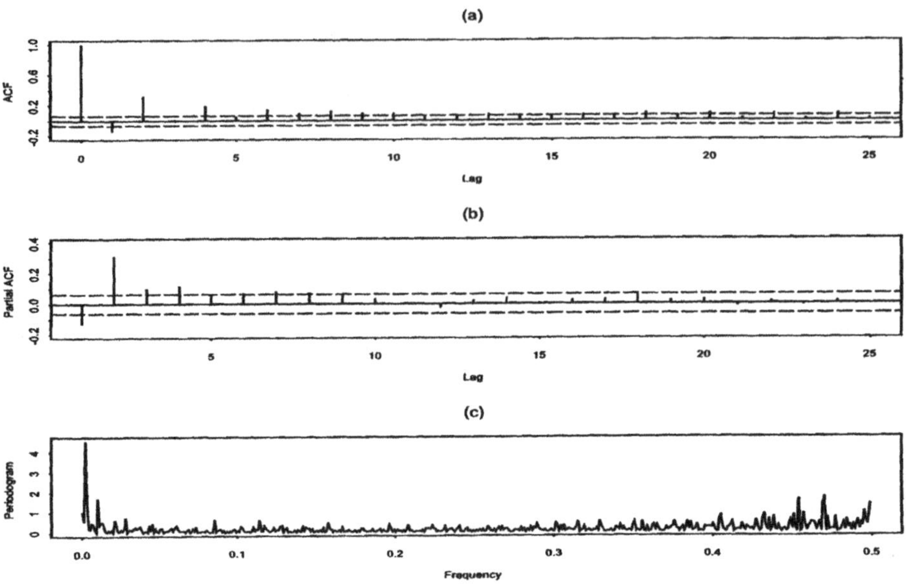

Figure 3.6 Example 3.3, ARFIMA$(1, d, 0)$ model with $\phi = 0.5$ and $d = 0.3$: (a) Sample autocorrelation function, (b) sample partial autocorrelation function, and (c) periodogram.

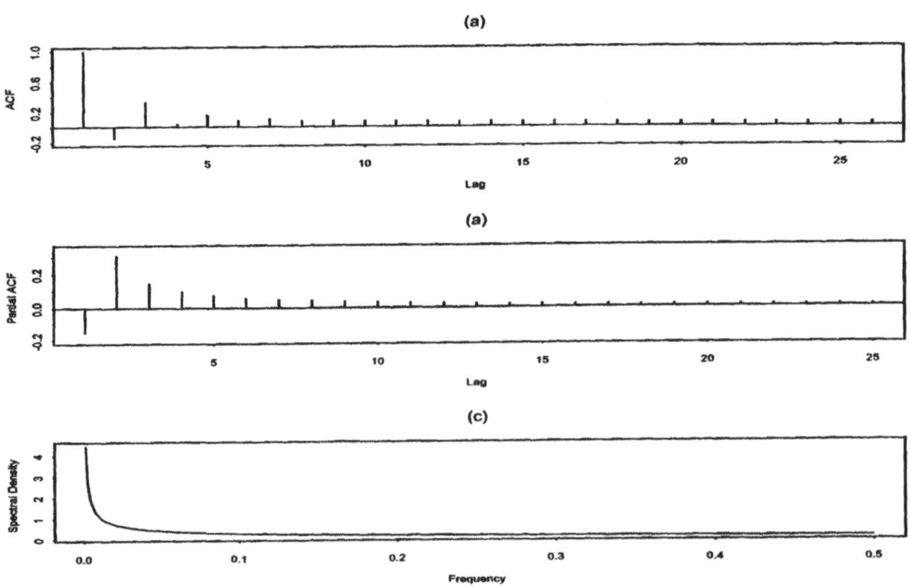

Figure 3.7 Example 3.3, ARFIMA$(1, d, 0)$ model with $\phi = 0.5$ and $d = 0.3$: (a) Theoretical autocorrelation function, (b) theoretical partial autocorrelation function, and (c) spectral density.

Figure 3.6(c) exhibits the periodogram. On the other hand, panels (a), (b), and (c) of Figure 3.7 exhibit the theoretical versions of the autocorrelation function, the partial autocorrelation function and the spectral density, respectively.

From Figures 3.6 and 3.7 note that in this case all the components of the sample ACF and the theoretical ACF are positive, excepting the component corresponding to lag 1. Besides, the empirical and the theoretical PACF behave similarly, exhibiting a negative component at lag 1 and positive components for most of the other lags. Finally, as in the previous examples, the periodogram displays a peak around the origin.

3.2.8 Approximation of Long-Memory Processes

The next theorem establishes that any long-memory process defined as in (3.11) can be approximated arbitrarily well by an ARFIMA model in the following sense.

Theorem 3.5. *Let* $\{y_t : t \in \mathbb{Z}\}$ *be a linear regular process satisfying* (1.1) *with strictly positive spectral density* f_y *satisfying* (3.11). *Then, there exists an ARFIMA process with spectral density* f *such that for any* $\varepsilon > 0$

$$\left| \frac{f_y(\lambda)}{f(\lambda)} - 1 \right| < \varepsilon,$$

uniformly for $\lambda \in [-\pi, \pi]$.

Proof. Since f_y satisfies (3.11) and according to Subsection 3.1.2 we assume that f_y is continuous outside a neighborhood of zero, we may define the function

$$g(\lambda) = \left[\frac{2\sin(\lambda/2)}{\lambda} \right]^{2d} f_0(\lambda),$$

where $f_0(\lambda) = 2\pi\sigma^{-2} f_y(\lambda)|\lambda|^{2d}$ is a continuous function. Since $g(\lambda)$ is continuous on $[-\pi, \pi]$, an application of the Stone-Weierstrass theorem establishes that for any $\eta > 0$ we can find polynomials $\theta(z)$ and $\phi(z)$ such that

$$\left| |\theta(e^{-i\lambda})|^2 - |\phi(e^{-i\lambda})|^2 g(\lambda) \right| < \eta,$$

uniformly in λ. Given that the polynomial $\theta(z)$ can be chosen such that all its roots are outside the closed closed unit disk $\{z\colon |z| \leq 1\}$—see, for example, Theorem 4.4.3 of Brockwell and Davis (1991)—we conclude that

$$\left| 1 - \frac{|\phi(e^{-i\lambda})|^2}{|\theta(e^{-i\lambda})|^2} g(\lambda) \right| < \frac{\eta}{|\theta(e^{-i\lambda})|^2}.$$

Now, by noting that

$$\frac{f_y(\lambda)}{f(\lambda)} = \frac{|\phi(e^{-i\lambda})|^2}{|\theta(e^{-i\lambda})|^2} g(\lambda),$$

and taking $\varepsilon = \eta \sup_\lambda |\theta(e^{-i\lambda})|^{-2}$, the result is obtained. □

3.3 FRACTIONAL GAUSSIAN NOISE

Another well-known long-range-dependent process is the so-called *fractional Gaussian noise* (fGn). This process may be defined as follows. Consider the fractional Brownian motion $B_d(t)$ introduced in Subsection 1.1.11 and let $\{y_t : t \in \mathbb{Z}\}$ be defined by the increments of $B_d(t)$:

$$y_t = B_d(t+1) - B_d(t). \tag{3.28}$$

The discrete-time process $\{y_t : t \in \mathbb{Z}\}$ is called *fractional Gaussian noise*. The following result—see Proposition 3.1 of Taqqu (2003)—describes some of the properties of this process.

Theorem 3.6. *Let* $\{y_t : t \in \mathbb{Z}\}$ *be the process defined by* (3.28). *Then*

(a) $\{y_t : t \in \mathbb{Z}\}$ *is stationary for* $d \in (-\frac{1}{2}, \frac{1}{2})$.

(b) $E(y_t) = 0$.

(c) $E(y_t^2) = E[B(1)^2]$.

(d) *The autocovariance function of* $\{y_t : t \in \mathbb{Z}\}$ *is*

$$\gamma(h) = \frac{\sigma^2}{2}\left(|h+1|^{2d+1} - 2|h|^{2d+1} + |h-1|^{2d+1}\right),$$

where $\sigma^2 = \text{Var}(y_t)$.

(e) *For* $d \neq 0$, *the asymptotic behavior of the ACF is given by*

$$\gamma(h) \sim \sigma^2 d(2d+1)\,|h|^{2d-1},$$

as $|h| \to \infty$.

3.3.1 Sample Mean

Let $\bar{y}_n$ be the sample mean of a fractional Gaussian noise described by (3.28). Then, by telescopic sum we have

$$\bar{y}_n = \frac{1}{n}[B_d(n+1) - B_d(1)].$$

Thus, an application of formula (1.13) yields

$$\text{Var}(\bar{y}_n) = \sigma^2 n^{2d-1}.$$

Consequently, since the process $\{B_d(t)\}$ is Gaussian, we conclude that

$$\bar{y}_n \sim N(0, \sigma^2 n^{2d-1}),$$

for all $n \in \mathbb{N}$.

3.4 TECHNICAL LEMMAS

In this section we present three technical results that are very useful for proving the theorems discussed in this chapter. Besides, these lemmas are also used in Chapter 8 about Bayesian methods for long-range dependent data and Chapter 12 dealing with seasonal long-memory processes.

The following lemma, due to Bondon and Palma (2007), establishes sufficient conditions for the existence of an autoregressive representation of a stationary process. Recall that a function $f : D \to \mathbb{R}$ is said to be *locally integrable* if its Lebesgue integral is finite over every compact subset of the domain D.

Lemma 3.1. *Let $\{y_t : t \in \mathbb{Z}\}$ be a stationary process with spectral density f and moving-average expansion convergent in $\mathcal{L}_2$*

$$y_t = \sum_{j=0}^{\infty} \psi_j \varepsilon_{t-j} = \psi(B)\varepsilon_t. \tag{3.29}$$

Define the operator $\pi(B) = 1/\psi(B)$. If the spectral density f satisfies the following three conditions

(a) $f(\lambda) \sim |\lambda|^{-2d}\ell(1/|\lambda|)$ when $\lambda \to 0$, where ℓ is a slowly varying function and $d \in (-1, \frac{1}{2})$,

(b) f is bounded on $[\epsilon, \pi]$ for every $\epsilon > 0$,

(c) f^{-1} is locally integrable on $(0, \pi]$,

then the expansion

$$\varepsilon_t = \sum_{j=0}^{\infty} \pi_j y_{t-j} = \pi(B)y_t, \tag{3.30}$$

converges in $\mathcal{L}_2$.

This next result is an immediate consequence of Lemma 4.3 of Inoue (1997).

Lemma 3.2. *Let $r, s \in \mathbb{R}$ with $s < 1 < r + s$, ℓ_1, ℓ_2 slowly varying functions and $\{f_k\}$, $\{g_k\}$ two real sequences with $\in \mathbb{N}$ satisfying $f_k \sim k^{-r}\ell_1(k)$, $g_k \sim k^{-s}\ell_2(k)$ as k tends to infinity. For any $n \in \mathbb{N}$, the sequence $\{f_{n+k}g_k\}$ is summable and*

$$\sum_{k=0}^{\infty} f_{n+k}g_k \sim n^{-(r+s-1)}\ell_1(n)\ell_2(n)B(r+s-1, 1-s),$$

as n tends to infinity, where $B(\cdot,\cdot)$ is the beta function.

Let $C \subseteq (0, \infty)$ be a compact set, the *total variation* of the real-valued function f on C is defined as

$$v(f, C) = \sup \sum_{j=1}^{n} |f(x_j) - f(x_{j-1})|,$$

where the supremum is over all finite sequences $x_0 \leq x_1 \leq \cdots \leq x_n$ in C. A function f is said to be of *locally bounded variation* on $(0, \infty)$ if $v(f, C) < \infty$ for each compact $C \subseteq (0, \infty)$. Now, a positive function f of *locally bounded variation* on $(0, \infty)$ is said to be quasi-monotone if for some $\delta > 0$,

$$\int_0^x t^\delta |df(t)| = O[x^\delta f(x)],$$

as x tends to infinity.

Parts (a), (b), and (c) of the next lemma follow directly from Proposition 1.3.6(v), Proposition 1.5.8, and Theorem 4.3.2 of Bingham, Goldie, and Teugels (1987), respectively.

Lemma 3.3. *(a) If $\ell(\cdot)$ is slowly varying and $\alpha > 0$, then $x^\alpha \ell(x) \to \infty$ as $x \to \infty$.*

(b) If $\ell(\cdot)$ is slowly varying, m is large enough such that $\ell(x)$ is locally bounded in $[m, \infty)$ and $\alpha > -1$, then

$$\sum_{k=m}^{n} k^\alpha \ell(k) \sim \frac{n^{\alpha+1}\ell(n)}{\alpha+1},$$

as $n \to \infty$.

(c) If $\ell(\cdot)$ is quasi-monotone slowly varying, then the following series is conditionally convergent and

$$\sum_{k=1}^{\infty} \ell(k)\cos(k\lambda)k^{-\mu} \sim \frac{\pi\lambda^{\mu-1}\ell(1/\lambda)}{2\Gamma(\mu)\cos(\pi\mu/2)},$$

for $\lambda \to 0_+$ and $0 < \mu < 1$.

3.5 BIBLIOGRAPHIC NOTES

Definitions of long-memory processes have been extensively discussed in the literature. Several technical aspects of our discussion about slowly varying functions can be found in Bingham, Goldie, and Teugels (1987), Feller (1971, Section VIII.8) and Zygmund (1959). Chapter 2 of Beran (1994b) and Taqqu (2003) address this issue in detail while the articles by Cox (1984) and Hall (1997) give overviews about different definitions of long-range dependence.

The paper by Hosking (1981) discusses several properties of ARFIMA models, including results about stationarity, invertibility, autocorrelations, and the like. The asymptotic expressions for the infinite autoregressive and moving-average coefficients of ARFIMA(p, d, q) processes reviewed in Subsection 3.2.2 are proved in Kokoszka and Taqqu (1995). Formulas for the exact autocovariance function of an ARFIMA process were established by Sowell (1992). A nice review of fractional Gaussian noise processes and their properties is given in Taqqu (2003). Finally, the books by Rangarajan and Ding (2003) and Teyssière and Kirman (2007) provide comprehensive discussions about long-memory time series methodologies and their applications to a wide range of fields.

Problems

3.1 Show that an ARFIMA model satisfies all definitions of a long-memory process discussed in this chapter.

3.2 Prove that $\ell(x) = \log(x)$ and $\ell(x) = \log[\log(x)]$ are slowly varying functions in Karamata's sense. Are these slowly varying functions in Zygmund's sense?

3.3 Prove the asymptotic expression (3.24) using Lemma 3.2.

3.4 Can we approximate arbitrarily well any regular linear process by an ARMA process? Discuss.

3.5 Calculate explicitly the autocovariance function of an ARFIMA$(1, d, 0)$ and ARFIMA$(0, d, 1)$. Compare your results with Hosking (1981).

3.6 Use the Stirling' approximation

$$\Gamma(x) \sim \sqrt{2\pi} e^{1-x}(x-1)^{x-1/2},$$

as $x \to \infty$, to show that

$$\frac{\Gamma(n+\alpha)}{\Gamma(n+\beta)} \sim n^{\alpha-\beta}, \tag{3.31}$$

as $n \to \infty$.

3.7 Applying the Stirling' approximation, show directly that for an ARFIMA$(0, d, 0)$ process the following asymptotic expressions hold

$$\begin{aligned} \psi_k &\sim \frac{k^{d-1}}{\Gamma(d)}, \\ \pi_k &\sim \frac{k^{-d-1}}{\Gamma(-d)}, \\ \rho(k) &\sim k^{2d-1}\frac{\Gamma(1-d)}{\Gamma(d)}, \end{aligned}$$

as $k \to \infty$.

3.8 Let $\{\lambda_t\}$ be a real sequence and let $\lambda_n(\delta)$ be the nth coefficient of the power expansion of $(1-B)^\delta \sum_{n=0}^{\infty} \lambda_n B^n$, that is,

$$\lambda_n(\delta) = \sum_{j=0}^{n} \lambda_j (-1)^{n-j} \binom{\delta}{n-j},$$

for $n \geq 0$. According to Lemma 2.1 of Inoue (2002), if the sequence $\{\lambda_t\}$ decays exponentially and $\delta \in (-1, \infty) \setminus \{0, 1, 2, \dots\}$, then

$$\lambda_n(\delta) = n^{-\delta-1} \frac{\sum_{j=0}^{\infty} \lambda_j}{\Gamma(-\delta)} + O(n^{-\delta-2}),$$

as $n \to \infty$.

Using the result above, show that the AR(∞) and MA(∞) coefficients of an ARFIMA(p, d, q) process satisfy

$$\begin{aligned} \psi_j &\sim \frac{\theta(1)}{\phi(1)} \frac{j^{d-1}}{\Gamma(d)}, \\ \pi_j &\sim \frac{\phi(1)}{\theta(1)} \frac{j^{-d-1}}{\Gamma(-d)}, \end{aligned}$$

as $j \to \infty$, cf. Corollary 3.1 of Kokoszka and Taqqu (1995).

3.9 Let $\psi_j(\boldsymbol{\theta})$ be the infinite moving-average coefficients of an ARFIMA(p, d, q) model defined by the parameter vector $\boldsymbol{\theta} = (d, \phi_1, \dots, \phi_p, \theta_1, \dots, \theta_q)$, that is,

$$y_t = \sum_{j=0}^{\infty} \psi_j(\boldsymbol{\theta}) \varepsilon_{t-j},$$

where $\{\varepsilon_t\}$ is a white noise sequence.

Show that

$$|\psi_j(\boldsymbol{\theta})| \leq K j^{d-1},$$

for $j = 0, 1, 2, \dots$ where $K > 0$ is a constant which does not depend on $\boldsymbol{\theta}$.

3.10 Let $\pi_j(\boldsymbol{\theta})$ be the infinite autoregressive coefficients of an ARFIMA(p, d, q) model defined by the parameter vector $\boldsymbol{\theta} = (d, \phi_1, \dots, \phi_p, \theta_1, \dots, \theta_q)$, that is,

$$\sum_{j=0}^{\infty} \pi_j(\boldsymbol{\theta}) y_{t-j} = \varepsilon_t,$$

where $\{\varepsilon_t\}$ is a white noise sequence. Verify that

$$|\pi_j(\boldsymbol{\theta})| \leq K j^{-d-1},$$

for $j = 0, 1, 2, \dots$ where $K > 0$ is a constant which does not depend on $\boldsymbol{\theta}$.

3.11 Prove that

$$f(\lambda) \sim \frac{\sigma^2}{2\pi} \frac{|\theta(1)|^2}{|\phi(1)|^2} |\lambda|^{-2d},$$

for $|\lambda| \to 0$. Hint: $\sin(x) \sim x$ for $|x| \to 0$.

3.12 Using the following formula involving the gamma function

$$\Gamma(1-d)\Gamma(d) = \frac{\pi}{\sin(\pi d)},$$

compare equation 8.334(3) of Gradshteyn and Ryzhik (2000), show that the constant c_γ appearing in expression (3.24) may be written as

$$c_\gamma = \sigma^2 \frac{|\theta(1)|^2}{|\phi(1)|^2} \frac{\Gamma(1-2d)}{\Gamma(1-d)\Gamma(d)}.$$

3.13 Show that the autocovariance function of a fractional Gaussian noise is positive for $d \in (0, \frac{1}{2})$ and negative for $d \in (-\frac{1}{2}, 0)$.

3.14 Let η_j be the MA(∞) coefficients of an ARFIMA$(0, d, 0)$ process with $d \in (0, \frac{1}{2})$ and define $\varphi_0 = 1$ and $\varphi_j = \eta_j - \eta_{j-1}$ for $j = 1, 2, \ldots$.

a) Verify that $\varphi_j = \frac{d-1}{j}\eta_{j-1}$.

b) Show that $\varphi_j \sim \frac{j^{d-2}}{\Gamma(d-1)}$ as $j \to \infty$.

c) Prove that

$$\sum_{j=0}^{\infty} \varphi_j^2 = \frac{\Gamma(3-2d)}{\Gamma(2-d)^2}.$$

3.15 Consider the linear process $\{t_t\}$ with Wold expansion

$$y_t = \varepsilon_t + \sum_{j=1}^{\infty} \frac{\varepsilon_{t-j}}{j},$$

where $\{\varepsilon_t\}$ is a white noise sequence with unit variance.

a) Show that the autocovariance function of this process is

$$\gamma(h) = \frac{1}{|h|} + \frac{1}{|h|} \sum_{j=1}^{|h|} \frac{1}{j},$$

for $|h| > 0$ and

$$\gamma(0) = 1 + \frac{\pi^2}{6}.$$

b) Show that

$$\gamma(h) \sim \frac{\log h}{h},$$

as $h \to \infty$.

c) Verify that for any $m > n > 0$,

$$\sum_{h=n}^{m} \gamma(h) > \sum_{h=n}^{m} \frac{1}{h}.$$

d) Is $\{y_t\}$ a long-memory process?
Hint: The following formulas may be useful,

$$\sum_{j=1}^{\infty} \frac{1}{j(j+h)} = \frac{1}{h} \sum_{j=1}^{h} \frac{1}{j},$$

for $h > 0$.

$$\sum_{j=1}^{h} \frac{1}{j} = C + \log h + O\left(\frac{1}{h}\right),$$

where C is the Euler's constant, $C = 0.577215664\cdots$

$$\sum_{j=1}^{\infty} \frac{1}{j^2} = \frac{\pi^2}{6}.$$

3.16 Let $\psi_j(\boldsymbol{\theta})$ be the MA(∞) coefficients of an ARFIMA(p, d, q) process where $\boldsymbol{\theta} = (d, \phi_1, \ldots, \phi_p, \theta_1, \ldots, \theta_q)$ and $d \in (0, \frac{1}{2})$. Define $\varphi_0(\boldsymbol{\theta}) = 1$ and $\varphi_j(\boldsymbol{\theta}) = \psi_j(\boldsymbol{\theta}) - \psi_{j-1}(\boldsymbol{\theta})$ for $j \geq 1$. Verify that for some $\epsilon > 0$ we have

$$\sup_{\boldsymbol{\theta}} |\varphi_j(\boldsymbol{\theta})| \leq K j^{d-2},$$

$$\sup_{\boldsymbol{\theta}} \left| \frac{\partial \varphi_j(\boldsymbol{\theta})}{\partial \boldsymbol{\theta}_i} \right| \leq K j^{d-2+\epsilon},$$

for $j \geq 1$ and $i = 1, \ldots, p+q+1$, where K is a positive constant; see Chan and Palma (1998, Lemma 3.1).

3.17 Suppose that $\gamma_j = \int_{-\pi}^{\pi} f(\lambda) e^{i\lambda j}\, d\lambda$ is the autocovariance function of a stationary process with spectral density f and $\sum_{j=-\infty}^{\infty} \gamma_j^2 < \infty$. Prove that for any $k \in \mathbb{Z}$,

$$\sum_{j=-\infty}^{\infty} \gamma_j \gamma_{j+k} = 2\pi \int_{-\pi}^{\pi} f^2(\lambda) e^{i\lambda k}\, d\lambda.$$

3.18 Let γ_j be the autocovariance function of a fractional noise FN(d) with $d < \frac{1}{4}$ and white noise variance σ^2.

a) Show that

$$\sum_{j=-\infty}^{\infty} \gamma_j^2 = \sigma^4 \frac{\Gamma(1-4d)}{\Gamma^2(1-2d)}.$$

b) Prove that

$$\sum_{j=0}^{\infty} \gamma_j \gamma_{j+1} = \sigma^4 \frac{d\,\Gamma(1-4d)}{(1-2d)\Gamma^2(1-2d)}.$$

CHAPTER 4

ESTIMATION METHODS

In this chapter we analyze several parameter estimation methods for long-memory models. Exact maximum-likelihood methods are reviewed in Section 4.1, including the Cholesky decomposition, the Durbin-Levinson algorithm, and state space methodologies. Section 4.2 discusses approximate maximum-likelihood methods based on truncations of the infinite autoregressive expansion of a long-memory process, including the Haslett-Raftery estimator and other similar procedures. Truncations of the infinite moving-average expansion of the process is discussed in Section 4.3 along with the corresponding Kalman filter recursions. The spectrum-based Whittle method and semiparametric procedures are studied in Section 4.4. Extensions to the non-Gaussian case are addressed in this section as well. On the other hand, a regression method for estimating the long-memory parameter is described in Section 4.5. This section also discusses several heuristic techniques for detecting long-range dependence and estimating the long-memory parameter including the so-called rescaled range statistics (R/S) method, variance plots, the detrended fluctuation analysis, and a wavelet-based estimation methodology. Numerical experiments that illustrate the finite sample performance of several estimates are discussed in Section 4.6 while Section 4.7 provides some bibliographic references. A list of proposed problems is presented at the end of the chapter.

Long-Memory Time Series. By Wilfredo Palma

4.1 MAXIMUM-LIKELIHOOD ESTIMATION

Assume that $\{y_t\}$ is a zero-mean stationary Gaussian process. Then, the log-likelihood function of this process is given by

$$\mathcal{L}(\theta) = -\tfrac{1}{2}\log\det\Gamma_\theta - \tfrac{1}{2}\boldsymbol{y}'\Gamma_\theta^{-1}\boldsymbol{y}, \tag{4.1}$$

where $\boldsymbol{y} = (y_1, \dots, y_n)'$, $\Gamma_\theta = \mathrm{Var}(\boldsymbol{y})$, and θ is the parameter vector. Consequently, the *maximum-likelihood* (ML) estimate $\widehat{\theta}$ is obtained by maximinzig $\mathcal{L}(\theta)$. The log-likelihood function (4.1) requires the calculation of the determinant and the inverse of the variance-covariance matrix Γ_θ. However, these calculations can be conducted by means of the Cholesky decomposition method. In the following subsections, we review this and other procedures for computing the function (4.1) such as the Durbin-Levinson algorithm and state space techniques.

4.1.1 Cholesky Decomposition Method

Given that the matrix Γ_θ is symmetric positive definite, it can be written as

$$\Gamma_\theta = U'U,$$

where U is an upper triangular matrix. According to this Cholesky decomposition, the determinant of Γ_θ is given by $\det\Gamma_\theta = (\det U)^2 = \prod_{j=1}^n u_{jj}^2$, where u_{jj} denotes the jth diagonal element of the matrix U. Besides, the inverse of Γ_θ can be obtained as $\Gamma_\theta^{-1} = U^{-1}(U^{-1})'$, where the inverse of U can be computed by means of a very simple procedure. As described in Press, Teukolsky, Vetterling, and Flannery (1992, p. 90) the operation count for the Cholesky algorithm is order $\mathcal{O}(n^3)$.

4.1.2 Durbin-Levinson Algorithm

The Cholesky decomposition could be inefficient for long time series. Thus, faster methods for calculating the log-likelihood function (4.1) have been developed. One of these algorithms designed to exploit the Toeplitz structure of the variance-covariance matrix Γ_θ, is known as the Durbin-Levinson algorithm.

Suppose that $\widehat{y}_1 = 0$ and $\widehat{y}_{t+1} = \phi_{t1}y_t + \cdots + \phi_{tt}y_1$, $t = 1, \dots, n-1$, are the one-step ahead forecasts of the process $\{y_t\}$ based on the finite past $\{y_1, \dots, y_{t-1}\}$, where the regression coefficients ϕ_{tj} are given by the equations

$$\begin{aligned}
\phi_{tt} &= [\nu_{t-1}]^{-1}\left[\gamma(t) - \sum_{i=1}^{t-1}\phi_{t-1,i}\gamma(t-i)\right], \\
\phi_{tj} &= \phi_{t-1,j} - \phi_{tt}\phi_{t-1,t-j}, \qquad j = 1, \dots, t-1, \\
\nu_0 &= \gamma(0), \\
\nu_t &= \nu_{t-1}[1 - \phi_{tt}^2], \qquad j = 1, \dots, t-1.
\end{aligned}$$

Furthermore, if $e_t = y_t - \widehat{y}_t$ is the prediction error and $e = (e_1, \ldots, e_n)'$, then $e = Ly$ where L is the lower triangular matrix:

$$L = \begin{pmatrix} 1 & & & & & \\ -\phi_{11} & 1 & & & & \\ -\phi_{22} & -\phi_{21} & 1 & & & \\ -\phi_{33} & -\phi_{32} & -\phi_{31} & & & \\ \vdots & \vdots & & & 1 & \\ -\phi_{n-1,n-1} & -\phi_{n-1,n-2} & -\phi_{n-1,n-3} & \cdots & -\phi_{n-1,1} & 1 \end{pmatrix}.$$

Hence, Γ_θ may be decomposed as $\Gamma_\theta = LDL'$, where $D = \text{diag}(\nu_0, \ldots, \nu_{n-1})$. Therefore, $\det \Gamma_\theta = \prod_{j=1}^n \nu_{j-1}$ and $y'\Gamma_\theta^{-1}y = e'D^{-1}e$. As a result, the log-likelihood function (4.1) may be expressed as

$$\mathcal{L}(\theta) = -\frac{1}{2}\sum_{t=1}^{n} \log \nu_{t-1} - \frac{1}{2}\sum_{t=1}^{n} \frac{e_t^2}{\nu_{t-1}}.$$

The numerical complexity of this algorithm is $\mathcal{O}(n^2)$ for a linear stationary process; see, for example, Ammar (1998). Nevertheless, for some Markovian processes such as the family of ARMA models, the Durbin-Levinson algorithm can be implemented in only $\mathcal{O}(n)$ operations. Unfortunately, this reduction in operations count does not apply to ARFIMA models since they are not Markovian.

4.1.3 Computation of Autocovariances

The calculation of the ACF of an ARFIMA process is a crucial aspect in the implementation of the Cholesky and the Durbin-Levinson algorithms. Recall that a closed form expression for the ACF of an ARFIMA model was discussed in Chapter 3; see, for example, equation (3.22).

Another approach for calculating the ACF is the so-called *splitting method*. This method is based on the decomposition of the ARFIMA model into its ARMA and its fractionally integrated (FI) parts. Let $\gamma_1(\cdot)$ be the ACF of the ARMA component and $\gamma_2(\cdot)$ be the ACF of the fractional noise given by (3.21). Then, the ACF of the corresponding ARFIMA process is given by the convolution of these two functions:

$$\gamma(h) = \sum_{j=-\infty}^{\infty} \gamma_1(j)\gamma_2(j-h).$$

If this infinite sum is truncated to m summands, then we obtain the approximation

$$\gamma(h) \approx \sum_{j=-m}^{m} \gamma_1(j)\gamma_2(j-h).$$

From this expression, the ACF $\gamma(\cdot)$ can be efficiently calculated with a great level of precision.

■ EXAMPLE 4.1

To illustrate the calculation of the ACF of a long-memory process consider the ARFIMA$(1, d, 1)$ model

$$(1+\phi B)y_t = (1+\theta B)(1-B)^{-d}\varepsilon_t,$$

with $\text{Var}(\varepsilon_t) = \sigma^2 = 1$.

Following Subsection 3.2.3 and Subsection 3.2.6, an exact formula for the ACF of this model is given by

$$\gamma(h) = \frac{\theta C(d,-h,-\phi) + (1+\theta^2)C(d,1-h,-\phi) + \theta C(d,2-h,-\phi)}{\phi(\phi^2-1)},$$

where the function $C(\cdot,\cdot,\cdot)$ is defined in (3.23).

On the other hand, an approximated ACF is obtained by the splitting algorithm

$$\gamma(k) \approx \sum_{h=-m}^{m} \gamma_0(h)\gamma_{\text{ARMA}}(k-h),$$

where

$$\gamma_0(h) = \sigma^2 \frac{\Gamma(1-2d)}{\Gamma(1-d)\Gamma(d)} \frac{\Gamma(h+d)}{\Gamma(1+h-d)},$$

and

$$\gamma_{\text{ARMA}}(h) = \begin{cases} \dfrac{1-2\phi\theta+\theta^2}{1-\phi^2} & h=0, \\[2ex] \dfrac{(1-\phi\theta)(\theta-\phi)}{1-\phi^2}(-\phi)^{|h|} & h \neq 0. \end{cases}$$

We consider two sets of parameters $d = 0.4$, $\phi = 0.5$, $\theta = 0.2$ and $d = 0.499$, $\phi = -0.9$, $\theta = -0.3$, and several lags between 0 and 999. The calculation of the exact and the approximate method has been carried out with $m = 200$ in a Pentium IV 2.8-GHz machine using an Splus program in a Windows XP platform. The central processing unit (CPU) time for computing the ACF at 1000 lags is less than 1 second for the splitting method and about 100 seconds for the exact method. The results are shown in Table 4.1.

Note that for the set of parameters $d = 0.4, \phi = 0.5, \theta = 0.2$, the accuracy of the splitting method is about six significant decimals while for the second set of parameters, $d = 0.499, \phi = -0.9, \theta = -0.3$, the accuracy drops to about three significant decimals for the range of lags studied.

Table 4.1 Calculation of the Autocorrelation Function of ARFIMA$(1, d, 1)$ Models[a]

		ACF	
Lag	Method	$d = 0.4, \phi = 0.5, \theta = 0.2$	$d = 0.499, \phi = -0.9, \theta = -0.3$
0	Exact	1.6230971100284379	7764.0440304632230
	Approx.	1.6230971200957560	7764.0441353477199
1	Exact	0.67605709850269124	7763.5195069108622
	Approx.	0.67605707826745276	7763.5196117952073
2	Exact	0.86835879142153161	7762.8534907771409
	Approx.	0.86835880133411103	7762.8535956613778
3	Exact	0.66265875439861421	7762.0404144912191
	Approx.	0.66265877143805063	7762.0405193753304
998	Exact	0.22351300800718499	7682.7366067938428
	Approx.	0.22351301379700828	7682.7367003641175
999	Exact	0.22346824274316196	7682.7212154918925
	Approx.	0.22346824853234257	7682.7213090555442

[a] Exact and approximated values for different lags and two sets of parameters.

4.1.4 State Space Approach

An alternative methodology for calculating exact ML estimates (MLE) is provided by the state space systems. In this section we address the application of Kalman filter techniques to long-memory processes. It is worth noting that since these processes are not Markovian, all the state space representations are infinite dimensional. Despite this fact, the Kalman filter equations may be used to calculate the exact log-likelihood (4.1) in a finite number of steps.

Recall that a causal ARFIMA(p, d, q) process $\{y_t\}$ has a linear process representation given by

$$y_t = \frac{\theta(B)}{\phi(B)}(1 - B)^{-d}\varepsilon_t = \sum_{j=0}^{\infty} \psi_j \varepsilon_{t-j}, \tag{4.2}$$

where ψ_j are the coefficients of $\psi(z) = \sum_{j=0}^{\infty} \psi_j z^j = \theta(z)\phi(z)^{-1}(1 - z)^{-d}$ and $\{\varepsilon_t\}$ is a Gaussian white noise sequence.

From equation (4.2), an infinite-dimensional state space system may be written as in (2.4)–(2.8). The Gaussian log-likelihood function can be evaluated by directly applying the Kalman recursive equations in (2.9)–(2.13) to the infinite-dimensional system.

Let $\Omega_t = (\omega_{ij}^{(t)})$ be the state estimation error covariance matrix at time t. The Kalman equations for the infinite-dimensional system are given by

$$\widehat{x}_1 = E[x_1], \tag{4.3}$$
$$\Omega_1 = E[x_1 x_1'] - E[\widehat{x}_1 \widehat{x}_1'], \tag{4.4}$$
$$\Delta_t = \omega_{11}^{(t)} + 1, \tag{4.5}$$
$$\omega_{ij}^{(t+1)} = \omega_{i+1,j+1}^{(t)} + \psi_i \psi_j - \frac{(\omega_{i+1,1}^{(t)} + \psi_i)(\omega_{j+1,1}^{(t)} + \psi_j)}{\omega_{11}^{(t)} + 1}, \tag{4.6}$$

and

$$\widehat{x}_{t+1} = [\widehat{x}_1^{(t+1)}, \widehat{x}_2^{(t+1)}, \ldots]' = [\widehat{x}_i^{(t+1)}]_{i=1,2,\ldots}',$$

where

$$\widehat{x}_i^{(t+1)} = \widehat{x}_{i+1}^{(t)} + \frac{(y_t - \widehat{x}_1^{(t)})(\omega_{i+1,1}^{(t)} + \psi_i)}{\omega_{11}^{(t)} + 1}, \tag{4.7}$$
$$\widehat{y}_{t+1} = G\widehat{x}_{t+1} = \widehat{x}_1^{(t+1)}, \tag{4.8}$$

and the log-likelihood function is given by

$$\mathcal{L}(\theta) = -\frac{1}{2}\left[n \log 2\pi + \sum_{t=1}^{n} \log \Delta_t + n \log \sigma^2 + \frac{1}{\sigma^2} \sum_{t=1}^{n} \frac{(y_t - \widehat{y}_t)^2}{\Delta_t} \right]. \tag{4.9}$$

According to Theorem 2.1, all the state space representations of an ARFIMA model are infinite-dimensional. Despite this fact, the exact likelihood function can be evaluated in a finite number of steps, as specified in the following result due to Chan and Palma (1998).

Theorem 4.1. *Let $\{y_1, \ldots, y_n\}$ be a finite sample of an ARFIMA(p,d,q) process. If Ω_1 is the variance of the initial state x_1 of the infinite-dimensional representation* (2.1)–(2.2), *then the computation of the exact likelihood function* (4.9) *depends only on the first n components of the Kalman equations* (4.3)–(4.8).

As a consequence of Theorem 4.1, for an ARFIMA time series with n observations, the calculation of the exact likelihood function can be based only on the first n components of the state vector and the remaining infinitely many components of the state vector may be omitted from the computations.

The numerical complexity of the exact state space method is $\mathcal{O}(n^3)$. Hence, this approach is comparable to the Cholesky decomposition but it is less efficient than the Durbin-Levinson procedure. The use of the exact Kalman algorithm is advisable for moderate sample sizes or for handling missing values. The state space systems provide a simple solution to this problem. On the contrary, the Durbin-Levinson method cannot be used when a time series has missing values since the variance-covariance matrix of the observed process does not have a Toeplitz structure.

The consistency, asymptotic normality, and efficiency of the exact MLE for long-memory process are analyzed in detail in Chapter 5. Here is a summary of the asymptotic properties; see Dahlhaus (1989).

Theorem 4.2. *Let $\widehat{\theta}_n$ be the value that maximizes the exact log-likelihood where*

$$\theta = (\phi_1, ..., \phi_p, \theta_1, ..., \theta_q, d)'$$

is a $p+q+1$ dimensional parameter vector and let θ_0 be the true *parameter. Under some regularity conditions we have*

(a) *Consistency: $\widehat{\theta}_n \to \theta_0$ in probability as $n \to \infty$.*

(b) *Central Limit Theorem: $\sqrt{n}(\widehat{\theta}_n - \theta_0) \to N(0, \Gamma^{-1}(\theta_0))$, as $n \to \infty$, where $\Gamma(\theta) = (\Gamma_{ij}(\theta))$ with*

$$\Gamma_{ij}(\theta) = \frac{1}{4\pi} \int_{-\pi}^{\pi} \left[\frac{\partial \log f_\theta(\lambda)}{\partial \theta_i} \right] \left[\frac{\partial \log f_\theta(\lambda)}{\partial \theta_j} \right] d\lambda, \tag{4.10}$$

where f_θ is the spectral density of the process.

(c) *Efficiency: $\widehat{\theta}_n$ is an efficient estimator of θ_0.*

The next two sections discuss maximum-likelihood procedures based on AR and MA approximations, respectively.

4.2 AUTOREGRESSIVE APPROXIMATIONS

Given that the computation of exact ML estimates is computationally demanding, many authors have considered the use of autoregressive approximations to speed up the calculation of parameter estimates. Let $\{y_t : t \in \mathbb{Z}\}$ be a long-memory process defined by the autoregressive expansion

$$y_t = \varepsilon_t + \pi_1(\theta) y_{t-1} + \pi_2(\theta) y_{t-2} + \pi_3(\theta) y_{t-3} + \cdots,$$

where $\pi_j(\theta)$ are the coefficients of $\phi(B)\theta^{-1}(B)(1-B)^d$. Since in practice only a finite number of observations is available, $\{y_1, \ldots, y_n\}$, the following truncated model is considered

$$y_t = \widetilde{\varepsilon}_t + \pi_1(\theta) y_{t-1} + \pi_2(\theta) y_{t-2} + \cdots + \pi_m(\theta) y_{t-m}, \tag{4.11}$$

for $m < t \leq n$. Then, the approximate maximum-likelihood estimate $\widehat{\theta}_n$ is obtained by minimizing the function

$$\mathcal{L}_1(\theta) = \sum_{t=m+1}^{n} [y_t - \pi_1(\theta) y_{t-1} - \pi_2(\theta) y_{t-2} - \cdots - \pi_m(\theta) y_{t-m}]^2. \tag{4.12}$$

Many improvements can be made on this basic framework to obtain better estimates. In the following subsections, we describe some of these refinements. For simplicity, an estimator produced by the maximization of an approximation of the Gaussian likelihood function (4.9) will be called *quasi-maximum-likelihood estimate* (QMLE).

4.2.1 Haslett-Raftery Method

Consider the ARFIMA process (4.2). An approximate one-step forecast of y_t is given by

$$\widehat{y}_t = \phi(B)\theta(B)^{-1}\sum_{j=1}^{t-1}\phi_{tj}y_{t-j}, \tag{4.13}$$

with prediction error variance

$$v_t = \mathrm{Var}(y_t - \widehat{y}_t) = \sigma_y^2\kappa\prod_{j=1}^{t-1}(1-\phi_{jj}^2),$$

where $\sigma_y^2 = \mathrm{Var}(y_t)$, κ is the ratio of the innovations variance to the variance of the ARMA(p,q) process as given by Equation (3.4.4) of Box, Jenkins, and Reinsel (1994), and

$$\phi_{tj} = -\binom{t}{j}\frac{\Gamma(j-d)\Gamma(t-d-j+1)}{\Gamma(-d)\Gamma(t-d+1)}, \tag{4.14}$$

for $j = 1,\ldots,t$.

To avoid the computation of a large number of coefficients ϕ_{tj}, the last term of the predictor (4.13) is approximated by

$$\sum_{j=1}^{t-1}\phi_{tj}y_{t-j} \approx \sum_{j=1}^{M}\phi_{tj}y_{t-j} - \sum_{j=M+1}^{t-1}\pi_j y_{t-j}, \tag{4.15}$$

since $\phi_{tj} \sim -\pi_j$ for large j, where for simplicity π_j denotes $\pi_j(\theta)$.

An additional approximation is made to the second term on the right-hand side of (4.15):

$$\sum_{j=M+1}^{t-1}\pi_j y_{t-j} \approx M\pi_M d^{-1}\left[1-\left(\frac{M}{t}\right)^d\right]\bar{y}_{M+1,t-1-M},$$

where $\bar{y}_{M+1,t-1-M} = \frac{1}{t-1-2M}\sum_{j=M+1}^{t-1-M}y_j$. Hence, a QMLE $\widehat{\theta}_n$ is obtained by maximizing

$$\mathcal{L}_2(\theta) = \text{constant} - \tfrac{1}{2}n\log[\widehat{\sigma}^2(\theta)],$$

with

$$\widehat{\sigma}^2(\theta) = \frac{1}{n}\sum_{t=1}^{n}\frac{(y_t-\widehat{y}_t)^2}{v_t}.$$

The Haslett-Raftery algorithm has numeric complexity of order $\mathcal{O}(nM)$. Therefore, if the truncation parameter M is fixed, then this method is order $\mathcal{O}(n)$. Thus, it is usually faster than the Cholesky Decomposition and the Durbin-Levinson method. It has been suggested that $M = 100$ works fine in most applications. Besides, by setting $M = n$ we get the exact ML estimator for the fractional noise process. But, the numerical complexity in situation is order $\mathcal{O}(n^2)$.

4.2.2 Beran Approach

A slightly different autoregressive approximation technique, proposed by Beran (1994a), proceeds as follows. Consider the following Gaussian innovation sequence:

$$\varepsilon_t = y_t - \sum_{j=1}^{\infty} \pi_j(\theta) y_{t-j}.$$

Since the values $\{y_t, t \leq 0\}$ are not observed, an approximate innovation sequence $\{u_t\}$ may be obtained by assuming that $y_t = 0$ for $t \leq 0$,

$$u_t = y_t - \sum_{j=1}^{t-1} \pi_j(\theta) y_{t-j},$$

for $j = 2, \ldots, n$. Let $r_t(\theta) = u_t(\theta)/\sigma$ and $\theta = (\sigma, \phi_1, \ldots, \phi_p, \theta_1, \ldots, \theta_q, d)$. Then, a QMLE for θ is provided by the minimization of

$$\mathcal{L}_2(\theta) = 2n \log(\sigma) + \sum_{t=2}^{n} r_t^2(\theta).$$

Now, by taking partial derivatives with respect to θ, the minimization problem is equivalent to solving the nonlinear equations

$$\sum_{t=2}^{n} \{r_t(\theta)\dot{r}_t(\theta) - E[r_t(\theta)\dot{r}_t(\theta)]\} = 0, \tag{4.16}$$

where $\dot{r}_t(\theta) = \left(\frac{\partial r_t(\theta)}{\partial \theta_1}, \ldots, \frac{\partial r_t(\theta)}{\partial \theta_r}\right)'$.

The arithmetic complexity of this method is $\mathcal{O}(n^2)$, so it is comparable to the Durbin-Levinson algorithm. Unlike the Haslett-Raftery method, this alternative approach uses the same variance for all the errors u_t. Hence, its performance may be poor for short time series.

The QMLE based on the autoregressive approximations share the same asymptotic properties with the exact MLE; see Beran (1994a).

Theorem 4.3. *Let $\widetilde{\theta}_n$ be the value that solves (4.16), then*

(a) *Consistency: $\widetilde{\theta}_n \to \theta_0$, in probability, as $n \to \infty$.*

(b) *Central Limit Theorem: $\sqrt{n}(\widetilde{\theta}_n - \theta_0) \to N(0, \Gamma^{-1}(\theta_0))$, as $n \to \infty$, with $\Gamma(\theta_0)$ is given in (4.10).*

(c) *Efficiency: $\widetilde{\theta}_n$ is an efficient estimator of θ_0.*

4.2.3 A State Space Method

The state space methodology may be also used for handling autoregressive approximations. For instance, starting from the AR(m) truncation (4.11) and dropping θ from the coefficients π_j and the tilde from ε_t, we have

$$y_t = \pi_1 y_{t-1} + \cdots + \pi_m y_{t-m} + \varepsilon_t.$$

Thus, we may write the following state space system:

$$\begin{aligned} x_{t+1} &= F x_t + H \varepsilon_{t+1}, \\ y_t &= G x_t, \end{aligned}$$

where the state is given by $x_t = [y_t \; y_{t-1} \; \cdots \; y_{t-m+2} \; y_{t-m+1}]'$, the state transition matrix is

$$F = \begin{pmatrix} \pi_1 & \pi_2 & \cdots & \pi_{m-1} & \pi_m \\ 1 & 0 & \cdots & 0 & 0 \\ 0 & 1 & \cdots & 0 & 0 \\ \vdots & & & & \vdots \\ 0 & 0 & \cdots & 0 & 0 \\ 0 & 0 & \cdots & 1 & 0 \end{pmatrix},$$

the observation matrix is

$$G = [1\,0\,0 \cdots 0],$$

and the state noise matrix is given by

$$H = [1\,0\,0 \cdots 0]'.$$

The variance of the observation noise is $R = 0$, the covariance between the state noise and the observation noise is $S = 0$ and the state noise variance-covariance matrix is given by

$$Q = \sigma^2 \begin{pmatrix} 1 & 0 & \cdots & 0 \\ 0 & 0 & \cdots & 0 \\ \vdots & & & \vdots \\ 0 & 0 & \cdots & 0 \\ 0 & 0 & \cdots & 0 \end{pmatrix}.$$

The Kalman recursions for this system are

$$
\begin{aligned}
\Delta_t &= \omega_t(1,1),\\
\Theta_t &= \left[\sum_{i=1}^{m}\pi_i\omega_t(i,1)\ \omega_t(1,1)\ \cdots\ \omega_t(m-1,1)\right]',\\
\omega_{t+1}(1,1) &= \sigma^2+\sum_{i,j=1}^{m}\pi_i\omega_t(i,j)\pi_j-\left[\sum_{i=1}^{m}\pi_i\omega_t(i,1)\right]^2/\omega_t(1,1),\\
\omega_{t+1}(1,j) &= \sum_{i=1}^{m}\pi_i\omega_t(i,j)-\left[\sum_{i=1}^{m}\pi_i\omega_t(i,1)\right]\frac{\omega_t(1,j)}{\omega_t(1,1)}\quad\text{for } j\geq 2,\\
\omega_{t+1}(i,j) &= \omega_t(i-1,j-1)-\omega_t(i-1,1)\omega_t(j-1,1)/\omega_t(1,1)\quad\text{for } i,j\geq 2,\\
\widehat{x}_{t+1}(1) &= \sum_{i=1}^{m}\pi_i\widehat{x}_t(i)+\left[\sum_{i=1}^{m}\pi_i\omega_t(i,1)\right]\frac{y_t-\widehat{x}_t(1)}{\omega_t(1,1)},\\
\widehat{x}_{t+1}(j) &= \widehat{x}_t(j-1)+\omega_t(j-1,1)\frac{y_t-\widehat{x}_t(1)}{\omega_t(1,1)}\quad\text{for } j\geq 2,\\
\widehat{y}_{t+1} &= \widehat{x}_{t+1}(1).
\end{aligned}
$$

The initial conditions for these iteration may be

$$\widehat{x}_0=0,$$

and

$$
\Omega_0=\begin{pmatrix}
\gamma(0) & \gamma(1) & \cdots & \gamma(m-1)\\
\gamma(1) & \gamma(0) & \cdots & \gamma(m-2)\\
\vdots & \vdots & & \vdots\\
\gamma(m-2) & \gamma(m-3) & \cdots & \gamma(1)\\
\gamma(m-1) & \gamma(m-2) & \cdots & \gamma(0)
\end{pmatrix}.
$$

The calculation of the log-likelihood proceeds analogously to the formula given in (4.9). This representation is particularly useful for interpolation of missing values since the state noise is uncorrelated with the observation noise; see Brockwell and Davis (1991, p. 487).

4.3 MOVING-AVERAGE APPROXIMATIONS

An alternative methodology to autoregressive approximations is the truncation of the Wold expansion of a long-memory process. Two advantages of this approach are the easy implementation of the Kalman filter recursions and the simplicity of the analysis of the theoretical properties of the ML estimates. Besides, if the long-memory time series is differenced, then the resulting moving-average truncation has smaller error variance than the autoregressive approximation.

A causal representation of an ARFIMA(p, d, q) process $\{y_t\}$ is given by

$$y_t = \sum_{j=0}^{\infty} \psi_j \varepsilon_{t-j}, \tag{4.17}$$

and we may consider an approximate model for (4.17) given by

$$y_t = \sum_{j=0}^{m} \psi_j \varepsilon_{t-j}, \tag{4.18}$$

which corresponds to a MA(m) process in contrast to the MA(∞) process (4.17). A canonical state space representation of the MA(m) model (4.18) is given by

$$\begin{aligned} x_{t+1} &= Fx_t + H\varepsilon_t, \\ y_t &= Gx_t + \varepsilon_t, \end{aligned}$$

with

$$x_t = [y(t|t-1)\ y(t+1|t-1)\ \cdots\ y(t+m-1|t-1)]',$$

where $y(t+j|t-1) = E[y_{t+j}|y_{t-1}, y_{t-2}, \cdots]$ and system matrices

$$F = \begin{bmatrix} 0 & I_{m-1} \\ 0 & 0 \end{bmatrix}, \qquad G = \begin{bmatrix} 1 & 0 & \cdots & 0 \end{bmatrix}, \quad H = [\psi_1 \cdots \psi_m]'.$$

The approximate representation of a causal ARFIMA(p, d, q) has computational advantages over the exact one. In particular, the order of the MLE algorithm is reduced from $\mathcal{O}(n^3)$ to $\mathcal{O}(n)$.

The log-likelihood function, excepting a constant, is given by

$$\mathcal{L}(\theta) = -\frac{1}{2}\left\{ \sum_{t=1}^{n} \log\Delta_t(\theta) + \sum_{t=1}^{n} \frac{[y_t - \widehat{y}_t(\theta)]^2}{\Delta_t(\theta)} \right\},$$

where $\theta = (\phi_1, \ldots, \phi_p, \theta_1, \ldots, \theta_q, d, \sigma^2)$ is the parameter vector associated to the ARFIMA representation (4.1).

In order to evaluate the log-likelihood function $\mathcal{L}(\theta)$ we may choose the initial conditions $\widehat{x}_1 = E[x_1] = 0$ and $\Omega_1 = E[x_1 x_1'] = [\omega(i,j)]_{i,j=1,2,\ldots}$ where $\omega(i,j) = \sum_{k=0}^{\infty} \psi_{i+k}\psi_{j+k}$.

The evolution of the state estimation and its variance, Ω_t, is given by the following recursive equations. Let $\delta_i = 1$ if $i \in \{0, 1, \ldots, m-1\}$ and $\delta_i = 0$ otherwise. Furthermore, let $\delta_{ij} = \delta_i \delta_j$. Then, the elements of Ω_{t+1} and x_{t+1} are as follows:

$$\Delta_t = \omega_t(1,1) + 1, \tag{4.19}$$

$$\begin{aligned} \omega_{t+1}(i,j) &= \omega_t(i+1, j+1)\delta_{ij} + \psi_i \psi_j \\ &\quad - \frac{[\omega_t(i+1,1)\delta_i + \psi_i][\omega_t(j+1,1)\delta_j + \psi_j]}{\omega_t(1,1) + 1}, \end{aligned} \tag{4.20}$$

the state estimation is

$$\widehat{x}_{t+1}(i) = \widehat{x}_t(i+1)\delta_i + \frac{[\omega_t(i+1,1)\delta_i + \psi_i][y_t - \widehat{x}_t(1)]}{\omega_t(1,1)+1}, \tag{4.21}$$

and the observation predictor is given by

$$\widehat{y}_t = G\widehat{x}_t = \widehat{x}_t(1).$$

A faster version of the previous algorithm can be obtained by differencing the series $\{y_t\}$ since the infinite MA representation of the differenced series converges more rapidly than the MA expansion of the original process. To illustrate this approach, consider the differenced process

$$z_t = (1-B)y_t = \sum_{j=0}^{\infty} \varphi_j \varepsilon_{t-j}, \tag{4.22}$$

where $\varphi_j = \psi_j - \psi_{j-1}$.

Remark 4.1. It is worth noting that by virtue of Theorem 3.4, the process $\{z_t\}$ is stationary and invertible for any $d \in (0, \frac{1}{2})$, provided that the AR(p) and MA(q) polynomials do not have common roots and all their roots are outside the closed unit disk.

By truncating the MA(∞) expansion (4.22) after m components, we get the approximate model

$$z_t = \sum_{j=0}^{m} \varphi_j \varepsilon_{t-j}. \tag{4.23}$$

An advantage of this approach is that, as shown in Problem 3.14, the coefficients φ_j converge faster to zero than the coefficients ψ_j. Consequently, a smaller truncation parameter m is necessary to achieve a good approximation level.

The truncated model (4.23) can be represented in terms of a state space system as

$$\begin{aligned} x_{t+1} &= \begin{bmatrix} 0 & I_{m-1} \\ 0 & 0 \end{bmatrix} x_t + \begin{bmatrix} \varphi_1 \\ \vdots \\ \varphi_m \end{bmatrix} \varepsilon_t, \\ z_t &= \begin{bmatrix} 1 & 0 & \cdots & 0 \end{bmatrix} x_t + \varepsilon_t. \end{aligned}$$

Under normality, the log-likelihood function of the truncated model (4.23) may be written as

$$\mathcal{L}_n(\theta) = -\frac{1}{2n} \log \det T_{n,m}(\theta) - \frac{1}{2n} z' T_{n,m}(\theta)^{-1} z, \tag{4.24}$$

where $[T_{n,m}(\theta)]_{r,s=1,\dots,n} = \int_{-\pi}^{\pi} f_{m,\theta}(\lambda) e^{i\lambda(r-s)} d\lambda$, is the covariance matrix of $z = (z_1, \dots, z_n)'$ with $f_{m,\theta}(\lambda) = (2\pi)^{-1}\sigma^2 |\varphi_m(e^{i\lambda})|^2$ and the polynomial $\varphi_m(\cdot)$ is given by

$$\varphi_m(e^{i\lambda}) = 1 + \varphi_1 e^{i\lambda} + \cdots + \varphi_m e^{mi\lambda}.$$

The matrices involved in the truncated Kalman equations are of size $m \times m$. Thus, only $\mathcal{O}(m^2)$ evaluations are required for each iteration and the algorithm has an order $\mathcal{O}(n \times m^2)$. For a fixed truncation parameter m, the calculation of the likelihood function is only of order $\mathcal{O}(n)$ for the approximate ML method. Therefore, for very large samples, it may be desirable to consider truncating the Kalman recursive equations after m components. With this truncation, the number of operations required for a single evaluation of the log-likelihood function is reduced to $\mathcal{O}(n)$. The following theorem, due to Chan and Palma (1998), establishes some large sample properties of the truncated maximum-likelihood estimate.

Theorem 4.4. *Let $\widetilde{\theta}_{n,m}$ be the value that maximizes* (4.24) *and let θ_0 be the true parameter. Then,*

(a) *Consistency: Assume that $m = n^\beta$ with $\beta > 0$, then $\widetilde{\theta}_{n,m} \to \theta_0$ in probability as $n \to \infty$.*

(b) *Normality: Suppose that $m = n^\beta$ with $\beta \geq \frac{1}{2}$, then $\sqrt{n}(\widetilde{\theta}_{n,m} - \theta_0) \to N(0, \Gamma(\theta_0)^{-1})$ as $n \to \infty$ where, $\Gamma(\theta)$ is given in* (4.10).

(c) *Efficiency: Assume that $m = n^\beta$ with $\beta \geq \frac{1}{2}$, then $\widetilde{\theta}_{n,m}$ is an efficient estimator of θ_0.*

It is worth noting that the autoregressive, AR(m), and the moving-average, MA(m), approximations produce algorithms with numerical complexity of order $\mathcal{O}(n)$, where n is the sample size. Nevertheless, the quality of these approximations is governed by the truncation parameter m. The variance of the truncation error for an AR(m) approximation is of order $\mathcal{O}(1/m)$ while this variance is of order $\mathcal{O}(m^{2d-1})$ in the MA(m) case. On the other hand, the truncation error variance is of order $\mathcal{O}(m^{2d-3})$ for the differenced approach.

4.4 WHITTLE ESTIMATION

A well-known methodology to obtain approximate maximum-likelihood estimates is based on the calculation of the periodogram—see equation (3.27)—by means of the *fast Fourier transform* (FFT) and the use of the so-called *Whittle approximation* of the Gaussian log-likelihood function. Since the calculation of the FFT has a numerical complexity of order $\mathcal{O}[n \log_2(n)]$, this approach produces very fast algorithms for computing parameter estimates.

Suppose that the sample vector $\boldsymbol{y} = (y_1, \ldots, y_n)'$ is normally distributed with zero mean and variance Γ_θ. Then, the log-likelihood function divided by the sample size is given by

$$\mathcal{L}(\theta) = -\frac{1}{2n} \log \det \Gamma_\theta - \frac{1}{2n} \boldsymbol{y}' \Gamma_\theta^{-1} \boldsymbol{y}. \tag{4.25}$$

Notice that the variance-covariance matrix Γ_θ may be expressed in terms of the spectral density of the process $f_\theta(\cdot)$ as follows:

$$(\Gamma_\theta)_{ij} = \gamma_\theta(i - j),$$

where

$$\gamma_\theta(k) = \int_{-\pi}^{\pi} f_\theta(\lambda) \exp(i\lambda k)\, d\lambda.$$

In order to obtain the Whittle method, two approximations are made. Since

$$\frac{1}{n} \log \det \Gamma_\theta \to \frac{1}{2\pi} \int_{-\pi}^{\pi} \log[2\pi f_\theta(\lambda)]\, d\lambda,$$

as $n \to \infty$, the first term in (4.25) is approximated by

$$\frac{1}{2n} \log \det \Gamma_\theta \approx \frac{1}{4\pi} \int_{-\pi}^{\pi} \log[2\pi f_\theta(\lambda)]\, d\lambda.$$

On the other hand, the second term in (4.25) is approximated by

$$\begin{aligned}
\frac{1}{2n} y' \Gamma_\theta^{-1} y &\approx \sum_{\ell=1}^{n} \sum_{j=1}^{n} y_\ell \left\{ \frac{1}{8\pi^2 n} \int_{-\pi}^{\pi} f_\theta^{-1}(\lambda) \exp[i\lambda(\ell - j)]\, d\lambda \right\} y_j \\
&= \frac{1}{8\pi^2 n} \int_{-\pi}^{\pi} f_\theta^{-1}(\lambda) \sum_{\ell=1}^{n} \sum_{j=1}^{n} y_\ell y_j \exp[i\lambda(\ell - j)]\, d\lambda \\
&= \frac{1}{8\pi^2 n} \int_{-\pi}^{\pi} f_\theta^{-1}(\lambda) \left| \sum_{j=1}^{n} y_j \exp(i\lambda j) \right|^2 d\lambda \\
&= \frac{1}{4\pi} \int_{-\pi}^{\pi} \frac{I(\lambda)}{f_\theta(\lambda)}\, d\lambda,
\end{aligned}$$

where

$$I(\lambda) = \frac{1}{2\pi n} \left| \sum_{j=1}^{n} y_j e^{i\lambda j} \right|^2$$

is the *periodogram* of the series $\{y_t\}$ defined in equation (3.27).

Thus, the log-likelihood function is approximated, up to a constant, by

$$\mathcal{L}_3(\theta) = -\frac{1}{4\pi} \left[\int_{-\pi}^{\pi} \log f_\theta(\lambda)\, d\lambda + \int_{-\pi}^{\pi} \frac{I(\lambda)}{f_\theta(\lambda)}\, d\lambda \right]. \tag{4.26}$$

The evaluation of the log-likelihood function (4.26) requires the calculation of integrals. To simplify this computation, the integrals can be substituted by Riemann sums as follows:

$$\int_{-\pi}^{\pi} \log f_\theta(\lambda)\, d\lambda \approx \frac{2\pi}{n} \sum_{j=1}^{n} \log f_\theta(\lambda_j),$$

and

$$\int_{-\pi}^{\pi} \frac{I(\lambda)}{f_\theta(\lambda)} d\lambda \approx \frac{2\pi}{n} \sum_{j=1}^{n} \frac{I(\lambda_j)}{f_\theta(\lambda_j)},$$

where $\lambda_j = 2\pi j/n$ are the Fourier frequencies. Thus, a discrete version of the log-likelihood function (4.26) is

$$\mathcal{L}_4(\theta) = -\frac{1}{2n} \left[\sum_{j=1}^{n} \log f_\theta(\lambda_j) + \sum_{j=1}^{n} \frac{I(\lambda_j)}{f_\theta(\lambda_j)} \right].$$

4.4.1 Other versions

Other versions of the Whittle likelihood function are obtained by making additional assumptions. For instance, if the spectral density is normalized as

$$\int_{-\pi}^{\pi} \log f_\theta(\lambda)\, d\lambda = 0, \tag{4.27}$$

then the Whittle log-likelihood function is reduced to

$$\mathcal{L}_5(\theta) = -\frac{1}{4\pi} \int_{-\pi}^{\pi} \frac{I(\lambda)}{f_\theta(\lambda)}\, d\lambda,$$

with the corresponding discrete version

$$\mathcal{L}_6(\theta) = -\frac{1}{2n} \sum_{j=1}^{n} \frac{I(\lambda_j)}{f_\theta(\lambda_j)}.$$

Note that according to the Szegö-Kolmogorov formula (1.6), the normalization (4.27) is equivalent to setting $\sigma^2 = 2\pi$. The Whittle estimator and the exact MLE share similar asymptotic properties, as established in the following result due to Fox and Taqqu (1986) and Dahlhaus (1989).

Theorem 4.5. *Let $\widehat{\theta}_n^{(i)}$ be the value that maximizes the log-likelihood function $\mathcal{L}_i(\theta)$ for $i = 3, \ldots, 6$ for a Gaussian process $\{y_t\}$. Then, under some regularity conditions, $\widehat{\theta}_n^{(i)}$ is consistent and $\sqrt{n}[\widehat{\theta}_n^{(i)} - \theta_0] \to N[0, \Gamma(\theta_0)^{-1}]$ as $n \to \infty$, where $\Gamma(\theta_0)$ is the matrix defined in* (4.10).

4.4.2 Non-Gaussian Data

The methods reviewed so far apply to Gaussian processes. However, if this assumption is dropped, we still can find well-behaved Whittle estimates. For example, let $\{y_t\}$ be a stationary process with Wold decomposition:

$$y_t = \sum_{j=0}^{\infty} \psi_j(\theta)\varepsilon_{t-j},$$

where ε_t is an independent and identically distributed sequence with finite four cumulant and $\sum_{j=0}^{\infty} \psi_j^2(\theta) < \infty$. The following result establishes the consistency and the asymptotic normality of the Whittle estimate under these circumstances; see Giraitis and Surgailis (1990).

Theorem 4.6. *Let $\widehat{\theta}_n$ be the value that maximizes the log-likelihood function $\mathcal{L}_5(\theta)$. Then, under some regularity conditions, $\widehat{\theta}_n$ is consistent and $\sqrt{n}(\widehat{\theta}_n - \theta_0) \to N[0, \Gamma(\theta_0)^{-1}]$ as $n \to \infty$, where $\Gamma(\theta_0)$ is the matrix defined in* (4.10).

It is important to emphasize that unlike Theorem 4.5, this result does not assume the normality of the process.

4.4.3 Semiparametric Methods

In this subsection we analyze a generalization of the Whittle approach called the *Gaussian semiparametric* estimation method. This technique does not require the specification of a parametric model for the data. It only relies on the specification of the shape of the spectral density of the time series.

Assume that $\{y_t\}$ is a stationary process with spectral density satisfying

$$f(\lambda) \sim G\lambda^{1-2H},$$

as $\lambda \to 0+$, with $G \in (0, \infty)$ and $H \in (0, 1)$. Observe that for an ARFIMA model, the terms G and H correspond to $\sigma^2\theta(1)^2/[2\pi\phi(1)^2]$ and $\frac{1}{2} + d$, respectively. Let us define $Q(G, H)$ as the objective function

$$Q(G, H) = \frac{1}{m} \sum_{j=1}^{m} \left[\log G\lambda_j^{1-2H} + \frac{\lambda_j^{2H-1}}{G} I(\lambda_j) \right],$$

where m is an integer satisfying $m < n/2$. If $(\widehat{G}, \widehat{H})$ is the value that minimizes $Q(G, H)$, then under some regularity conditions such as

$$\frac{1}{m} + \frac{m}{n} \to 0,$$

as $n \to \infty$, Robinson (1995a) established the following result.

Theorem 4.7. *If H_0 is the true value of the self-similarity parameter, then the estimator $\widehat{H}$ is consistent and $\sqrt{m}(\widehat{H} - H_0) \to N\left(0, \frac{1}{4}\right)$ as $n \to \infty$.*

4.5 OTHER METHODS

In this section we review four additional methodologies for estimating the long-memory parameter d. Among these techniques we study a log-periodogram regression, the so-called *rescaled range statistic* (R/S), the variance plots, the *detrended fluctuation analysis*, and a wavelet-based approach.

4.5.1 A Regression Method

Under the assumption that the spectral density of a stationary process may be written as

$$f(\lambda) = f_0(\lambda)[2\sin(\lambda/2)]^{-2d}, \tag{4.28}$$

we may consider the following regression method for parameter estimation proposed by Geweke and Porter-Hudak (1983).

Taking logarithms on both sides of (4.28) and evaluating the spectral density at the Fourier frequencies $\lambda_j = 2\pi j/n$, we have that

$$\log f(\lambda_j) = \log f_0(0) - d\log\left[2\sin\frac{\lambda_j}{2}\right]^2 + \log\left[\frac{f_0(\lambda_j)}{f_0(0)}\right]. \tag{4.29}$$

On the other hand, the logarithm of the periodogram $I(\lambda_j)$ may be written as

$$\log I(\lambda_j) = \log\left[\frac{I(\lambda_j)}{f(\lambda_j)}\right] + \log f(\lambda_j). \tag{4.30}$$

Now, combining (4.29) and (4.30) we have

$$\log I(\lambda_j) = \log f_0(0) - d\log\left[2\sin\frac{\lambda_j}{2}\right]^2 + \log\left\{\frac{I(\lambda_j)[2\sin(\lambda/2)]^{2d}}{f_0(0)}\right\}.$$

By defining $y_j = \log I(\lambda_j)$, $\alpha = \log f_0(0)$, $\beta = -d$, $x_j = \log[2\sin(\lambda_j/2)]^2$, and

$$\varepsilon_j = \log\left\{\frac{I(\lambda_j)[2\sin(\lambda/2)]^{2d}}{f_0(0)}\right\},$$

we obtain the regression equation

$$y_j = \alpha + \beta x_j + \varepsilon_j.$$

In theory, one could expect that for frequencies near zero (that is, for $j = 1, \dots, m$ with $m \ll n$)

$$f(\lambda_j) \sim f_0(0)[2\sin(\lambda_j/2)]^{-2d},$$

so that

$$\varepsilon_j \sim \log\left[\frac{I(\lambda_j)}{f(\lambda_j)}\right].$$

The least squares estimate of the long-memory parameter d is given by

$$\widehat{d}_m = -\frac{\sum_{j=1}^{m}(x_j - \bar{x})(y_j - \bar{y})}{\sum_{j=1}^{m}(x_j - \bar{x})^2},$$

where $\bar{x} = \sum_{j=1}^{m} x_j/m$ and $\bar{y} = \sum_{j=1}^{m} y_j/m$. The asymptotic properties of this and other related estimates have been analyzed by Robinson (1995b) and Hurvich, Deo, and Brodsky (1998), among others. The following result corresponds to Theorem 2 of Hurvich, Deo, and Brodsky (1998).

Theorem 4.8. *Let $\{y_t\}$ be a Gaussian process with spectral density satisfying* (4.28) *where f_0 is an even, positive, continuous function on $[-\pi, \pi]$. Furthermore, assume that f_0 has null first derivative at the origin and that it has bounded second and third derivative in a neighborhood of zero. Let $\widehat{d}_m$ be the least squares estimate of the parameter d. If $m = o(n^{4/5})$ and $\log^2 n = o(m)$, then*

$$\sqrt{m}(\widehat{d}_m - d) \to N\left(0, \frac{\pi^2}{24}\right),$$

as $m \to \infty$.

4.5.2 Rescaled Range Method

Consider the sample $\{y_1, \dots, y_n\}$ from a stationary long-memory process and let x_t be the partial sums of $\{y_t\}$, that is, $x_t = \sum_{j=1}^{t} y_j$ for $t = 1, \dots, n$ and let $s_n^2 = \sum_{t=1}^{n}(y_t - \bar{y})^2/(n-1)$ be the sample variance where $\bar{y} = x_n/n$.

The *rescaled range statistic* (R/S) introduced by Hurst (1951) is defined by

$$R_n = \frac{1}{s_n}\left[\max_{1\le t\le n}\left(x_t - \frac{t}{n}x_n\right) - \min_{1\le t\le n}\left(x_t - \frac{t}{n}x_n\right)\right].$$

This statistic satisfies the following asymptotic property as shown by Mandelbrot (1975, 1976).

Theorem 4.9. *Let $\{y_t : t \in \mathbb{Z}\}$ be a zero mean stationary process such that y_t^2 is ergodic and*

$$n^{-\frac{1}{2}-d} x_{[tn]} \to B_d(t),$$

in distribution, as $n \to \infty$, where $B_d(t)$ is the fractional Brownian motion defined in Subsection 1.1.11. *Define $Q_n = n^{-1/2-d}R_n$, then*

$$Q_n \to Q,$$

in distribution, as $n \to \infty$, where

$$Q = \sup_{0\le t\le 1}\left[B_d(t) - tB_d(1)\right] - \inf_{0\le t\le 1}\left[B_d(t) - tB_d(1)\right].$$

Note that $\log R_n = E\,Q_n + (d + \frac{1}{2})\log n + (\log Q_n - E\,Q_n)$, so that we can obtain an estimator of the long-memory parameter d by a least squares technique similar to the one studied in Subsection 4.5.1. For instance, if $R_{t,k}$ is the R/S statistic based on the sample of size k, $\{y_t, \dots, y_{t+k-1}\}$ for $1 \le t \le n-k+1$, then an estimator of d can be obtained by regressing $\log R_{t,k}$ on $\log k$ for $1 \le t \le n-k+1$.

EXAMPLE 4.2

Figures 4.1 and 4.2 illustrate how the R/S method works in a practical context. For comparison purpose, we first analyze a simulated series of independent observations (white noise) and then apply the R/S technique to a real-life data exhibiting long-memory behavior.

Figure 4.1 exhibits the values of the statistics $\log R_{t,k}$ plotted against $\log k$ with $k = 10, 20, \ldots, 200$, $t = 50j$, $j = 1, 2, \ldots$ for 1000 observations of a Gaussian white noise with zero mean and unit variance.

The slope of the least squares regression line in this case is 0.4858 and the intercept is 0.0256. Thus, for this white noise sequence the estimated value of the long-memory parameter is $\widehat{d} = -0.014$, reflecting the lack of memory of these data.

On the other hand, Figure 4.2 displays the values of the statistics $\log R_{t,k}$ plotted against $\log k$ with $k = 10, 20, \ldots, 200$, $t = 50j$, $j = 1, 2, \ldots$ for the White Mountains tree ring data for the period 964–1963. This long-range-dependent series, available from StatLib, is described in Hipel and McLeod (1994).

The slope of the least squares regression line in this case is 0.658 and the intercept is -0.315. Therefore, the estimated value of the long-memory parameter for these data is $\widehat{d} = 0.158$. On the other hand, the estimate of the long-memory parameter reported by Hipel and McLeod (1994, p. 368) is 0.195.

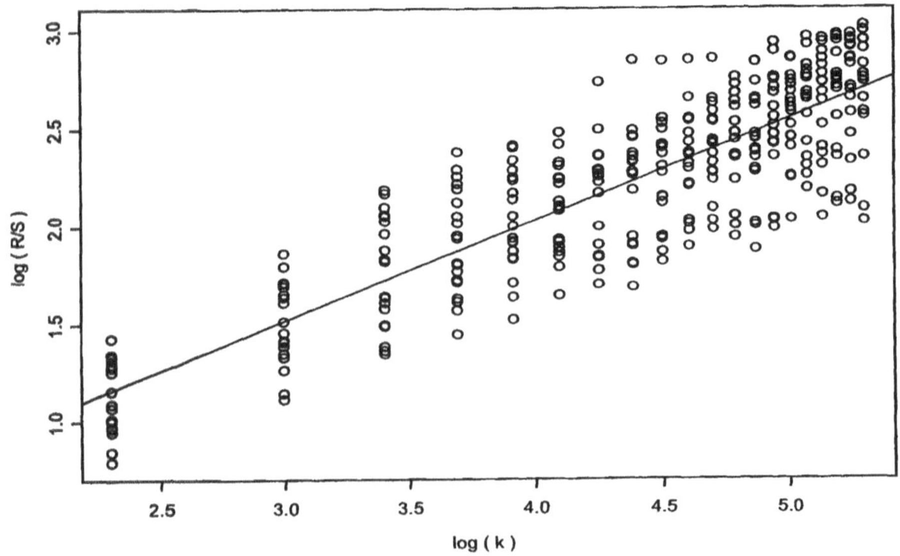

Figure 4.1 Rescaled range method: Simulated white noise process $N(0, 1)$.

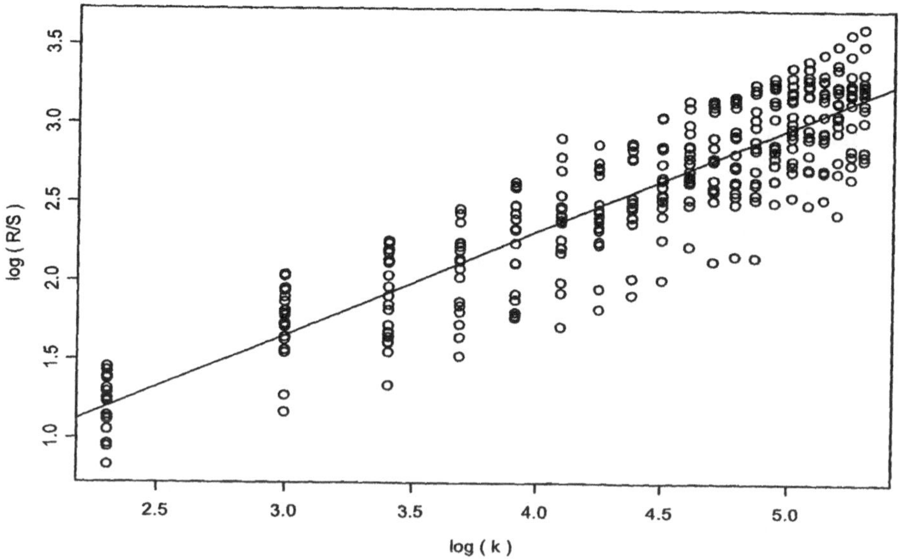

Figure 4.2 Rescaled range method: White Mountains tree ring data.

4.5.3 Variance Plots

According to (3.25), the variance of the sample mean of a long-memory process based on m observations behaves like

$$\operatorname{Var}(\bar{y}_m) \sim c\ m^{2d-1},$$

for large m, where c is a positive constant. Consequently, by dividing a sample of size n, $\{y_1, \ldots, y_n\}$, into k blocks of size m each with $n = k \times m$, we have

$$\log \operatorname{Var}(\bar{y}_j) \sim c + (2d-1)\log j, \tag{4.31}$$

for $j = 1, \ldots, k$, where $\bar{y}_j$ is the average of the jth block, that is,

$$\bar{y}_j = \frac{1}{m} \sum_{t=(j-1)\times m+1}^{j\times m} y_t.$$

From (4.31), a heuristic least squares estimator of d is

$$\widehat{d} = \frac{1}{2} - \frac{\sum_{j=1}^{k} (\log j - a)[\log \operatorname{Var}(\bar{y}_j) - b]}{2\sum_{j=1}^{k} (\log j - a)^2},$$

where $a = (1/k)\sum_{j=1}^{k} \log j$ and $b = (1/k)\sum_{j=1}^{k} \log \operatorname{Var}(\bar{y}_j)$.

Thus, for a short-memory process, $d = 0$, and then the slope of the line described by equation (4.31) should be -1. On the other hand, for a long-memory process with parameter d, the slope is $2d - 1$.

■ EXAMPLE 4.3

Figure 4.3 shows an example of a variance plot for a simulated fractional noise process with $d = 0.4$ and sample size $n = 1000$. We have taken $k = 20$ and $m = 50$. The heavy line corresponds to expression (4.31) where the intercept and the slope have been calculated by their respective least squares estimates.

The broken line indicates the slope -1. Observe that in this case both lines are quite different suggesting the presence of a long-memory component. On the other hand, Figure 4.4 displays a variance plot for a Gaussian ARMA$(1,1)$ process

$$(1 + \phi B)y_t = (1 + \theta B)\varepsilon_t,$$

$t = 1, \ldots, n$, with $\phi = 0.3$, $\theta = 0.7$, $\text{Var}(\varepsilon_t) = 1$, $n = 1000$, $k = 20$, and $m = 50$. In this case, both lines are very close suggesting that the process does not have long-range dependence.

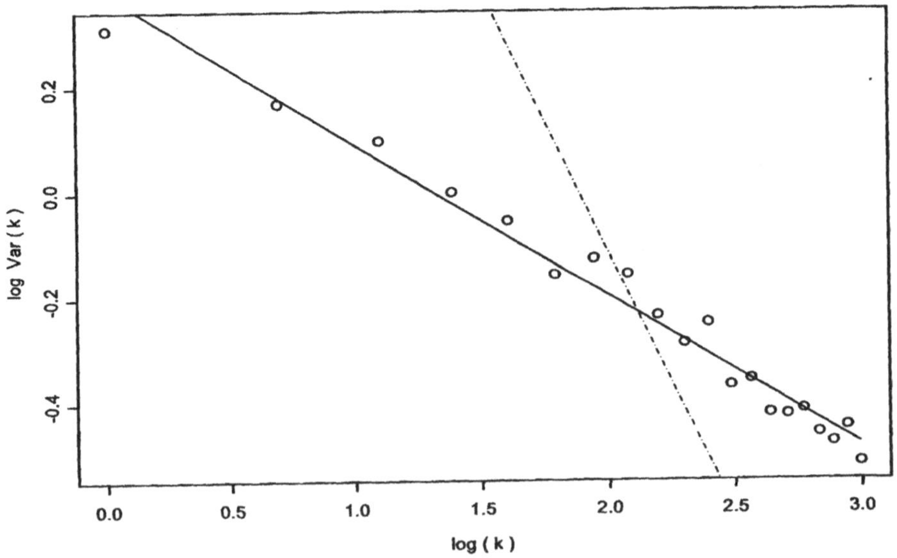

Figure 4.3 Variance plot: Simulated fractional noise process with $d = 0.4$. Estimated intercept and slope from (4.31) with $k = 20$ and $m = 50$ (heavy line). Line with null intercept and slope -1 (broken line).

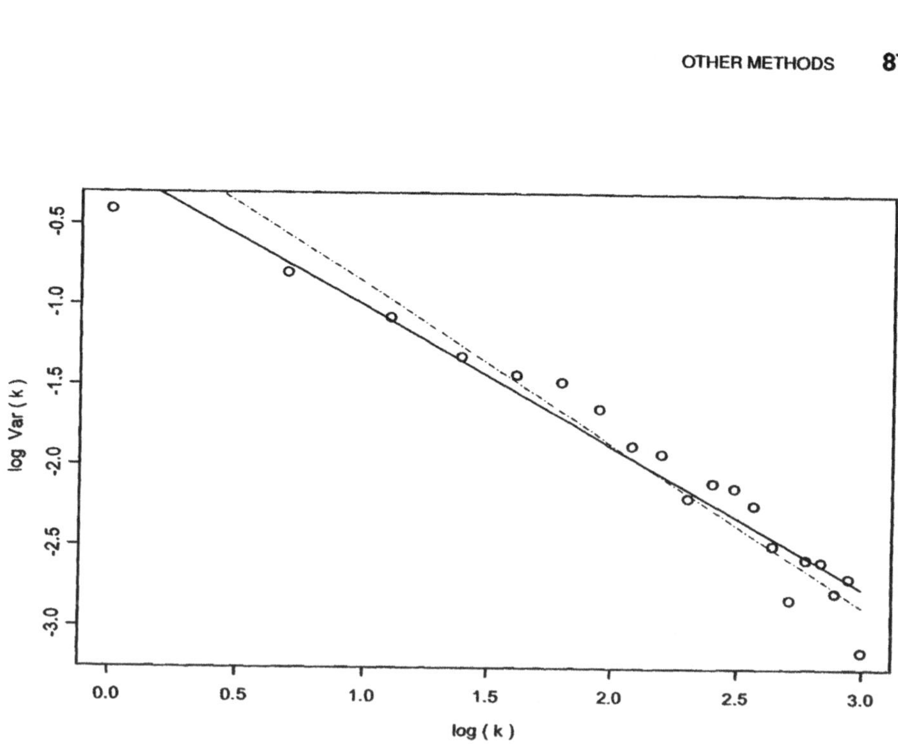

Figure 4.4 Variance plot: Simulated ARMA(1,1) process with parameters $\phi = 0.3$ and $\theta = 0.7$. Estimated intercept and slope from (4.31) with $k = 20$ and $m = 50$ (heavy line). Line with null intercept and slope -1 (broken line).

4.5.4 Detrended Fluctuation Analysis

Let $\{y_1, \ldots, y_n\}$ be a sample from a stationary long-memory process and let $\{x_t\}$ be the sequence of partial sums of $\{y_t\}$, that is, $x_t = \sum_{j=1}^{t} y_j$ for $t = 1, \ldots, n$. The so-called *detrended fluctuation analysis* (DFA) method, introduced by Peng et al. (1994) for estimating the long-memory parameter d of the process $\{y_t : t \in \mathbb{Z}\}$, proceeds as follows. The sample $\{y_1, \ldots, y_n\}$ is divided into k nonoverlapping blocks, each containing $m = n/k$ observations. Within each block, we fit a linear regression model to x_t versus $t = 1, \ldots, m$. Let σ_k^2 be the estimated residual variance from the regression within block k,

$$\sigma_k^2 = \frac{1}{m} \sum_{t=1}^{m} (x_t - \widehat{\alpha}_k - \widehat{\beta}_k t)^2,$$

where $\widehat{\alpha}_k$ and $\widehat{\beta}_k$ are the least squares estimators of the intercept and the slope of the regression line.

Let $F^2(k)$ be the average of these variances

$$F^2(k) = \frac{1}{k} \sum_{j=1}^{k} \sigma_j^2.$$

As described by Peng et al. (1994), for a random walk this term behaves like

$$F(k) \sim c\, k^{1/2},$$

while for a long-range sequence,

$$F(k) \sim c\, k^{d+1/2}.$$

Thus, by taking logarithms, we have

$$\log F(k) \sim \log c + (d + \tfrac{1}{2}) \log k.$$

Therefore, by fitting a least squares regression model to

$$\log F(k) = \alpha + \beta \log k + \varepsilon_k, \tag{4.32}$$

for $k \in K$, we may obtain an estimate of d as

$$\widehat{d} = \widehat{\beta} - \tfrac{1}{2},$$

where $\widehat{\beta}$ is the least squares estimator of the parameter β.

There are several ways to select the set of indexes K. If $k_0 = \min\{K\}$ and $k_1 = \max\{K\}$, then, for example, some authors choose $k_0 = 4$ and k_1 a fraction of the sample size n. Of course, for $k_0 = 2$ the regression error variance is zero since only two observations are fitted by the straight line. On the other hand, for $k_1 = n$, there is only one block and therefore the average of error variances is taken over only one sample.

Mathematical support for this methodology derives from the following theoretical result about the behavior of the residual variances $\{\sigma_k^2\}$, due to Taqqu, Teverovsky, and Willinger (1995).

Theorem 4.10. *Let $\{y_t\}$ be a fractional Gaussian noise process—see definition* (3.28)*— and let σ_k^2 be the residual variance from the least squares fitting in block k. Then,*

$$E[\sigma_k^2] \sim c(d)\, m^{2d+1},$$

as $m \to \infty$ where the constant $c(d)$ is given by the formula

$$c(d) = \frac{1 - 2d}{(d+1)(2d+3)(2d+5)}.$$

Remark 4.2. The numerical results reported in Example 4.4 are based on a Splus implementation of the DFA method. Additionally, the R package RandomFields provides a computational implementation of this method through the function *hurst*; see Schlather (2006) for further details.

■ **EXAMPLE 4.4**

In order to illustrate the DFA technique, consider a trajectory $\{y_1, \ldots, y_n\}$ of a simulated fractional noise process $\{y_t : t \in \mathbb{Z}\}$ with long-memory parameter $d = 0.4$, mean $\mu = 0$, noise variance $\sigma^2 = 1$, and sample size $n = 1000$ displayed in Figure 4.5.

On the other hand, Figure 4.6 shows the cumulative sum of this fractional noise process, $\{x_1, \ldots, x_n\}$, where

$$x_t = \sum_{j=1}^{t} y_j,$$

for $t = 1, 2, \ldots, n$. In this plot, we have divided the sample into 5 blocks containing 200 observations each. Within every block, we have fitted a *local trend* through linear regression models and the resulting fitted line have been plotted along the series $\{x_t\}$.

The regression line from (4.32) has been plotted in Figure 4.7 for $\log F(k)$ versus $\log k$ for $k = 4, \ldots, 125$. The least squares estimates are $\widehat{\alpha} = -2.0883$ and $\widehat{\beta} = 0.8957$. Thus, an estimate of d is $\widehat{d} = 0.3957$.

Finally, Figure 4.8 compares the behavior of $\log F(k)$ for the fractional noise process with $d = 0.4$ and a white noise $\{\varepsilon_1, \ldots, \varepsilon_n\}$ following a standard

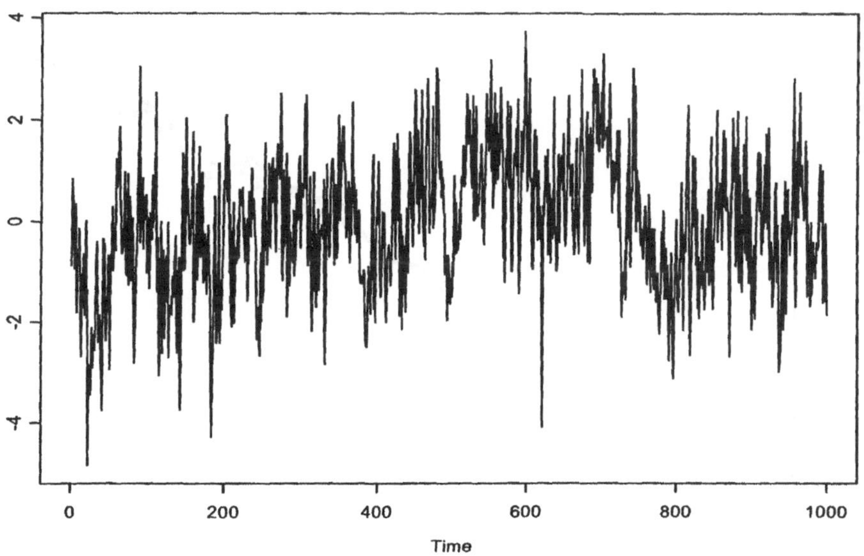

Figure 4.5 Detrended fluctuation analysis: Simulated fractional noise process with long-memory paramter $d = 0.4$.

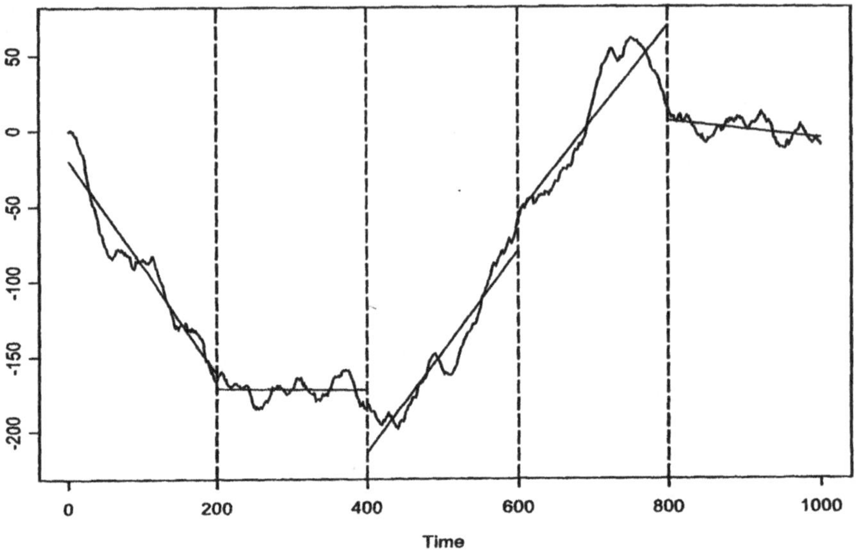

Figure 4.6 Detrended fluctuation analysis: Cumulative sum $\{x_t\}$ and local trends for the simulated fractional noise process with $d = 0.4$.

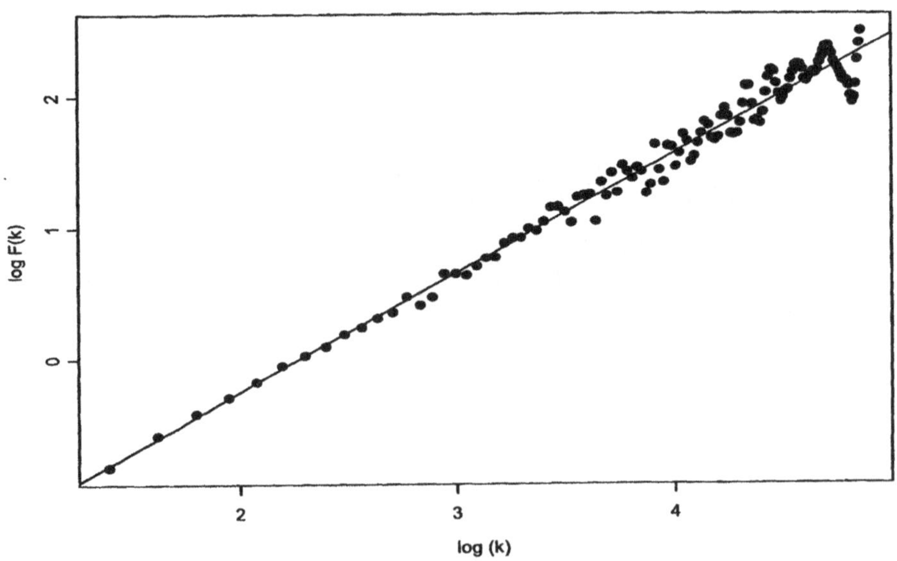

Figure 4.7 Detrended fluctuation analysis: $\log F(k)$ versus $k = 1, 2, \ldots, 125$, for the simulated fractional noise process with $d = 0.4$.

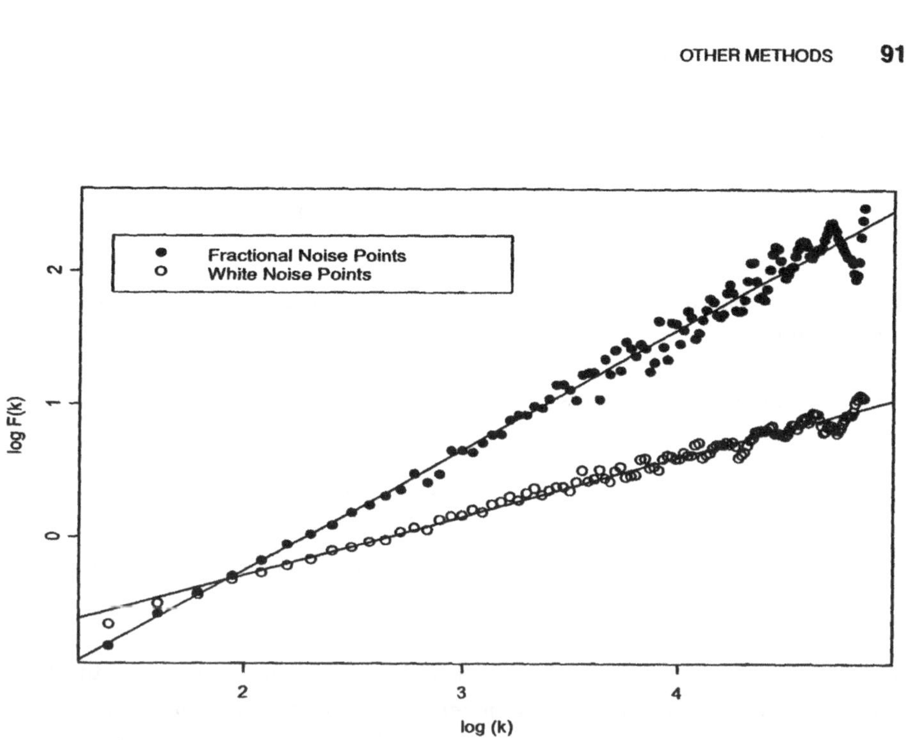

Figure 4.8 Detrended fluctuation analysis: Comparative $\log F(k)$ versus k graphs for the fractional noise process with $d = 0.4$ and a white noise process.

normal distribution $N(0,1)$. The estimated parameters for the random walk are $\widehat{\alpha} = -1.2397$ and $\widehat{\beta} = 0.4718$, which lead to the estimate $\widehat{d} = -0.0282$.

4.5.5 A Wavelet-Based Method

Consider the discrete wavelet transform coefficients d_{jk} discussed in Chapter 1 and define the statistics

$$\widehat{\mu}_j = \frac{1}{n_j} \sum_{k=1}^{n_j} \widehat{d}_{jk}^2,$$

where n_j is the number of coefficients at octave j available to be calculated. As shown by Veitch and Abry (1999),

$$\widehat{\mu}_j \sim \frac{z_j}{n_j} \mathcal{X}_{n_j},$$

where $z_j = 2^{2dj}c$, $c > 0$, and $\mathcal{X}_{n_j}$ is a chi-squared random variable with n_j degrees of freedom. Thus, by taking logarithms we may write

$$\log_2 \widehat{\mu}_j \sim 2dj + \log_2 c + \log \mathcal{X}_{n_j} / \log 2 - \log_2 n_j.$$

Recall that the expected value and the variance of the random variable $\log \mathcal{X}_n$ are given by

$$\begin{aligned} E(\log \mathcal{X}_n) &= \psi(n/2) + \log 2, \\ \mathrm{Var}(\log \mathcal{X}_n) &= \zeta(2, n/2), \end{aligned}$$

where $\psi(z)$ is the psi function, $\psi(z) = d/dz \log \Gamma(z)$, and $\zeta(2, n/2)$ is the Riemann zeta function.

By defining $\varepsilon_j = \log_2 \log \mathcal{X}_{n_j} - \log_2 n_j - g_j$, where $g_j = \psi(n_j/2) - \log(n_j/2)$, we conclude that this sequence satisfies

$$\begin{aligned} E(\varepsilon_j) &= 0, \\ \mathrm{Var}(\varepsilon_j) &= \frac{\zeta(2, n_j/2)}{(\log 2)^2}. \end{aligned}$$

Therefore, we could write the following *heteroskedastic regression equation*:

$$y_j = \alpha + \beta x_j + \varepsilon_j,$$

where $y_j = \log_2 \widehat{\mu}_j - g_j$, $\alpha = \log c$, and $\beta = 2d$. Thus, once the estimate $\widehat{\beta}$ is obtained, an estimate for the long-memory parameter d is given by $\widehat{d} = \widehat{\beta}/2$. Furthermore, and estimate of the variance of $\widehat{d}$ is provided by the estimate of the variance of $\widehat{\beta}$, $\mathrm{Var}(\widehat{d}) = \mathrm{Var}(\widehat{\beta})/4$.

As an illustration of this methodology, consider the fractional noise process FN(d) with $d = 0.4$, discussed in the context of the DFA approach. In this case, an application of the function *waveletFit* of the *R* package *fSeries* yields the estimates $\widehat{d} = 0.4447$ and $\mathrm{Var}(\widehat{d}) = 0.0525$ ($t_d = 17.9955$).

4.6 NUMERICAL EXPERIMENTS

Table 4.2 displays the results from several simulations comparing five ML estimation methods for Gaussian processes: Exact MLE, Haslett and Raftery's approach, AR(40) approximation, MA(40) approximation, and the Whittle method.

The process considered is a fractional noise ARFIMA$(0, d, 0)$ with three values of the long-memory parameter: $d = 0.1, 0.25, 0.4$, Gaussian innovations with zero mean and unit variance, and sample sizes $n = 200$ and $n = 400$. The mean and standard deviations of the estimates are based on 1000 repetitions. All the simulations reported in Table 4.2 were carried out by means of Splus programs.

From Table 4.2, it seems that all estimates are somewhat downward biased for the three values of d and the two sample sizes considered. All the estimators, excepting the Whittle; seem to behave similarly in terms of bias and standard deviation. The sample standard deviations of all the methods considered are relatively close to its theoretical value 0.05513 for $n = 200$ and 0.03898 for $n = 400$. Observe that the Whittle method exhibits less bias for $d = 0.4$ but greater bias for $d = 0.1$. Besides, this procedure seems to have greater standard deviations than the other estimators, for the three values of d and the two sample sizes under study.

Table 4.2 Finite Sample Behavior of Maximum Likelihood Estimates[a]

d		Exact	HR	AR	MA	Whittle
				$n = 200$		
0.40	Mean	0.3652	0.3665	0.3719	0.3670	0.3874
	SD	0.0531	0.0537	0.0654	0.0560	0.0672
0.25	Mean	0.2212	0.2219	0.2224	0.2220	0.2156
	SD	0.0612	0.0613	0.0692	0.0610	0.0706
0.10	Mean	0.0780	0.0784	0.0808	0.0798	0.0585
	SD	0.0527	0.0529	0.0561	0.0525	0.0522
				$n = 400$		
0.40	Mean	0.3799	0.3808	0.3837	0.3768	0.3993
	SD	0.0393	0.0396	0.0444	0.0402	0.0466
0.25	Mean	0.2336	0.2343	0.2330	0.2330	0.2350
	SD	0.0397	0.0397	0.0421	0.0394	0.0440
0.10	Mean	0.0862	0.0865	0.0875	0.0874	0.0753
	SD	0.0394	0.0395	0.0410	0.0390	0.0413

[a]Sample sizes $n = 200, 400$ with truncation $m = 40$ for AR and MA approximations.

4.7 BIBLIOGRAPHIC NOTES

Estimation of long-memory models have been considered by a large number of authors. An overview of the technique discussed in this chapter can be found in Chan and Palma (2006). Most of the estimation methodologies proposed in the literature can be classified into the time domain and the spectral domain procedures. In the first group, we have the exact maximum-likelihood estimators (MLE) and the quasi-maximum-likelihood estimators (QMLE); see, for example, the works by Granger and Joyeux (1980), Sowell (1992), and Beran (1994a). In the second group, we have, for instance, the Whittle and the semiparametric estimators; see Fox and Taqqu (1986), Giraitis and Surgailis (1990), and Robinson (1995a), among others.

Autoregressive approximations have been studied by Granger and Joyeux (1980), Li and McLeod (1986), Hasslett and Raftery (1989), Beran (1994b), Shumway and Stoffer (2000), and Bhansali and Kokoszka (2003), among others. The Haslett-Raftery method discussed in Subsection 4.2.1 was introduced by Hasslett and Raftery (1989). In particular, they used $M = 100$ in their implementation of the method in Splus. The alternative method addressed in Subsection 4.2.2 was suggested by Beran

(1994b). Additionally, an estimation method based on a multivariate central limit theorem was proposed by Beran and Terrin (1994).

Details about the computation of the ACF of an ARFIMA process and the application of the Choslesky decomposition method to long-memory models can be found in Sowell (1992) while the inversion of the lower triangular matrix L is discussed in Press, Teukolsky, Vetterling, and Flannery (1992, p. 89ff). This text also discusses the arithmetic complexity of the Cholesky algorithm.

The Durbin-Levinson algorithm is based on the seminal works by Levinson (1947) and Durbin (1960). The arithmetic complexity of this algorithm for a linear stationary process have been discussed, for instance, by Ammar (1998). The Durbin-Levinson algorithm can be implemented for an ARMA process in only $\mathcal{O}(n)$ operations; see, for example, Section 5.3 of Brockwell and Davis (1991).

The splitting algorithm has been applied to the calculation of the ACF of long-memory processes; see, for example, the numerical experiments reported by Bertelli and Caporin (2002). Besides, several computational aspects of parameter estimation are discussed by Doornik and Ooms (2003).

The asymptotic properties of the MLE have been established by Yajima (1985) for the fractional noise process and by Dahlhaus (1989, 2006) for a general class of long-memory processes including the ARFIMA model.

The approximation $\phi_{tj} \sim -\pi_j$ for large j appears in Hosking (1981). The autoregressive approximation approach described in Section 4.2.2 and Theorem 4.3 are due to Beran (1994a). The convergence rate $\mathcal{O}(1/m)$ for the AR(m) truncation has been proved by Bondon and Palma (2006).

The so-called Whittle method was proposed by Whittle (1951) while Theorem 4.5 combines results by Fox and Taqqu (1986) and Dahlhaus (1989). Estimates based on the maximization of the log-likelihood function $\mathcal{L}_5(\theta)$ for a general class of linear processes with independent innovations has been studied, for example, by Giraitis and Surgailis (1990) who proved Theorem 4.6. Furthermore, Theorem 4.7 was proved by Robinson (1995a) who proposed the Gaussian semiparametric estimation method presented in Subsection 4.4.3. A nice review of spectral methods for long-memory process is given in Moulines and Soulier (2003).

A study comparing the properties of the R/S with other estimators can be found in Giraitis, Kokoszka, Leipus, and Teyssière (2003). The large sample behavior of the periodogram of long-range-dependent processes has been extensively studied; see, for example, Fox and Taqqu (1987), Yajima (1989), and Terrin and Hurvich (1994), among others.

Finally, various estimators of the long-range dependence parameter including the R/S, DFA, and the Whittle methods are studied in the article by Taqqu, Teverovsky, and Willinger (1995).

Problems

4.1 Give another state space representation based on the infinite autoregressive expansion of an ARFIMA process. Discuss the advantages or disadvantages of this AR(∞) with respect to the MA(∞) representation.

4.2 Consider following the state transition matrix associated to the AR(m) state space approximation:

$$F = \begin{pmatrix} \pi_1 & \pi_2 & \cdots & \pi_{m-1} & \pi_m \\ 1 & 0 & \cdots & 0 & 0 \\ 0 & 1 & \cdots & 0 & 0 \\ \vdots & & & & \vdots \\ 0 & 0 & \cdots & 0 & 0 \\ 0 & 0 & \cdots & 1 & 0 \end{pmatrix}.$$

a) Show that the eigenvalues of this matrix are the roots of the polynomial

$$\lambda^m - \pi_1 \lambda^{m-1} - \cdots - \pi_{m-1}\lambda - \pi_m = 0.$$

b) Verify that space of eigenvectors is given by the $\text{sp}\{(\lambda^{m-1}, \ldots, \lambda, 1)'\}$.

4.3 Find an expression for the maximum-likelihood estimator of σ^2.

4.4 Implement computationally the convolution algorithm for estimating the ACF of an ARFIMA process. What numerical difficulties display this approach?

4.5 Show that if the spectral density of a stationary process satisfies $f_\theta(\lambda) > c$ for all $\lambda \in (-\pi, \pi]$ where c is a positive constant, then $x'\Gamma_\theta x > 0$ for all $x \in \mathbb{R}^n$, $x \neq 0$.

4.6 Using (4.14) show that $\phi_{tj} \sim -\pi_j$ for large j.

4.7 Explain why the Haslett-Raftery method yields exact maximum-likelihood estimates in the fractional noise case when $M = n$, where n is the sample size.

4.8 Consider the following formula for the coefficients of the best linear predictor from the Durbin-Levinson algorithm:

$$\phi_{tj} = \phi_{t-1,j} - \phi_{tt}\phi_{t-1,t-j},$$

for $j = 1, \ldots, n$. Let $\alpha_t = 1 - \sum_{j=1}^{t} \phi_{tj}$.

a) Show that α_t satisfies the following recursive equation:

$$\alpha_t = (1 - \phi_{tt})\alpha_{t-1}.$$

b) Verify that a solution to the above equation is

$$\alpha_t = \prod_{j=1}^{t} (1 - \phi_{jj}),$$

so that $\sum_{j=1}^{t} \phi_{tj} = 1 - \prod_{j=1}^{t} (1 - \phi_{jj})$.

4.9 [cf. Fox and Taqqu (1986)] Consider the periodogram defined in (3.27).

a) Show that the periodogram satisfies

$$\int_{-\pi}^{\pi} e^{ik\lambda} I(\lambda)\, d\lambda = \begin{cases} w(k,n), & |k| < n, \\ 0, & |k| > n, \end{cases}$$

where

$$w(k,n) = \frac{1}{n}\sum_{t=1}^{n-k}(y_t - \bar{y}_n)(y_{t+k} - \bar{y}_n).$$

b) Prove that if the process y_t is stationary and ergodic with mean μ we have that

$$\lim_{n\to\infty} w(k,n) = \lim_{n\to\infty} \frac{1}{n}\sum_{t=1}^{n-k}(y_t - \mu)(y_{t+k} - \mu) = \gamma(k),$$

where $\gamma(k)$ is the autocovariance at lag k.

4.10 Suppose that $\{y_1, \dots, y_n\}$ follows an ARFIMA model and $I_n(\lambda)$ is its periodogram. Based on the previous problem, prove that if $f(\lambda)$ is the spectral density of the ARFIMA process and $g(\lambda)$ is a continuous function on $[-\pi, \pi]$ then

$$\lim_{n\to\infty} \int_{-\pi}^{\pi} g(\lambda) I_n(\lambda)\, d\lambda = \int_{-\pi}^{\pi} g(\lambda) f(\lambda)\, d\lambda.$$

CHAPTER 5

ASYMPTOTIC THEORY

In the previous chapter we have discussed several parameter estimation methodologies for long-memory models. Additionally, we stated without proof some large sample properties of these estimators. In particular, Theorem 4.2 summarized some asymptotic results for the exact maximum-likelihood estimates. In this chapter we present the proof of this theorem in the context of Gaussian long-range-dependent processes. To simplify the exposition, we have divided the analysis into three separate theorems establishing the consistency, asymptotic normality, and efficiency, respectively. The results given in this chapter are applicable to a large class of long-range-dependent processes which includes as a particular case the ARFIMA family. However, other long-memory models such as those including seasonal effects, for example, Gegenbauer autoregressive moving average (GARMA) or seasonal fractional integrated autoregressive moving average (SARFIMA), possess more complex spectral density structures and consequently they are not covered by these theorems. The large sample properties of these seasonal models are analyzed in Chapter 12. Notational issues and basic definitions are discussed in Section 5.1. The consistency of the MLE, a central limit theorem and the efficiency is addressed in Section 5.2. Several examples illustrating the application of these asymptotic results are discussed in Sections 5.3 and 5.4. Two technical lemmas used in the proofs are listed in Section 5.5. Bibliographic notes are given in Section 5.6 while a list of problems is presented at the end of this chapter.

5.1 NOTATION AND DEFINITIONS

Given the zero mean Gaussian process $y = (y_1, \dots, y_n)'$, the exact maximum-likelihood estimate $\widehat{\theta}_n$ is obtained by minimizing the function

$$\mathcal{L}_n(\theta) = \frac{1}{2n} \log \det T_n(\theta) + \frac{1}{2n} Y_n' T_n^{-1}(\theta) y,$$

where $T_n(\theta) = T_n(f_\theta)$ is the variance of y and f_θ its spectral density. We assume that the following conditions on the process, parameter space, and the spectral density hold.

(A0) $\{y_t : t \in \mathbb{Z}\}$ is a stationary Gaussian sequence with mean zero and spectral density $f_\theta(\lambda), \theta \in \Theta \subset \mathbb{R}^p$, where θ is an unknown parameter. Let θ_0 be the true parameter of the process such that θ_0 is in the interior of Θ, where Θ is assumed to be compact. Suppose that the set $\{\lambda | f_\theta(\lambda) \neq f_{\theta'}(\lambda)\}$ has positive Lebesgue measure, whenever $\theta \neq \theta'$.

(A1) $g(\theta) = \int_{-\pi}^{\pi} \log f_\theta(\lambda) d\lambda$ can be differenced twice under the integral sign.

The following assumptions require the existence of a function $\alpha : \Theta \to (0,1)$ such that for each $\delta > 0$:

(A2) $f_\theta(\lambda)$ is continuous at all $(\lambda, \theta), \lambda \neq 0$, $f_\theta^{-1}(\lambda)$ is continuous at all (λ, θ), and

$$f_\theta(\lambda) = \mathcal{O}(|\lambda|^{-\alpha(\theta)-\delta}).$$

(A3) $\partial/\partial\theta_j f_\theta^{-1}(\lambda), \partial^2/\partial\theta_j\partial\theta_k f_\theta^{-1}(\lambda)$, and $\partial^3/\partial\theta_j\partial\theta_k\partial\theta_\ell f_\theta^{-1}(\lambda)$ are continuous at all (λ, θ):

$$\frac{\partial}{\partial\theta_j} f_\theta^{-1}(\lambda) = \mathcal{O}(|\lambda|^{\alpha(\theta)-\delta}), \qquad 1 \leq j \leq p,$$

$$\frac{\partial^2}{\partial\theta_j\,\partial\theta_k} f_\theta^{-1}(\lambda) = \mathcal{O}(|\lambda|^{\alpha(\theta)-\delta}), \qquad 1 \leq j,k \leq p,$$

$$\frac{\partial^3}{\partial\theta_j\,\partial\theta_k\,\partial\theta_\ell} f_\theta^{-1}(\lambda) = \mathcal{O}(|\lambda|^{\alpha(\theta)-\delta}), \qquad 1 \leq j,k,\ell \leq p.$$

(A4) $\partial/\partial\lambda\, f_\theta(\lambda)$ is continuous at all $(\lambda, \theta), \lambda \neq 0$, and

$$\frac{\partial}{\partial\lambda} f_\theta^{-1}(\lambda) = \mathcal{O}(|\lambda|^{-\alpha(\theta)-1-\delta}).$$

(A5) $\partial^2/\partial\lambda\,\partial\theta_j\, f_\theta^{-1}(\lambda)$ is continuous at all $(\lambda, \theta), \lambda \neq 0$, and

$$\frac{\partial^2}{\partial\lambda\,\partial\theta_j} f_\theta^{-1}(\lambda) = \mathcal{O}(|\lambda|^{\alpha(\theta)-1-\delta}), \qquad 1 \leq j \leq p.$$

(A6) $\partial^3/\partial^2\lambda\,\partial\theta_j\, f_\theta^{-1}(\lambda)$ is continuous at all (λ, θ), $\lambda \neq 0$, and

$$\frac{\partial}{\partial\lambda} f_\theta^{-1}(\lambda) = \mathcal{O}(|\lambda|^{-\alpha(\theta)-2-\delta}).$$

(A7) $(\partial/\partial\lambda) f_\theta^{-1}(\lambda)$ and $\partial^2/(\partial\lambda)^2 f_\theta^{-1}(\lambda)$ are continuous at all (λ, θ) $\lambda \neq 0$ and

$$\left(\frac{\partial}{\partial\lambda}\right)^k f_\theta^{-1}(\lambda) = \mathcal{O}(|\lambda|^{\alpha(\theta)-k-\delta}),$$

for $k = 0, 1, 2$.

(A8) The above constants can be chosen independently of θ (not of δ).

(A9) α is assumed to be continuous.

Furthermore, define the partial derivatives

$$\nabla f(\theta) = \left(\frac{\partial}{\partial\theta_j} f(\theta)\right)_{j=1,\ldots,p}, \qquad \nabla^2 f(\theta) = \left(\frac{\partial^2}{\partial\theta_j \partial\theta_k} f(\theta)\right)_{j,k=1,\ldots,p},$$

and the matrices

$$\begin{aligned} A^{(0)} &= T^{-1}, & A^{(1)} &= T^{-1}T_\nabla T^{-1} T_\nabla T^{-1}, \\ A^{(2)} &= T^{-1}T_{\nabla^2}T^{-1}, & A^{(3)} &= T^{-1}T_\nabla T^{-1}, \end{aligned}$$

where $T = T_n(f_\theta)$, $T_\nabla = T_n(\nabla f_\theta)$ and $T_{\nabla^2} = T_n(\nabla^2 f_\theta)$.

Let $\mathcal{U}(\theta, \epsilon) = \{\theta_1 : |\theta_1 - \theta| < \epsilon\}$ be a neighborhood of θ and let $z_n^{(i)}$ be the quadratic form

$$z_n^{(i)} = \frac{1}{n} y' A^{(i)} y - \frac{1}{n}\,\mathrm{tr}[A^{(i)}T],$$

for $i = 0, \ldots, 3$.

5.2 THEOREMS

In the proof of the following theorems K denotes a generic positive constant that may change from line to line, $T_0 = T(f_{\theta_0})$, and $T_1 = T(f_{\theta_1})$. These theorems and their proofs are based on Dahlhaus (1989, 2006).

5.2.1 Consistency

Theorem 5.1. *Let θ_0 be the true parameter of a Gaussian long-memory model. Under assumptions (A0), (A2), (A3) and (A7)–(A9) we have that*

$$\widehat{\theta}_n \to \theta_0,$$

in probability as $n \to \infty$.

Proof. Let $\theta \neq \theta_0$, $x_n(\theta) = \mathcal{L}_n(\theta) - \mathcal{L}_n(\theta_0)$, $c = c(\theta) > 0$, and $A = T_0 T_1^{-1} - I$. This theorem may be proved proceeding through the following three steps: (i) $\lim_{n\to\infty} E[x_n(\theta)] \geq c$, (ii) $\lim_{n\to\infty} P(x_n(\theta) < c/2) = 0$, and (iii)

$$\lim_{n\to\infty} P\left(\inf_{\theta_1 \in \mathcal{U}(\theta,\delta)} x_n(\theta_1) \geq c/4\right) = 1,$$

where the expectation and the probability are evaluated at θ_0.

Part (i). $E[x_n(\theta)] = [\operatorname{tr}(A) - \log\det(I+A)]/2n$. Define the real-valued function

$$f(t) = \log\det(I + tA) = \sum_{j=1}^{n} \log(1 + t\lambda_j),$$

where λ_j are the eigenvalues of A. Applying Taylor's theorem to $f(t)$ around $t = 0$, we get

$$f(1) = \operatorname{tr}(A) - \frac{1}{2}\sum_{j=1}^{n}\left(\frac{\lambda_j}{1+\tau\lambda_j}\right)^2, \tag{5.1}$$

for some $\tau \in [0, 1]$. Hence,

$$E[x_n(\theta)] = \frac{1}{4n}\sum_{j=1}^{n}\left(\frac{\lambda_j}{1+\tau\lambda_j}\right)^2.$$

For $\alpha(\theta) \geq \alpha(\theta_0)$, $f_{\theta_0}(\lambda) \leq K f_\theta(\lambda)$ and thus $T_0 \leq K T_1$. Therefore, $0 \leq \lambda_j < C$ for all $j = 1, \dots, n$ with $C = 1 + \tau(K-1)$ where K can be chosen arbitrarily large. Hence,

$$E[x_n(\theta)] \geq \frac{1}{4nC^2}\sum_{j=1}^{n}(\lambda_j - 1)^2 = \frac{1}{4nC^2}\operatorname{tr}(A^2).$$

Since by Lemma 5.1 in Subsection 5.5 the last term of this expression converges to

$$c = \frac{1}{8\pi C^2}\int_{-\pi}^{\pi}\left(\frac{f_{\theta_0}(\lambda)}{f_\theta(\lambda)} - 1\right)^2 d\lambda > 0,$$

the result is obtained. The argument proceeds analogously for $\alpha(\theta) < \alpha(\theta_0)$.

Part (ii). Observe that

$$\operatorname{Var}[x_n(\theta)] = \frac{1}{2n^2}\operatorname{tr}(A^2).$$

Therefore, by Lemma 5.1, $\lim_{n\to\infty} \operatorname{Var}[x_n(\theta)] = 0$. From this result and part (i) we obtain (ii).

Part (iii). As a consequence of the equicontinuity of $z_n^{(0)}$ given by Lemma 5.2; see Subsection 5.5, we have

$$\lim_{n\to\infty} P\left(\sup_{\theta_1 \in \mathcal{U}(\theta,\delta)} |x_n(\theta_1) - x_n(\theta)| \geq c/4\right) = 0. \tag{5.2}$$

Since $x_n(\theta_1) = x_n(\theta) + [x_n(\theta_1) - x_n(\theta)]$, we have

$$P\left(\inf_{\theta_1 \in \mathcal{U}(\theta,\delta)} x_n(\theta_1) \le c/4\right) = P\left(x_n(\theta) + \inf_{\theta_1 \in \mathcal{U}(\theta,\delta)} [x_n(\theta_1) - x_n(\theta)] \le c/4\right)$$

$$\begin{aligned}
&= P\left(x_n(\theta) - c/2 \le -\inf_{\theta_1 \in \mathcal{U}(\theta,\delta)} [x_n(\theta_1) - x_n(\theta)] - c/4\right) \\
&\le P\left(x_n(\theta) - c/2 \le \sup_{\theta_1 \in \mathcal{U}(\theta,\delta)} |x_n(\theta) - x_n(\theta_1)| - c/4\right) \\
&\le P(x_n(\theta) \le c/2) + P\left(\sup_{\theta_1 \in \mathcal{U}(\theta,\delta)} |x_n(\theta) - x_n(\theta_1)| \ge c/4\right). \qquad (5.3)
\end{aligned}$$

From part (ii) the first summand on the right-hand side of (5.3) tends to zero while by (5.2) the second summand vanishes as $n \to \infty$. Thus, part (iii) has been proved.

Finally, given that the parameter space is compact, there is a finite covering of open neighborhoods $\{\mathcal{U}(\theta,\delta), \theta \neq \theta_0\}$ and $\mathcal{U}(\theta_0, \delta_0)$ so that

$$\lim_{n\to\infty} P\left(\inf_{\theta} x_n(\theta) = \inf_{\theta \in \mathcal{U}(\theta_0,\delta_0)} x_n(\theta)\right) = 1.$$

Since $\delta_0 > 0$ is arbitrary, we conclude $\lim_{n\to\infty} P(|\widehat{\theta}_n - \theta_0| \ge \delta_0) = 0$. □

5.2.2 Central Limit Theorem

Theorem 5.2. *Let θ_0 be the true parameter of a Gaussian long-memory model. Assume that conditions* (A0), (A2), (A3), *and* (A7)–(A9) *hold. Then,*

$$\sqrt{n}(\widehat{\theta}_n - \theta_0) \to N[0, \Gamma^{-1}(\theta_0)],$$

in distribution, as $n \to \infty$, where

$$\Gamma(\theta) = \frac{1}{4\pi} \int_{-\pi}^{\pi} [\nabla \log f_\theta(\lambda)][\nabla \log f_\theta(\lambda)]' \, d\lambda.$$

Proof. Notice that we may write

$$\sqrt{n}[\nabla \mathcal{L}_n(\widehat{\theta}_n) - \nabla \mathcal{L}_n(\widehat{\theta}_n)] = \nabla^2 \mathcal{L}_n(\bar{\theta}_n)\sqrt{n}(\widehat{\theta}_n - \theta_0),$$

for some $\bar{\theta}_n \in \mathcal{U}(\theta_0, \epsilon)$. From this, it suffices to prove

(i) $\sqrt{n}\mathcal{L}_n(\theta_0) \to N(0, \Gamma(\theta_0))$ in distribution as $n \to \infty$,

(ii) $|\nabla^2 \mathcal{L}_n(\bar{\theta}_n) - \nabla^2 \mathcal{L}_n(\theta_0)| \to 0$ as $\bar{\theta}_n \to \theta_0$ in probability as $n \to \infty$, and

(iii) $\nabla^2 \mathcal{L}_n(\theta_0) \to \Gamma(\theta_0)$ in probability as $n \to \infty$.

Part (i). First, note that the gradient of $\mathcal{L}_n$ is given by

$$\nabla \mathcal{L}_n(\theta) = \frac{1}{2n} \operatorname{tr}[T^{-1}T_\nabla] - \frac{1}{2n} y' A^{(3)} y.$$

This part may be proved by the cumulant method since $E[\sqrt{n}\nabla \mathcal{L}_n(\theta_0)] = 0$,

$$n \operatorname{Var}[\nabla \mathcal{L}_n(\theta_0)] = \frac{1}{2n} \operatorname{tr}(T_0^{-1}T_{\nabla,0}T_0^{-1}T_{\nabla,0}),$$

tends to $\Gamma(\theta_0)$ by Lemma 5.1 and the cumulants of order k:

$$n^{k/2} \operatorname{cum}(\nabla \mathcal{L}_n(\theta_0)_{i_1}, \ldots, \nabla \mathcal{L}_n(\theta_0)_{i_k}) = \frac{(-1)^k}{2kn^{k/2}} \sum_{j_1,\ldots,j_k} \operatorname{tr}\left[\prod_{l=1}^{k} T^{-1}T_0\left(\frac{\partial f_{\theta_0}}{\partial \theta_{j_l}}\right)\right],$$

where $j_1, \ldots, j_k$ is a permutation of $i_1, \ldots, i_k$, converge to zero by Lemma 5.1.

Part (ii). The second derivatives of $\mathcal{L}_n(\theta)$ may be written as

$$\begin{aligned} \nabla^2 \mathcal{L}_n(\theta) &= \frac{1}{2n}\left[-\operatorname{tr}\left(T^{-1}T_\nabla T^{-1}T_\nabla\right) + \operatorname{tr}\left(T^{-1}T_{\nabla^2}\right) + 2y'A^{(1)}y - y'A^{(2)}y\right] \\ &= z_n^{(1)}(\theta) - \tfrac{1}{2} z_n^{(1)}(\theta) + r_n(\theta), \end{aligned}$$

where $r_n(\theta) = \operatorname{tr}(T^{-1}T_\nabla T^{-1}T_\nabla)/2n$. Since by Lemma 5.2 the quadratic forms $z_n^{(1)}$ and $z_n^{(2)}$ are equicontinuous, it suffices to prove that $|r_n(\bar{\theta}_n) - r_n(\theta_0)|$ converges to zero in probability as $n \to \infty$.

By the mean value theorem we have that

$$|r_n(\bar{\theta}_n) - r_n(\theta_0)| \le b_n(\theta)|\bar{\theta}_n - \theta_0|,$$

for some $\theta \in \mathcal{U}(\theta_0, \epsilon)$ with

$$b_n(\theta) = \frac{1}{n}|\operatorname{tr}(T^{-1}T_\nabla)^3| + \frac{1}{n}|\operatorname{tr}(A^{(2)}T_\nabla)|. \tag{5.4}$$

In what follows, we prove that b_n is uniformly bounded on $\mathcal{U}(\theta_0, \epsilon)$.

Let $g^+ = (\partial f/\partial\theta)_+$ and $g^- = (\partial f/\partial\theta)_-$. Then, the first summand on (5.4) is bounded by

$$\frac{1}{n}|\operatorname{tr}(T^{-1}T_\nabla)^3| \le 3\sum_{j=0}^{3} |\operatorname{tr}[T^{-1}T(g^+)]^{3-j}||\operatorname{tr}[T^{-1}T(g^-)]^j|.$$

Observe that since $\alpha(\cdot)$ is continuous by (A9), we have $|\alpha(\theta) - \alpha(\theta_0)| < \eta$ for $\eta > 0$ and $\theta \in \mathcal{U}(\theta_0, \epsilon)$. Hence,

$$T(g^+) \le T(K|\lambda|^{-\alpha(\theta_0)-\eta-\delta}),$$

and then for $i = 1, 2, 3$ we have

$$\begin{aligned} \frac{1}{n} | \operatorname{tr}[T^{-1}T(g^+)]^i | &\leq \frac{1}{n} \left| \operatorname{tr} \left[T^{-1} T \left(\left| \frac{\partial f}{\partial \theta} \right| \right) \right]^i \right| \\ &\leq \frac{1}{n} | \operatorname{tr}[T^{-1}T(K|\lambda|^{-\alpha(\theta_0)-\eta-\delta})]^i |. \end{aligned}$$

On the other hand, from (A7) we have that

$$T^{-1} \leq T(K|\lambda|^{-\alpha(\theta_0)+\eta+\delta})^{-1}.$$

Therefore, the following inequality holds:

$$\frac{1}{n} | \operatorname{tr}[T^{-1}T(g^+)]^i | \leq \frac{1}{n} | \operatorname{tr}[T(K|\lambda|^{-\alpha(\theta_0)+\eta+\delta})^{-1} T(K|\lambda|^{-\alpha(\theta_0)-\eta-\delta})]^i |.$$

Hence, by Lemma 5.1

$$\limsup_{n \to \infty} \frac{1}{n} | \operatorname{tr}[T^{-1}T(g^+)]^i | \leq K \int_{-\pi}^{\pi} |\lambda|^{-i(\eta+\delta)} d\lambda \leq K_i,$$

for η and δ sufficiently small and some positive constants K_i, $i = 1, 2, 3$. Analogously, we have

$$\limsup_{n \to \infty} \frac{1}{n} | \operatorname{tr}[T^{-1}T(g^-)]^i | \leq K_i,$$

for $i = 4, 5, 6$, and then

$$\limsup_{n \to \infty} \frac{1}{n} | \operatorname{tr}[T^{-1}T_\nabla] | \leq K,$$

with $K = \max\{K_1, \ldots, K_6\}$. Since the second summand in (5.4) can be bounded by a similar argument, the result is obtained.

Part (iii). Note that

$$\begin{aligned} E[\nabla^2 \mathcal{L}_n(\theta_0)] &= \frac{1}{2n} \{ -\operatorname{tr} \left(T^{-1}T_\nabla T^{-1}T_\nabla \right) + \operatorname{tr} \left(T^{-1}T_{\nabla^2} \right) \\ &\quad + 2\, E[y'A^{(1)}y] - E[y'A^{(2)}y] \}. \end{aligned}$$

Thus, by Problem 5.11 we have that

$$E[\nabla^2 \mathcal{L}_n(\theta_0)] = \frac{1}{2n} \left\{ -\operatorname{tr} \left(T^{-1}T_\nabla T^{-1}T_\nabla \right) \right\},$$

and then Lemma 5.1 yields

$$\lim_{n \to \infty} E[\nabla^2 \mathcal{L}_n(\theta_0)] = \frac{1}{4\pi} \int_{-\pi}^{\pi} \left[\frac{\nabla f(\lambda)}{f(\lambda)} \right] \left[\frac{\nabla f(\lambda)}{f(\lambda)} \right]' d\lambda = \Gamma(\theta_0). \qquad (5.5)$$

On the other hand,

$$\mathrm{Var}[\nabla^2\mathcal{L}_n(\theta_0)] = \frac{1}{4n^2}\,\mathrm{Var}\left\{y'[2A^{(1)} - A^{(2)}]y\right\}.$$

Hence, again by Problem 5.11 we obtain

$$\mathrm{Var}[\nabla^2\mathcal{L}_n(\theta_0)] = \frac{1}{4n^2}\,\mathrm{Var}\left\{T^{-2}[2T_\nabla T^{-1}T_\nabla - T_{\nabla^2}]^2\right\}.$$

Consequently, by Lemma 5.1 we conclude that

$$\lim_{n\to\infty}\mathrm{Var}[\nabla^2\mathcal{L}_n(\theta_0)] = 0. \tag{5.6}$$

Finally, from (5.5) and (5.6) we obtain part (iii). □

5.2.3 Efficiency

Theorem 5.3. *Under assumptions* (A0)–(A7), $\widehat{\theta}_n$ *is an efficient estimator of* θ_0.

Proof. Let $I_n(\theta_0)$ be the Fisher information matrix of the model evaluated at θ_0, respectively. Then,

$$\begin{aligned}\frac{1}{n}I_n(\theta_0) &= n\,E[\nabla\mathcal{L}_n(\theta_0)\nabla\mathcal{L}_n(\theta_0)'] \\ &= \frac{1}{4n}\,\mathrm{Cov}[y'T^{-1}T_\nabla T^{-1}y] \\ &= \frac{1}{2n}\,\mathrm{tr}\{T^{-1}T_\nabla T^{-1}T_\nabla\}.\end{aligned}$$

Finally, an application of Lemma 5.1 yields

$$\lim_{n\to\infty}\frac{1}{n}I_n(\theta_0) = \Gamma(\theta_0).$$

□

5.3 EXAMPLES

In this section we discuss the application of the previous theoretical results to the analysis of the large sample properties of MLE for some well-known long-memory models.

EXAMPLE 5.1

For a fractional noise process with long-memory parameter d, FN(d), the maximum-likelihood estimate $\widehat{d}_n$ satisfies the following limiting distribution:

$$\sqrt{n}(\widehat{d}_n - d) \to N\left(d, \frac{6}{\pi^2}\right),$$

as $n \to \infty$. Observe that the asymptotic variance of this estimate does not depend on the value of d.

EXAMPLE 5.2

Consider the ARFIMA$(1, d, 1)$ model

$$(1 + \phi B)y_t = (1 + \theta B)(1 - B)^{-d}\varepsilon_t,$$

where $\{\varepsilon_t\}$ is independent and identically distributed $N(0, \sigma^2)$. The parameter variance-covariance matrix $\Gamma(d, \phi, \theta)$ may be calculated as follows. The spectral density of this process is given by

$$f(\lambda) = \frac{\sigma^2}{2\pi}[2(1 - \cos\lambda)]^{-d}\frac{1 + \theta^2 + 2\theta\cos\lambda}{1 + \phi^2 + 2\phi\cos\lambda}.$$

Hence,

$$\begin{aligned}\log f(\lambda) &= \log\left(\frac{\sigma^2}{2\pi}\right) - d\log[2(1 - \cos\lambda)] \\ &\quad + \log[1 + \theta^2 + 2\theta\cos\lambda] - \log[1 + \phi^2 + 2\phi\cos\lambda],\end{aligned}$$

and the gradient is

$$\nabla \log f(\lambda) = \begin{bmatrix} -\log[2(1 - \cos\lambda)] \\ -\dfrac{2[\phi + \cos\lambda]}{1 + \phi^2 + 2\phi\cos\lambda} \\ \dfrac{2[\theta + \cos\lambda]}{1 + \theta^2 + 2\theta\cos\lambda} \end{bmatrix}.$$

Thus, by dropping the parameters d, ϕ, and θ from the 3×3 matrix $\Gamma(d, \phi, \theta)$ we have

$$\Gamma_{11} = \frac{1}{4\pi}\int_{-\pi}^{\pi}\{\log[2(1 - \cos\lambda)]\}^2\, d\lambda = \frac{\pi^2}{6},$$

$$\begin{aligned}\Gamma_{12} &= \frac{1}{4\pi}\int_{-\pi}^{\pi}\{\log[2(1 - \cos\lambda)]\}\frac{2[\phi + \cos\lambda]}{1 + \phi^2 + 2\phi\cos\lambda}d\lambda \\ &= \frac{1}{2\pi}\left\{\frac{\phi^2 - 1}{\phi}\int_0^{\pi}\frac{\log[2(1 - \cos\lambda)]}{1 + \phi^2 + 2\phi\cos\lambda}d\lambda \right.\\ &\quad \left. + \frac{1}{\phi}\int_0^{\pi}\log[2(1 - \cos\lambda)]d\lambda\right\}.\end{aligned}$$

Now, by Gradshteyn and Ryzhik (2000, p. 586) we have

$$\Gamma_{12} = -\frac{\log(1 + \phi)}{\phi}.$$

Analogously,

$$\Gamma_{13} = \frac{\log(1+\theta)}{\theta}.$$

In addition, for the two ARMA parameters we have

$$\begin{aligned}\Gamma_{22} &= \frac{1}{1-\phi^2},\\ \Gamma_{23} &= -\frac{1}{1-\phi\theta},\\ \Gamma_{33} &= \frac{1}{1-\theta^2}.\end{aligned}$$

Finally,

$$\Gamma(d,\phi,\theta) = \begin{bmatrix} \frac{\pi^2}{6} & -\frac{\log(1+\phi)}{\phi} & \frac{\log(1+\theta)}{\theta} \\ -\frac{\log(1+\phi)}{\phi} & \frac{1}{1-\phi^2} & -\frac{1}{1-\phi\theta} \\ \frac{\log(1+\theta)}{\theta} & -\frac{1}{1-\phi\theta} & \frac{1}{1-\theta^2} \end{bmatrix}. \tag{5.7}$$

Observe that similarly to the fractional noise case, the asymptotic variance of the MLE of an ARFIMA$(1,d,1)$ model does not depend on the value of the long-memory parameter.

■ **EXAMPLE 5.3**

The asymptotic variance of the MLE of ARFIMA$(1,d,0)$ and ARFIMA$(0,d,1)$ may be derived analogously to Example 5.2.

For the ARFIMA$(1,d,0)$ model we have that

$$\Gamma(d,\phi) = \begin{bmatrix} \frac{\pi^2}{6} & -\frac{\log(1+\phi)}{\phi} \\ -\frac{\log(1+\phi)}{\phi} & \frac{1}{1-\phi^2} \end{bmatrix},$$

and for the ARFIMA$(0,d,1)$ model

$$\Gamma(d,\theta) = \begin{bmatrix} \frac{\pi^2}{6} & \frac{\log(1+\theta)}{\theta} \\ \frac{\log(1+\theta)}{\theta} & \frac{1}{1-\theta^2} \end{bmatrix}.$$

From these expressions, we conclude that the asymptotic correlation between the maximum-likelihood estimates $\widehat{d}_n$ and $\widehat{\phi}_n$ of the ARFIMA$(1,d,0)$ model is

$$\lim_{n\to\infty} \operatorname{corr}(\widehat{d}_n, \widehat{\phi}_n) = \frac{\sqrt{6}}{\pi}\sqrt{1-\phi^2}\,\frac{\log(1+\phi)}{\phi}, \tag{5.8}$$

which is always positive for $\phi \in (-1, 1)$. On the other hand, the asymptotic correlation of the maximum-likelihood estimates $\widehat{d}_n$ and $\widehat{\theta}_n$ of an ARFIMA$(0, d, 1)$ model is given by

$$\lim_{n\to\infty} \operatorname{corr}(\widehat{d}_n, \widehat{\theta}_n) = -\frac{\sqrt{6}}{\pi}\sqrt{1-\theta^2}\,\frac{\log(1+\theta)}{\theta}, \tag{5.9}$$

which is always negative for $\theta \in (-1, 1)$. Observe that the asymptotic correlation formulas (5.8) and (5.9) do not depend on the value of the long-memory parameter d.

The limiting correlation between the maximum-likelihood estimates $\widehat{d}_n$ and $\widehat{\phi}_n$ of an ARFIMA$(1, d, 0)$ model provided by formula (5.8) is displayed in Figure 5.1 for $\phi \in (-1, 1)$. Additionally, Figure 5.2 exhibits the theoretical asymptotic correlation between the maximum-likelihood estimates $\widehat{d}_n$ and $\widehat{\theta}_n$ of an ARFIMA$(0, d, 1)$ model given by formula (5.9) for $\theta \in (-1, 1)$. Notice from these figures that the correlation between the estimators tends to 0 as $\phi \to \pm 1$ or $\theta \to \pm 1$. The maximum (minimum) value of the correlation is reached near $\phi = -0.68$ ($\theta = -0.68$).

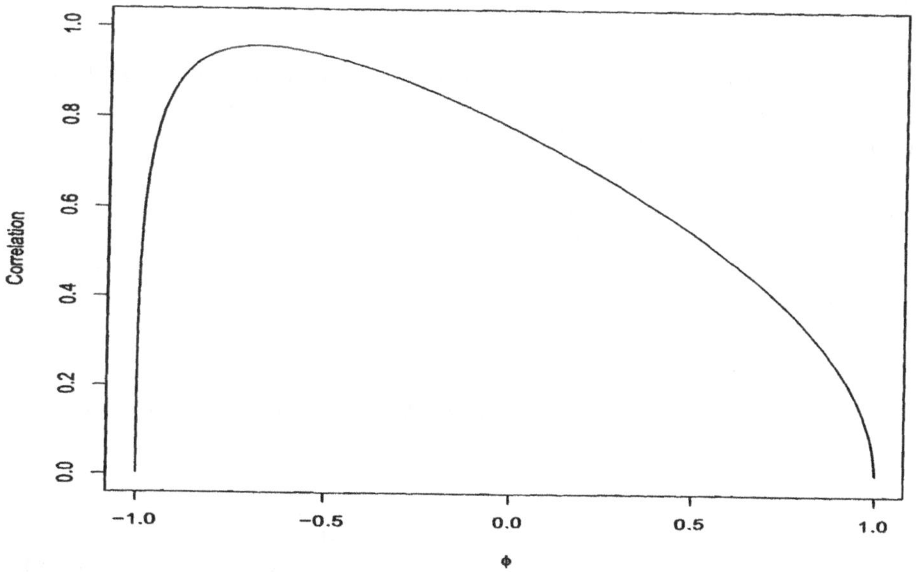

Figure 5.1 ARFIMA$(1, d, 0)$ example: Asymptotic correlation between the maximum-likelihood estimates $\widehat{d}_n$ and $\widehat{\phi}_n$.

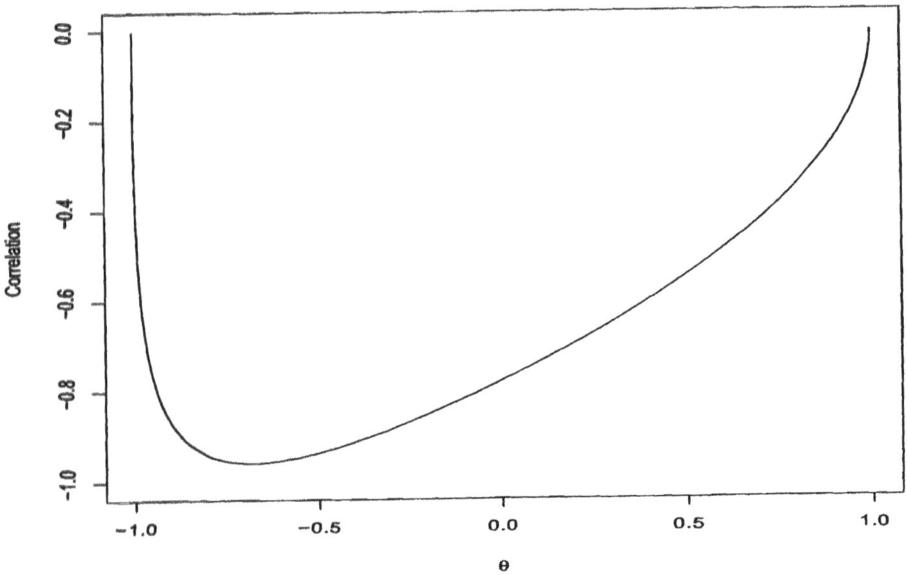

Figure 5.2 ARFIMA$(0, d, 1)$ example: Asymptotic correlation between the maximum-likelihood estimates $\widehat{d}_n$ and $\widehat{\theta}_n$.

5.4 ILLUSTRATION

To illustrate how the finite sample performance of the maximum-likelihood estimates of ARFIMA models compare to the theoretical results revised in this chapter consider the following Monte Carlo experiments.

Table 5.1 exhibits the maximum-likelihood parameter estimations from simulated ARFIMA$(1, d, 1)$ processes with sample size $n = 1000$ and parameters $d = 0.3$, $\phi = -0.5$, and $\theta = 0.2$. The results are based on 1000 replications.

Notice that the sample mean and standard deviations are close to their theoretical counterparts. The theoretical standard deviations are calculated from formula (5.7).

Table 5.1 MLE Simulations for an ARFIMA$(1, d, 1)$ Model with $d = 0.3$, $\phi = -0.5$, and $\theta = 0.2$

	d	ϕ	θ
Sample mean	0.2775	-0.5054	0.1733
Sample SD	0.0514	0.0469	0.0843
Theoretical SD	0.0487	0.0472	0.0834

5.5 TECHNICAL LEMMAS

In this chapter we have used the following lemmas whose proofs can be found in Dahlhaus (1989).

Lemma 5.1. *Let f_j and g_j, $j = 1, ..., p$, be real-valued symmetric functions, where all f_j are nonnegative and satisfy* (A2) *and* (A7) *for $k = 0, 1$ and g_j are continuous at all $\lambda \neq 0$ and*

$$g_j(\lambda) = \mathcal{O}(\lambda^{-\beta-\delta}),$$

for all $\delta > 0$. Assuming that $0 < \alpha, \beta < 1$ and $p(\beta - \alpha) < \frac{1}{2}$, we have

$$\lim_{n\to\infty} \frac{1}{n} \operatorname{tr} \left[\prod_{j=1}^{q} T(f_j)^{-1} T(g_j) \right] = \frac{1}{2\pi} \int_{-\pi}^{\pi} \left[\prod_{j=1}^{q} \frac{g_j(\lambda)}{f_j(\lambda)} \right] d\lambda. \tag{5.10}$$

Lemma 5.2. *Under assumptions* (A2)–(A9), *the quadratic forms $z_n^{(i)}(\theta)$ are equicontinuous in probability for $i = 0, \ldots, 3$. That is, for each $\eta > 0$ and $\epsilon > 0$ there exists $\delta > 0$ such that*

$$\limsup_{n\to\infty} P\left(\sup_{|\theta_1-\theta_2|\leq\delta} |z_n^{(i)}(\theta_1) - z_n^{(i)}(\theta_2)| > \eta \right) < \epsilon. \tag{5.11}$$

5.6 BIBLIOGRAPHIC NOTES

Yajima (1985) showed the large sample properties discussed in this chapter for fractional noise processes. The proofs presented in this chapter correspond to the extension to the general case of a long-memory process established by Dahlhaus (1989, 2006), with some minor changes for readability. Note that the work by Fox and Taqqu (1986) is fundamental for obtaining some of these results; see Dahlhaus (1989).

Following Dahlhaus (2006), assumption (A9) has been slightly changed to include ARFIMA processes. Another difference is that for simplicity we have assumed that the process has zero mean. If the mean is not zero, μ say, the proof is almost identical by assuming that we have a consistent estimate $\widehat{\mu}_n$; see Dahlhaus (1989) for further details. Extensions of the theorems discussed in this chapter to the case $-\frac{1}{2} < d < 0$ can be found in Möhring (1990).

Problems

5.1 Show that the property

$$\frac{\partial f_\theta(\lambda)}{\partial \theta_j} = \mathcal{O}(|\lambda|^{-\alpha(\theta)-\delta}),$$

for $\lambda \neq 0$ can be derived from assumptions (A2) and (A3).

5.2 Show that a Gaussian ARFIMA(p, d, q) process with compact parameter space Θ satisfies assumption (A0)–(A9).

5.3 Let A be the $n \times n$ matrix defined in the proof of Theorem 5.1, $A = T_0 T_1^{-1} - I$.

a) Show that if λ is an eigenvalue of A, then $1 + t\lambda$ is an eigenvalue of $I + tA$ for any $t \in \mathbb{R}$.

b) Consider the function

$$f(t) = \sum_{j=1}^{n} \log(1 + t\lambda_j),$$

where $\{\lambda_j\}$ are the eigenvalues of A. Verify that $f'(0) = \operatorname{tr}(A)$ and that

$$f''(t) = -\sum_{j=1}^{n} \left(\frac{\lambda_j}{1 + t\lambda_j} \right)^2.$$

c) Verify expression (5.1) using Taylor's theorem.

d) Show that

$$\log\det(I + tA) = \sum_{j=1}^{\infty} (-1)^{j+1} \operatorname{tr}(A^j) \frac{t^j}{j}.$$

For what values of t and A does the above series converge?

5.4 Prove that a spectral density f_θ satisfying (A0)–(A9) is uniformly bounded from below, that is, there exists a positive constant $c > 0$ such that $f_\theta(\lambda) > c$.

5.5 Show that if the spectral density f_θ is uniformly bounded from below there is a constant $K > 0$ such that $\|T(f_\theta)^{-1}\| \leq K$, where $\|A\|$ is the spectrum norm of the $n \times n$ matrix:

$$\|A\| = \sup_x \left(\frac{x' A' A x}{x' x} \right)^{1/2}.$$

5.6 Let $A(f_\theta)$ be a $n \times n$ matrix with elements

$$a_{rs}(\theta) = \int_{-\pi}^{\pi} f_\theta(\lambda) e^{i(r-s)\lambda} d\lambda,$$

where f_θ is the spectral density of a Gaussian process and $\theta = (\theta_1, \ldots, \theta_p)$.

a) Let $\lambda_j(\theta)$ be an eigenvalue of $A(\theta)$. Show that $\partial\lambda_j(\theta)/\partial\theta_k$ is an eigenvalue of $A\,(\partial f_\theta / \partial\theta_k)$.

b) Let $g(\theta) = \log\det[A(f_\theta)]$. Verify that

$$\frac{\partial g(\theta)}{\partial\theta_k} = \operatorname{tr}\left[A(f_\theta)^{-1} A \left(\frac{\partial f_\theta}{\partial\theta_k} \right) \right].$$

5.7 Suppose that X and Y are real-valued random variables.

a) Show that $P(X \leq Y) \leq P(X \leq 0) + P(Y \geq 0)$.

b) Use part (a) to verify (5.3).

5.8 Consider the following positive symmetric functions such that

$$\begin{aligned} f^{-1}(\lambda) &= \mathcal{O}(|\lambda|^{\alpha}), \\ g(\lambda) &= \mathcal{O}(|\lambda|^{-\beta}), \end{aligned}$$

with $0 < \alpha, \beta < 1$.

a) Show that

$$\|T(g)^{1/2}T(f)^{-1/2}\|^2 \leq K \sup_{h \in \mathcal{P}_n} \frac{\int_{-\pi}^{\pi} |\lambda|^{-\beta} h(\lambda)\, d\lambda}{\int_{-\pi}^{\pi} |\lambda|^{-\alpha} h(\lambda)\, d\lambda},$$

where

$$\mathcal{P}_n = \{h(\lambda) : h(\lambda) \text{ is a probability density on } [-\pi, \pi] \text{ with } h(\lambda) \leq n\}.$$

b) Show that if $\beta \leq \alpha$, then $\|T(g)^{1/2}T(f)^{-1/2}\|$ is bounded.

c) Show that if $\beta > \alpha$, then the sup in part (a) is reached by $h(\lambda) = n\mathbf{1}\,(|\lambda| < 1/2n)$.

d) Verify that $\|T(g)^{1/2}T(f)^{-1/2}\|^2 = \mathcal{O}(n^{\max\{\beta-\alpha, 0\}})$.

5.9 Show that for an ARFIMA$(1, d, 0)$ model,

$$\text{corr}(\widehat{d}, \widehat{\phi}) \to \frac{\sqrt{6}}{\pi},$$

as $\phi \to 0$.

5.10 Show that for an ARFIMA$(0, d, 1)$ model,

$$\text{corr}(\widehat{d}, \widehat{\theta}) \to -\frac{\sqrt{6}}{\pi},$$

as $\theta \to 0$.

5.11 (Quadratic forms). Assume that $y = (y_1, \ldots, y_n)'$ follows a multivariate normal distribution $N(0, T)$ and consider the *quadratic form*

$$Q = y'Ay,$$

where A is a real symmetric $n \times n$ matrix.

a) Verify that the characteristic function of Q is

$$\phi(t) = \det(I - 2itAT)^{-1/2}.$$

b) Verify that for two squared matrices A and B

$$\det(I - AB) = \det(I - BA).$$

c) Let S be a nonsingular matrix such that $T = SS'$. Show that

$$\det(I - 2itAT) = \det(I - 2itS'AS).$$

d) Since $S'AS$ is symmetric, verify that

$$\det(I - 2itAT) = \det(I - 2itD),$$

where D is a diagonal matrix, $D = \text{diag}(\lambda_1, \cdots, \lambda_n)$, such that $\{\lambda_j\}$ are the eigenvalues of AT.

e) Prove that

$$\det(I - 2itAT) = \prod_{j=1}^{n}(1 - 2it\lambda_j).$$

f) By using the fact that the characteristic function of a χ^2 random variable with one degree of freedom is given by

$$\phi_{\chi^2}(t) = \frac{1}{(1 - 2it)^{1/2}},$$

show that the quadratic form Q may be written as

$$Q = \sum_{j=1}^{n} \lambda_j x_j,$$

where λ_j are the eigenvalues of AT and x_j are independent χ^2 random variables with one degree of freedom.

g) Show that $E(Q) = \text{tr}\{AT\}$.

h) Show that $\text{Var}(Q) = 2\,\text{tr}\{(AT)^2\}$.

Hint: Recall that

$$\int_{\mathbb{R}^n} \exp\left[-\frac{1}{2}y'T^{-1}y\right] dy = (2\pi)^{n/2}[\det T]^{1/2}.$$

5.12 Let $\{y_t : t \in \mathbb{Z}\}$ be a stationary process with spectral density $f(\lambda)$ and let $y = (y_1, y_2, \ldots, y_n)' \sim N(0, T_{\theta_0})$ where the elements of the $n \times n$ variance-covariance matrix $T_{\theta_0} = (T_{ij})$ are given by

$$T_{ij} = \int_{-\pi}^{\pi} f(\lambda)e^{i\lambda(i-j)}d\lambda.$$

Consider the function

$$\mathcal{L}_n(\theta) = \frac{1}{n}\log\det T_\theta + \frac{1}{n}y'T_\theta^{-1}y.$$

a) Prove that

$$E[\mathcal{L}_n(\theta_0)] = 1 + \log\det T_{\theta_0}.$$

b) Show that

$$\lim_{n\to\infty} E[\mathcal{L}_n(\theta_0)] = \log(2\pi) + \frac{1}{2\pi}\int_{-\pi}^{\pi} \log f(\lambda)\, d\lambda.$$

c) Verify that

$$\lim_{n\to\infty} E[\mathcal{L}_n(\theta_0)] = 1 + \log\sigma^2.$$

d) Verify that

$$\mathrm{Var}[\mathcal{L}_n(\theta_0)] = \frac{2}{n},$$

and prove that $\mathcal{L}_n(\theta_0)$ converges to $1 + \log\sigma^2$ in probability as n tends to infinity.

CHAPTER 6

HETEROSKEDASTIC MODELS

The time series models discussed so far are governed by a Wold expansion or a linear state space system, where the innovations are usually assumed to have constant conditional variance. However, there is strong empirical evidence that a large number of time series from fields as diverse as economics and physics exhibit some *stylized facts* such as clusters of highly variable observations followed by clusters of observations with low variability and strong autocorrelations either in the series or its squares. In this chapter we examine some of the models proposed to account for these features. In particular, we consider *heteroskedastic* time series models where the conditional variance given the past is no longer constant. To motivate the introduction of the concept of heteroskedastic time series we present an analysis of copper prices in Section 6.1. Three well-known classes of heteroskedastic models are discussed in Sections 6.2–6.4. Section 6.2 is devoted to the analysis of ARFIMA-GARCH models, Section 6.3 focuses on ARCH(∞) processes and Section 6.4 addresses long-memory stochastic volatility models. The performance of these methodologies is assessed in Section 6.5 in the context of a simulated time series setup. Applications of these heteroskedastic models to the analysis of a real-life data is discussed in Section 6.6. Bibliographic notes with further readings on this topic are given in Section 6.7 while a list of proposed problems is given at the end of this chapter.

Long-Memory Time Series. By Wilfredo Palma

6.1 INTRODUCTION

As an illustration of the stylized facts frequently found in economic time series, consider the daily log-returns of the copper prices from January 1, 1996 to May 11, 2005, defined by $r_t = \log(P_t) - \log(P_{t-1})$, where P_t is the price of a pound of copper at the London Metal Exchange at time t. These data are available from the Chilean Copper Commission (www.cochilco.cl/english). The series $\{r_t\}$ is displayed in panel (a) of Figure 6.1 while its squares are shown in panel (b).

From Figure 6.1(a) we note a period of high volatility during 1996. Besides, Figure 6.1(b) suggests that the squared series suffers from bursts of high volatility followed by periods of low volatility. On the other hand, the sample autocorrelation function, Figure 6.2(a), shows some significant autocorrelations in the returns while the sample autocorrelation of the squares exhibits a strong level of dependence; see Figure 6.2(b).

Several models have been proposed to account for these features. Most of these models specify an ARMA or an ARFIMA process for the returns and specify some parametric model for the conditional variance of the series given its infinite past. In some cases this model resembles an ARMA in the form of a *generalized autoregressive conditionally heteroskedastic* (GARCH) process or it resembles an AR(∞) process in the form of an ARCH(∞) model.

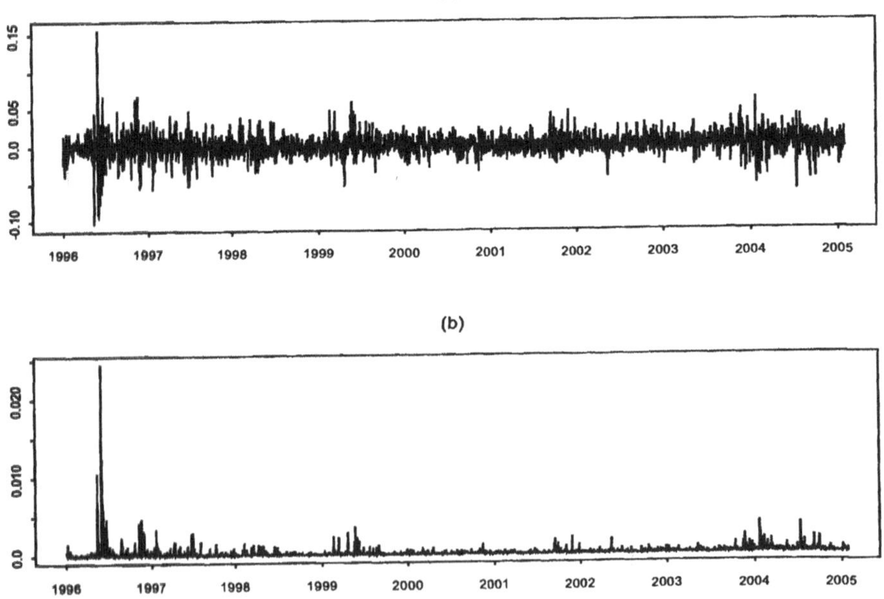

Figure 6.1 Copper price data (1996–2005): (a) Daily log-returns and (b) squared log-returns.

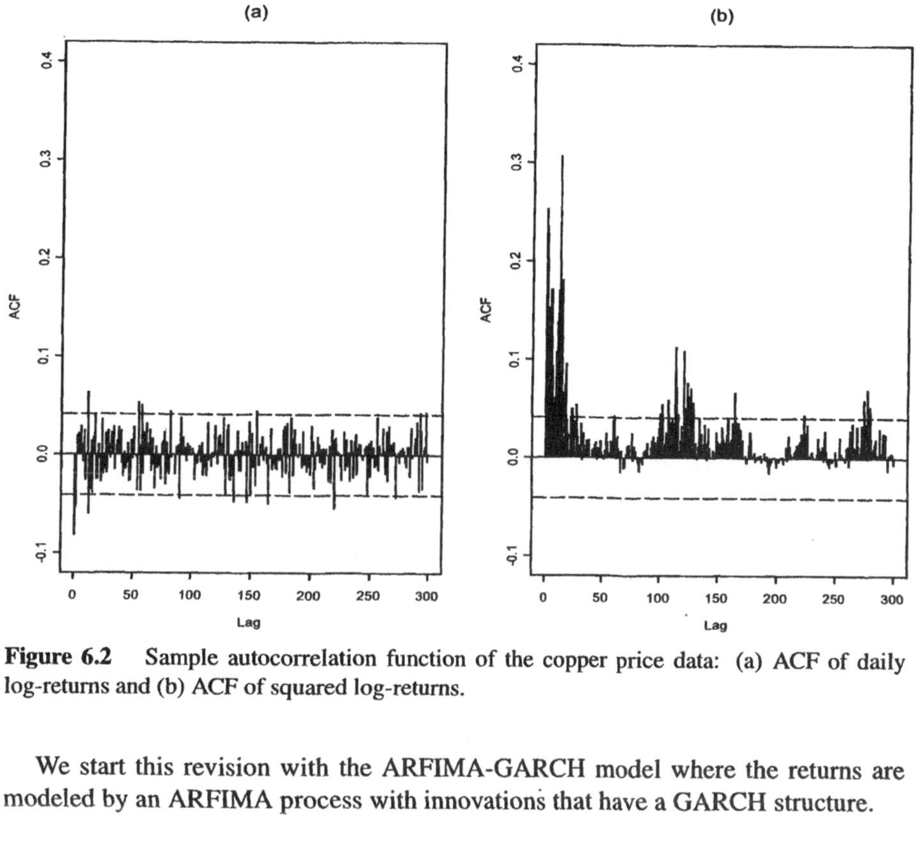

Figure 6.2 Sample autocorrelation function of the copper price data: (a) ACF of daily log-returns and (b) ACF of squared log-returns.

We start this revision with the ARFIMA-GARCH model where the returns are modeled by an ARFIMA process with innovations that have a GARCH structure.

6.2 ARFIMA-GARCH MODEL

An ARFIMA(p, d, q)-GARCH(r, s) process is defined by the discrete-time equation

$$\phi(B)y_t = \theta(B)(1-B)^{-d}\varepsilon_t, \tag{6.1}$$

$$\varepsilon_t = \epsilon_t\sigma_t, \tag{6.2}$$

$$\sigma_t^2 = \alpha_0 + \sum_{j=1}^{r}\alpha_j\varepsilon_{t-j}^2 + \sum_{j=1}^{s}\beta_j\sigma_{t-j}^2, \tag{6.3}$$

where $\mathcal{F}_{t-1}$ is the σ-algebra generated by the past observations $y_{t-1}, y_{t-2}, \ldots$, $\sigma_t^2 = E[y_t^2|\mathcal{F}_{t-1}]$ is the conditional variance of the process $\{y_t\}$, the GARCH coefficients $\alpha_1, \ldots, \alpha_r$ and $\beta_1, \ldots, \beta_s$ are positive, $\sum_{j=1}^{r}\alpha_j + \sum_{j=1}^{s}\beta_j < 1$, and $\{\epsilon_t\}$ is sequence of independent and identically distributed zero mean and unit variance random variables. Although ϵ_t is often assumed to be Gaussian, in some cases it may be specified by a t-distribution or a double exponential distribution, among others. These distributions have a greater flexibility to accommodate a possible heavy tail behavior of some financial time series.

■ **EXAMPLE 6.1**

In order to explore the structure of the model described by (6.1)–(6.3), consider the ARFIMA(p, d, q)-GARCH$(1, 1)$ process:

$$\begin{aligned} y_t &= \sum_{j=0}^{\infty} \psi_j \varepsilon_{t-j}, \\ \varepsilon_t &= \epsilon_t \sigma_t, \\ \sigma_t^2 &= \alpha_0 + \alpha_1 \epsilon_{t-1}^2 + \beta_1 \sigma_{t-1}^2, \end{aligned}$$

where $\psi(B) = \phi(B)^{-1}\theta(B)(1-B)^{-d}$ and ϵ_t follows a standard normal distribution. Thus, we may write

$$\begin{aligned} \sigma_t^2 &= \alpha_0 + (\alpha_1 \epsilon_{t-1}^2 + \beta_1)\sigma_{t-1}^2, \\ &= \left[\prod_{k=1}^{n} (\alpha_1 \epsilon_{t-k}^2 + \beta_1)\right] \sigma_{t-1-n}^2 + \alpha_0 \left[1 + \sum_{k=0}^{n-1} \prod_{j=1}^{k+1} (\alpha_1 \epsilon_{t-j}^2 + \beta_1)\right]. \end{aligned}$$

Define the random variable $z_n = \sum_{k=1}^{n} \log(\alpha_1 \epsilon_{t-k}^2 + \beta_1)$ and let $\gamma_n = z_n/n$. By the strong law of the large numbers, $\gamma_n \to E[\log(\alpha_1 \epsilon_1^2 + \beta_1]$ almost surely as $n \to \infty$. This limit is called the *top Lyapunov exponent* of the process, γ.

If $\gamma < 0$, then we may write

$$\sigma_t^2 = \alpha_0 \left[1 + \sum_{k=0}^{\infty} \prod_{j=1}^{k+1} (\alpha_1 \epsilon_{t-j}^2 + \beta_1)\right]. \tag{6.4}$$

Consequently, the process y_t may be expressed as

$$y_t = \sqrt{\alpha_0} \sum_{j=0}^{\infty} \psi_j \epsilon_{t-j} \left[1 + \sum_{k=0}^{\infty} \prod_{j=1}^{k+1} (\alpha_1 \epsilon_{t-j}^2 + \beta_1)\right]^{1/2}.$$

Thus, since $\{y_t\}$ corresponds to a measurable transformation of the independent and identically distributed sequence $\{\epsilon_t\}$, by Theorem 1.7, the process $\{y_t\}$ is stationary and ergodic. This result can be readily extended to the general model ARFIMA(p, d, q)-GARCH(r, s).

Observe that the conditional variance σ_t^2 specified by a GARCH(r, s) process may be expressed as an ARMA(p, r) model with $p = \max\{r, s\}$ as follows. Let $u_t = \sigma_t^2(\epsilon_t^2 - 1)$. This sequence is white noise, since $E[u_t] = 0$, $E[u_t^2] = E[\sigma_t^4(\epsilon_t^2 - 1)^2] = E[\sigma_t^4]\,E[(\epsilon_t^2 - 1)^2]$, and for $k > 0$ we have

$$E[u_t u_{t+k}] = E[E(u_t u_{t+k} | \mathcal{F}_{t+k-1})] = E[u_t \sigma_{t+k}^2\, E(\epsilon_{t+k}^2 - 1)] = 0.$$

Thus, σ_t^2 may be written as

$$(1 - \lambda_1 B - \cdots - \lambda_p B^p)\sigma_t^2 = \alpha_0 + \sum_{j=1}^{r} \alpha_j u_{t-j},$$

where $\lambda_j = \alpha_j \mathbf{1}_{\{1,\ldots,r\}}(j) + \beta_j \mathbf{1}_{\{1,\ldots,s\}}(j)$.

6.2.1 Estimation

An approximate MLE $\widehat{\theta}$ for the ARFIMA-GARCH model is obtained by maximizing the conditional log-likelihood

$$\mathcal{L}(\theta) = -\frac{1}{2n} \sum_{t=1}^{n} \left[\log \sigma_t^2 + \frac{\varepsilon_t^2}{\sigma_t^2} \right]. \tag{6.5}$$

Let $\theta = (\theta_1, \theta_2)'$, where $\theta_1 = (\phi_1, \ldots, \phi_p, \theta_1, \ldots, \theta_q, d)'$ is the parameter vector involving the ARFIMA components and $\theta_2 = (\alpha_0, \ldots, \alpha_r, \beta_1, \ldots, \beta_s)'$ is the parameter vector containing the GARCH component. The following result establishes some asymptotic properties of this estimate; see Ling and Li (1997b).

Theorem 6.1. *Let $\widehat{\theta}_n$ be the value that maximizes the conditional log-likelihood function (6.5). Then, under some regularity conditions, $\widehat{\theta}_n$ is a consistent estimate and $\sqrt{n}(\widehat{\theta}_n - \theta_0) \to N(0, \Omega^{-1})$, as $n \to \infty$, where $\Omega = diag(\Omega_1, \Omega_2)$ with*

$$\Omega_1 = E\left[\frac{1}{\sigma_t^2} \frac{\partial \varepsilon_t}{\partial \theta_1} \frac{\partial \varepsilon_t}{\partial \theta_1'} + \frac{1}{2\sigma_t^4} \frac{\partial \sigma_t^2}{\partial \theta_1} \frac{\partial \sigma_t^2}{\partial \theta_1'} \right],$$

and

$$\Omega_2 = E\left[\frac{1}{2\sigma_t^4} \frac{\partial \sigma_t^2}{\partial \theta_2} \frac{\partial \sigma_t^2}{\partial \theta_2'} \right].$$

At this point, it is necessary to introduce the concept of *intermediate memory* which will be used in the next section. We say that a second-order stationary process has intermediate memory if for a large lag h its ACF behaves like $\gamma(h) \sim \ell(h)|h|^{2d-1}$ with $d < 0$, where $\ell(\cdot)$ is a slowly varying function. Thus, the ACF decays to zero at an hyperbolic rate but it is summable, that is,

$$\sum_{h=0}^{\infty} |\gamma(h)| < \infty.$$

6.3 OTHER MODELS

The ARFIMA-GARCH process introduced in Section 6.2 has been employed for modeling many financial series exhibiting long-range dependence. Nevertheless, as shown in Chapter 7, the squares of an ARFIMA-GARCH process have only intermediate memory for $d \in (0, \frac{1}{4})$. In fact, for any $d \in (0, \frac{1}{2})$, the ACF of the squared series behaves like $k^{2\widetilde{d}-1}$, where k denotes the kth lag and $\widetilde{d} = 2d - \frac{1}{2}$. As a consequence, the long-memory parameter of the squared series, $\widetilde{d}$, is always smaller than the corresponding parameter of the original series d, that is, $\widetilde{d} < d$ for $d < \frac{1}{2}$.

Given that the squares of many financial series have similar or greater level of autocorrelation than their returns, the memory reduction that affects the squares of an ARFIMA-GARCH process may not be adequate in practice. This circumstance leads us to explore other classes of processes to model the strong dependence of the squared returns directly. As an important example of this approach, consider the following ARCH(∞) model:

$$y_t = \sigma_t \epsilon_t, \tag{6.6}$$

$$\sigma_t^2 = \alpha_0 + \sum_{j=1}^{\infty} \alpha_j y_{t-j}^2, \tag{6.7}$$

where $\{\epsilon_t\}$ is a sequence of independent and identically distributed random variables with zero mean and unit variance, α_0 is a positive constant, and $\alpha_j \geq 0$ for $j \geq 1$. This model can be formally written as

$$y_t^2 = \alpha_0 + \nu_t + \sum_{j=1}^{\infty} \alpha_j y_{t-j}^2, \tag{6.8}$$

where $\sigma_n^2 = E[y_t^2 | y_{t-1}, y_{t-2}, \ldots]$, $\nu_t = y_t^2 - \sigma_t^2$ is a martingale difference sequence.

If $E[\epsilon_0^2 \sum_{j=0}^{\infty} \alpha_j] < 1$, then the conditional variance may be written in terms of a Volterra expansion

$$\sigma_t^2 = \alpha_0 \sum_{k=0}^{\infty} \sum_{j_1,\ldots j_k=1}^{\infty} \alpha_{j_1}\alpha_{j_2}\cdots\alpha_{j_k}\epsilon_{t-j_1}^2\epsilon_{t-j_2}^2\cdots\epsilon_{t-j_k}^2, \tag{6.9}$$

and the process $\{y_t\}$ may be expressed as

$$y_t = \epsilon_t \left[\alpha_0 \sum_{k=0}^{\infty} \sum_{j_1,\ldots j_k=1}^{\infty} \alpha_{j_1}\alpha_{j_2}\cdots\alpha_{j_k}\epsilon_{t-j_1}^2\epsilon_{t-j_2}^2\cdots\epsilon_{t-j_k}^2,\right]^{1/2}.$$

In particular, when the coefficients $\{\alpha_j\}$ of the expansion (6.8) are specified by an ARFIMA(p,d,q) model, the resulting expression defines the FIGARCH(p,d,q) model. If $\pi(B) = \phi(B)(1-B)^d\theta(B)^{-1}$, then from (6.8) we get

$$\pi(B) y_t^2 = \alpha_0 + \nu_t.$$

Therefore, by multiplying both sides by $\theta(B)$ we conclude that

$$\phi(B)(1-B)^d y_t^2 = \omega + \theta(B)\nu_t,$$

where $\omega = \theta(B)\alpha_0$. This process is strictly stationary and ergodic but not second-order stationary. On the other hand, writing this model in terms of the conditional variance as in (6.7) we have

$$\sigma_t^2 = \alpha_0 + [1 - \pi(B)] y_t^2. \tag{6.10}$$

From (6.6) we may write $\log(y_t^2) = \log(\sigma_t^2) + 2\log(|\epsilon_t|)$. Thus, by considering $\log(y_t^2)$ as the *observed* returns, it may seem natural to attempt to specify a long-memory model directly to the term $\log(\sigma_t^2)$ instead of σ_t^2. An advantage of this formulation is that $\log(\sigma_t^2)$ is allowed to be negative. Therefore, unlike the FIGARCH model, no additional conditions on the parameters are needed to ensure the positivity of σ_t^2.

An example of this type of processes is the *fractionally integrated exponential* GARCH (FIEGARCH) model specified by

$$\phi(B)(1-B)^d \log(\sigma_t^2) = \alpha + \theta(B)|\epsilon_{t-1}| + \lambda(B)\epsilon_{t-1}, \tag{6.11}$$

where $\phi(B) = 1 + \phi_1 B + \cdots + \phi_p B^p$, $\alpha \in \mathbb{R}$, $\theta(B) = \theta_1 + \cdots + \theta_q B^{q-1}$, and the polynomial $\lambda(B) = \lambda_1 + \cdots + \lambda_q B^{q-1}$ accounts for the *leverage effect*, that is, conditional variances may react distinctly to negative or positive shocks.

6.3.1 Estimation

Consider the quasi-log-likelihood function

$$\mathcal{L}(\theta) = -\frac{1}{2}\log(2\pi) - \frac{1}{2}\sum_{t=1}^{n}\left[\log\sigma_t^2 + \frac{\varepsilon_t^2}{\sigma_t^2}\right], \tag{6.12}$$

where $\theta = (\omega, d, \phi_1, \ldots, \phi_p, , \theta_1, \ldots, \theta_q)$. A QMLE $\widehat{\theta}_n$ can be obtained by maximizing (6.12). But, even though this estimation approach has been widely used in many practical applications, to the best of our knowledge, asymptotic results for these estimators remain an open issue.

6.4 STOCHASTIC VOLATILITY

A *stochastic volatility* (SV) process is defined by the equations

$$r_t = \sigma_t \epsilon_t, \tag{6.13}$$

$$\sigma_t = \sigma \exp(v_t/2), \tag{6.14}$$

where $\{\epsilon_t\}$ is an independent and identically distributed sequence with zero mean and unit variance, and $\{v_t\}$ is a stationary process independent of $\{\epsilon_t\}$. In particular, $\{v_t\}$ may be specified as a long-memory ARFIMA(p, d, q) process. The resulting process is called *long-memory stochastic volatility* (LMSV) model.

From (6.13), we may write

$$\log(r_t^2) = \log(\sigma_t^2) + \log(\epsilon_t^2),$$

$$\log(\sigma_t^2) = \log(\sigma^2) + v_t.$$

Let $y_t = \log(r_t^2)$, $\mu = \log(\sigma^2) + E[\log(\epsilon_t^2)]$ and $\varepsilon_t = \log(\epsilon_t^2) - E[\log(\epsilon_t^2)]$. Then,

$$y_t = \mu + v_t + \varepsilon_t. \tag{6.15}$$

Consequently, the transformed process $\{y_t\}$ corresponds to a stationary long-memory process plus an additive noise.

On the other hand, an application of Theorem 1.7 yields the following result.

Theorem 6.2. *If $\{v_t\}$ is strictly stationary and ergodic, then the LMSV process $\{r_t\}$ and the transformed process $\{y_t\}$ are strictly stationary and ergodic.*

The ACF of (6.15) is given by

$$\gamma_y(h) = \gamma_v(k) + \sigma_\varepsilon^2 \delta_0(h),$$

where $\delta_0(h) = 1$ for $h = 0$ and $\delta_0(h) = 0$ otherwise. Furthermore, the spectral density of $\{y_t\}$, f_y, is given by

$$f_y(\lambda) = f_v(\lambda) + \frac{\sigma_\varepsilon^2}{2\pi},$$

where f_v is the spectral density of the long-memory process $\{v_t\}$.

In particular, if the process $\{v_t\}$ is an ARFIMA(p, d, q) model

$$\phi(B)v_t = \theta(B)(1-B)^{-d}\eta_t, \tag{6.16}$$

and $\theta = (d, \sigma_\eta^2, \sigma_\varepsilon^2, \phi_1, \ldots, \phi_p, \theta_1, \ldots, \theta_q)'$ is the parameter vector that specifies model (6.16), then the spectral density is given by

$$f_\theta(\lambda) = \frac{\sigma_\eta^2}{2\pi} \frac{|\theta(e^{i\lambda})|^2}{|1 - e^{i\lambda}|^{2d}|\phi(e^{i\lambda})|^2} + \frac{\sigma_\varepsilon^2}{2\pi}.$$

6.4.1 Estimation

The parameter θ can be estimated by minimizing the spectral likelihood

$$\mathcal{L}(\theta) = \frac{2\pi}{n} \sum_{j=1}^{n/2} \left[\log f_\theta(\lambda_j) + \frac{I(\lambda_j)}{f_\theta(\lambda_j)} \right]. \tag{6.17}$$

Let $\widehat{\theta}$ be the value that minimizes $\mathcal{L}(\theta)$ over the parameter space Θ. This estimator satisfies the following result; see Breidt, Crato, and de Lima (1998).

Theorem 6.3. *Assume that the parameter vector θ is an element of the compact parameter space Θ and assume that $f_{\theta_1} = f_{\theta_2}$ implies that $\theta_1 = \theta_2$. Let θ_0 be the true parameter value. Then, $\widehat{\theta}_n \to \theta_0$ in probability as $n \to \infty$.*

6.5 NUMERICAL EXPERIMENTS

The finite sample performance of the spectral-likelihood estimator based on (6.17) is analyzed in this section by means of Monte Carlo simulations. The model investigated

Table 6.1 Estimation of Long-Memory Stochastic Volatility Models

d	$\widehat{d}$	$\widehat{\sigma}_\eta$	SD($\widehat{d}$)	SD($\widehat{\sigma}_\eta$)
0.10	0.0868	9.9344	0.0405	0.4021
0.25	0.2539	10.0593	0.0400	0.4199
0.40	0.4139	10.1198	0.0415	0.3773

is the LMSV with an ARFIMA$(0, d, 0)$ structure, $\sigma_\varepsilon = \pi/\sqrt{2}$, ϵ_t follows a standard normal distribution, $\sigma_\eta = 10$, and the sample size is $n = 400$. Observe that $\varepsilon_t = \log \epsilon_t^2 - E[\log \epsilon_t^2]$. Thus, if $\epsilon_t \sim N(0,1)$, then $\text{Var}(\varepsilon_t) = \pi^2/2$.

The results displayed in Table 6.1 are based on 1000 replications. From this table, observe that the estimates of both the long-memory parameter d and the scale parameter σ_η are close to their true values. On the other hand, the standard deviations of $\widehat{d}$ and $\widehat{\sigma}_\eta$ seem to be similar for all the values of d simulated. However, to the best of our knowledge there are no formally established results for the asymptotic distribution of these QMLEs yet.

6.6 APPLICATION

Consider the copper prices series discussed in the introduction of this chapter. From Figure 6.2(a) it seems that the returns of this series have a short-memory structure. On the other hand, from Figure 6.2(b), the squared series seems to display a long-memory behavior. In order to account for these two features, we fitted an ARMA-FIEGARCH class of models with ARMA(p, q) specification for the mean and FIEGARCH(r, s) for the conditional variance. We considered two cases, models without leverage and models with leverage.

6.6.1 Model without Leverage

The model selected by the Akaike's information criterion (AIC)

$$\text{AIC} = -2 \log \mathcal{L}(\boldsymbol{\theta}) + 2r,$$

where $\mathcal{L}(\boldsymbol{\theta})$ is the likelihood of the data and r is the number of estimated parameters of the model, is the ARMA$(0, 1)$-FIEGARCH$(2, 1)$. The results from the quasi-maximum-likelihood estimation of this model without leverage effects are presented in Table 6.2.

All the parameters of the model are statistically significant at the 5% level. Furthermore, the estimated fractional differencing parameter belongs to the stationary region.

Table 6.2 Copper Data: ARMA-FIEGARCH Quasi-Maximum-Likelihood Estimation without Leverage Effect

Parameter	Estimate	SD	t-stat	$P(>\|t\|)$
MA(1)	-0.06892	0.02156	-3.196	0.0007
α	-0.09314	0.03632	-2.564	0.0052
GARCH(1)	0.89362	0.03910	22.855	0.0000
ARCH(1)	0.26953	0.02148	12.549	0.0000
ARCH(2)	-0.18475	0.03317	-5.570	0.0000
d	0.43554	0.10305	4.227	0.0000

6.6.2 Model with Leverage

Table 6.3 shows the results from a quasi-likelihood estimation when the leverage effects are incorporated into the model. All the parameters of the model are significant at the 5% level. However, the long-memory parameter d is greater than 0.5. Thus, the fitted model is nonstationary.

6.6.3 Model Comparison

The AIC of the six-parameter ARMA-FIEGARCH model without leverage effects is -13571.1 and the AIC of the eight-parameter ARMA-FIEGARCH model with leverage effects is -13579.5. These values are very close, even though formally we should select the second one. Despite the fact that the model with leverage is more complex, the leverage coefficients seem to make a significant contribution to the model fitting. However, the model with leverage effects leads to a nonstationary behavior since the estimated fractional differencing parameter is outside the stationarity region.

Table 6.3 Copper Data: ARMA-FIEGARCH Quasi-Maximum-Likelihood Estimation with Leverage Effect

Parameter	Estimate	SD	t-stat	$P(>\|t\|)$
MA(1)	-0.06393	0.02204	-2.901	0.0019
α	-0.03738	0.01183	-3.159	0.0008
GARCH(1)	0.93063	0.03029	30.722	0.0000
ARCH(1)	0.27820	0.02589	10.746	0.0000
ARCH(2)	-0.24043	0.02789	-8.622	0.0000
LEV(1)	-0.07070	0.01686	-4.192	0.0000
LEV(2)	0.07935	0.01706	4.652	0.0000
d	0.52601	0.09143	5.753	0.0000

Table 6.4 Copper Data: Ljung-Box Whiteness Tests for the ARMA-FIEGARCH Fitted Models

Model	Q_{12}	$P(> Q_{12})$
	Without Leverage	
Residuals	10.80	0.5461
Square residuals	11.57	0.4808
	With Leverage	
Residuals	10.21	0.5972
Square residuals	13.28	0.3487

In terms of model adequacy measured by the whiteness of the residuals or the squared residuals, the Ljung-Box tests with 12 degrees of freedom, Q_{12}, indicates that the residuals from both models behave adequately; see Table 6.4.

6.7 BIBLIOGRAPHIC NOTES

Several studies report evidence of long-memory behavior in returns or empirical volatilities; see, for example, Robinson (1991), Shephard (1996), Lobato and Savin (1998), and Baillie (1996). Similar features have been observed in data from other fields as well. In physics, for instance, the presence of strong autocorrelation in the squares of differences in velocity of the mean wind direction has been explored by Barndorff-Nielsen and Shephard (2001) and Mantegna and Stanley (2000).

Engle (1982) proposed the ARCH models to account for the stylized facts exhibited by many economic and financial time series. Based on this seminal work, a plethora of related models have been introduced. Among these methodologies we find the GARCH models [see, for example, Bollerslev (1986) and Taylor (1986)], the EGARCH models [see, for instance, Nelson (1991)], the *stochastic volatility processes* (SV) [see, for example, Harvey, Ruiz, and Shephard (1994)], the FIGARCH and FIEGARCH models [see, for instance, Baillie, Bollerslev, and Mikkelsen (1996) and Bollerslev and Mikkelsen (1996)], and the *long-memory generalized autoregressive conditionally heteroskedastic* (LMGARCH) models [see, for example, Robinson (1991), Robinson and Henry (1999) and Henry (2001)].

Most econometric models dealing with long-memory and heteroskedastic behaviors are nonlinear in the sense that the noise sequence is not necessarily independent. In particular, in the context of ARFIMA-GARCH models, the returns have long-memory and the noise has a conditional heteroskedasticity structure. These processes have received considerable attention; see, for example, Ling and Li (1997b) and references therein.

A related class of interesting models is the extension of the ARCH(p) processes to the ARCH(∞) models to encompass the longer dependence observed in many squared financial series. The ARCH(∞) class was first introduced by Robinson (1991).

On the other hand, extensions of the stochastic volatility processes to the long-memory case have produced the LMSV models; see Harvey, Ruiz, and Shephard (1994), Ghysels, Harvey, and Renault (1996), Breidt, Crato, and de Lima (1998), and Deo and Hurvich (2003). Other estimation procedures for LMSV using state space systems can be found in Chan and Petris (2000) and Section 11 of Chan (2002). Furthermore, exact likelihood-based Bayesian estimation of LMSV is discussed in Section 4 of Brockwell (2004).

Theorem 6.1 corresponds to Theorem 3.2 of Ling and Li (1997b) while Theorem 6.3 is due to Breidt, Crato, and de Lima (1998). The diagnostic tests employed in Section 6.6 are based on the Ljung-Box statistic and its extension to the analysis of squares; see, for example, the book by Li (2004) which gives a detailed account of this topic.

Problems

6.1 Consider the stochastic volatility process $\{r_t\}$ given by

$$\begin{aligned} r_t &= \varepsilon_t \sigma_t, \\ \sigma_t &= \sigma \exp(\nu_t/2), \end{aligned}$$

where $\{\varepsilon_t\}$ is an independent and identically distributed sequence with zero mean and unit variance and $\{\nu_t\}$ is a regular linear process satisfying

$$\nu_t = \sum_{j=0}^{\infty} \psi_j \eta_{t-j},$$

with $\sum_{j=0}^{\infty} \psi_j^2 < \infty$ and $\{\eta_t\}$ an independent and identically distributed sequence with zero mean and unit variance, independent of the sequence $\{\varepsilon_t\}$. Show that the process r_t is strictly stationary and ergodic.

6.2 Assume that $\pi(B) = (1 - B)^d$. Show that $\pi(B)\alpha_0 = 0$, where α_0 is any real constant and $d > 0$.

6.3 Show that the FIGARCH process may be written as

$$\theta(B)\sigma_t^2 = \omega + [\theta(B) - \phi(B)(1 - B)^d] y_t^2, \tag{6.18}$$

where $\omega = \theta(B)\alpha_0$. What conditions must satisfy the polynomial $\theta(B)$ in order to ensure that ω is a positive constant?

6.4 Let $\lambda(d) = (1 - B)^{-d}$ for $|d| < \frac{1}{2}$.

a) Verify that the ARFIMA(0, d, 0)-GARCH model may be written as

$$\lambda(d)\varepsilon_t(d) = c,$$

where c is a constant with respect to d. Note that the data $\{y_t\}$ do not depend on d.

b) Let $\psi(B) = \sum_{j=0}^{\infty} \psi_j(d) B^j = (1-B)^{-d} = \lambda(d)^{-1}$. Show that

$$\sum_{j=0}^{\infty} \left[\frac{\partial}{\partial d} \psi_j(d) \right] \varepsilon_{t-j}(d) + \sum_{j=0}^{\infty} \psi_j(d) \left[\frac{\partial}{\partial d} \varepsilon_{t-j}(d) \right] = 0.$$

c) Show that

$$\begin{aligned} \frac{\partial}{\partial d} \varepsilon_t(d) &= -\lambda(d) \sum_{j=0}^{\infty} \left[\frac{\partial}{\partial d} \psi_j(d) \right] \varepsilon_{t-j}(d) \\ &= -\lambda(d) \left[\frac{\partial}{\partial d} \lambda(d) \right] \varepsilon_t(d) \\ &= -\left[\frac{\partial}{\partial d} \log \lambda(d) \right] \varepsilon_t(d). \end{aligned}$$

d) Verify that

$$\frac{\partial}{\partial d} \varepsilon_t(d) = -\sum_{j=1}^{\infty} \frac{1}{j} \varepsilon_{t-j}(d),$$

and prove that this expansion is well-defined in $\mathcal{L}_2$.

6.5 Assume that the sequence $\{\epsilon_t\}$ in (6.2) corresponds to independent and identically distributed uniform random variables $U(-\sqrt{3}, \sqrt{3})$.

a) Verify that $\{\epsilon_t\}$ is a sequence of zero mean and unit variance random variables.

b) Show that for this specification of $\{\epsilon_t\}$, the top Lyapunov exponent of the model in Example 6.1 is given by

$$\gamma = E[\log(\alpha_1 \epsilon^2 + \beta_1)] = 2\left[\log \sqrt{3\alpha_1 + \beta_1} + \sqrt{\frac{\beta_1}{3\alpha_1}} \arctan \sqrt{\frac{3\alpha_1}{\beta_1}} - 1 \right].$$

Hint: The following formula could be useful:

$$\int \log(x^2 + a^2)\, dx = x \log(x^2 + a^2) + 2a \arctan \frac{x}{a} - 2x;$$

see Gradshteyn and Ryzhik (2000, p. 236).

6.6 Consider the ARFIMA(p, d, q)-GARCH$(1, 1)$ process:

$$\begin{aligned} y_t &= \sum_{j=0}^{\infty} \psi_j \varepsilon_{t-j}, \\ \varepsilon_t &= \epsilon_t \sigma_t, \\ \sigma_t^2 &= \alpha_0 + \alpha \varepsilon_{t-1}^2 + \beta \sigma_{t-1}^2, \end{aligned}$$

where ϵ_t is a random variable with density

$$f(\epsilon) = \frac{2}{\pi(1+\epsilon^2)^2},$$

for $-\infty < \epsilon < \infty$.

a) Verify that the random variable ϵ satisfies

$$E(\epsilon) = 0,$$
$$\mathrm{Var}(\epsilon) = 1.$$

b) Show that the top Lyapunov exponent in this case is given by

$$\gamma = 2\left[\log(\sqrt{\alpha}+\sqrt{\beta}) - \frac{\sqrt{\alpha}}{\sqrt{\alpha}+\sqrt{\beta}}\right]$$

c) Verify whether the Lyapunov exponent γ is negative for $\alpha > 0, \beta > 0$, and $\alpha + \beta < 1$.

Hint: The following formula could be useful:

$$\int_0^\infty \log(a^2 + b^2x^2)\frac{dx}{(1+x^2)^2} = \frac{\pi}{2}\left[\log(a+b) - \frac{b}{a+b}\right],$$

for $a, b > 0$; see Gradshteyn and Ryzhik (2000, p. 557).

6.7 Consider the ARFIMA-GARCH process defined in Problem 6.6 where ϵ_t is a random variable with density

$$f(\epsilon) = \frac{1}{\pi\sqrt{1-\epsilon^2}},$$

for $\epsilon \in (-1, 1)$.

a) Verify that ϵ is a zero mean and unit variance random variable.

b) Prove that the the top Lyapunov exponent in this case is

$$\gamma = 2\log\frac{\sqrt{\beta}+\sqrt{\alpha+\beta}}{2}.$$

c) Show that the Lyapunov exponent γ is negative for $\alpha > 0$, $\beta > 0$, and $\alpha + \beta < 1$.

Hint: The following integrals could be useful:

$$\int \frac{dx}{\sqrt{1-x^2}} = \arcsin x,$$
$$\int \frac{x^2dx}{\sqrt{1-x^2}} = \frac{1}{2}(\arcsin x - x\sqrt{1-x^2}),$$

and

$$\int_0^1 \log(1+ax^2)\frac{dx}{\sqrt{1-x^2}} = \pi\log\frac{1+\sqrt{1+a}}{2},$$

for $a \geq -1$; see Gradshteyn and Ryzhik (2000, p. 558).

6.8 The definition of the FIEGARCH model given by Equation (11) of Bollerslev and Mikkelsen (1996) is

$$\log(\sigma_t^2) = \omega + \phi(B)^{-1}(1-B)^{-d}\psi(B)g(\epsilon_{t-1}), \tag{6.19}$$

where $\phi(B) = 1 + \phi_1 B + \cdots + \phi_p B^p$, $\psi(B) = 1 + \psi_1 B + \cdots + \psi_q B^q$, and

$$g(\epsilon_t) = \theta\epsilon_t + \gamma[|\epsilon_t| - E(|\epsilon_t|)].$$

On the other hand, Zivot and Wang (2003) provide the following definition of a FIEGARCH process in their Equation (8.18):

$$\phi(B)(1-B)^d \log(\sigma_t^2) = a + \sum_{j=1}^{q}(b_j|\epsilon_{t-j}| + \gamma_j\epsilon_{t-j}). \tag{6.20}$$

a) Show that $\phi(B)(1-B)^d\omega = 0$.

b) Starting from definition (6.19), prove that

$$\phi(B)(1-B)^d \log(\sigma_t^2) = -\psi(B)\gamma\mu + \theta\psi(B)\epsilon_{t-1} + \gamma\psi(B)|\epsilon_{t-1}|,$$

where $\mu = E(|\epsilon_1|)$.

c) Show that by taking $b_j = \theta\psi_j$, $\gamma_j = \gamma\psi_j$, and $a = -\mu\gamma\sum_{j=1}^q \psi_j$, we obtain definition (6.20). Observe, however, that in model (6.20) we could release the parameters b_j and γ_j from the restriction $b_j\gamma = \gamma_j\theta$.

d) Verify that by taking $\theta_j = b_j$, $\lambda_j = \gamma_j$, and $\alpha = a$, definition (6.11) is obtained.

6.9 Consider the FIEGARCH model described by equation (6.11) and assume that $\alpha = -\sum_{j=1}^q \theta_j E(|\epsilon_1|)$.

a) Verify that

$$E[\alpha + \theta(B)|\epsilon_{t-1}| + \lambda(B)\epsilon_{t-1}] = 0.$$

b) Show that conditional variance σ_t^2 may be formally written as

$$\sigma_t^2 = \exp\left\{\phi(B)^{-1}(1-B)^{-d}[\theta(B)(|\epsilon_{t-1}| - E|\epsilon_{t-1}|) + \lambda(B)\epsilon_{t-1}]\right\}.$$

c) Show that the FIEGARCH process y_t may be formally written as

$$y_t = \epsilon_t \exp\left\{\tfrac{1}{2}\phi(B)^{-1}(1-B)^{-d}[\theta(B)(|\epsilon_{t-1}| - E|\epsilon_{t-1}|) + \lambda(B)\epsilon_{t-1}]\right\}.$$

d) Consider a FIEGARCH$(0, d, 1)$ where $\phi(B) = 1$, $(1-B)^{-d} = \sum_{j=0}^\infty \psi_j B^j$, $\theta(B) = \theta$, and $\lambda(B) = \lambda$. Show that σ_t^2 may be formally expressed as

$$\sigma_t^2 = \prod_{j=0}^{\infty} \exp\left[\theta\psi_j(|\epsilon_{t-1}| - E|\epsilon_{t-1}|) + \lambda\psi_j\epsilon_{t-1}\right].$$

Under what conditions is the above infinite product well defined?

CHAPTER 7

TRANSFORMATIONS

In this chapter we examine the autocorrelation of nonlinear transformations of a stationary process with a Wold expansion. The analysis of such transformations, including squares or higher powers of a time series, may give valuable clues about crucial aspects such as linearity, normality, or memory of the process. For instance, these issues are particularly important when studying the behavior of heteroskedastic processes since in this context the series usually represents the return of a financial instrument and the squared series is a rough empirical measure of its volatility. In this case, since we are interested in predicting both the returns and the volatility, we must analyze the dependence structure of a time series and its squares. Our analysis begins with general transformations of Gaussian processes through Hermite polynomials expansions; see Section 7.1. Subsequently, Section 7.2 presents an explicit expression for the autocorrelation of the squares for any regular process, not necessarily Gaussian. This formula is used in Section 7.3 to study the asymptotic behavior of the autocorrelation of a time series and its squares, as the lag increases to infinity. Furthermore, these results are applied to the analysis of several models including combinations of ARMA or ARFIMA with GARCH, EGARCH or ARCH(∞) errors, among others. Illustrations of these ideas through simulated time series are discussed in Section 7.4. Bibliographic notes are given in Section 7.5 while several problems are proposed at the end of the chapter.

7.1 TRANSFORMATIONS OF GAUSSIAN PROCESSES

Let $\{y_t : t \in \mathbb{Z}\}$ be a Gaussian process and suppose that f is a measurable transformation with Hermite expansion

$$f(y) = \sum_{j=0}^{\infty} \frac{\alpha_j H_j(y)}{j!}, \tag{7.1}$$

where $H_j(y)$ are the Hermite polynomials

$$H_j(y) = (-1)^j e^{y^2/2} \frac{d^j}{dy^j} e^{-y^2/2},$$

which form an orthonormal basis for Gaussian random variables, that is,

$$\langle H_i(y_t), H_j(y_t) \rangle = j!\delta(i-j),$$

where

$$\langle H_i(y_t), H_j(y_t) \rangle = \int_{-\infty}^{\infty} H_i(y) H_j(y) \frac{1}{\sqrt{2\pi}} e^{-y^2/2} dy,$$

and $\delta(h) = 1$ for $h = 0$ and $\delta(h) = 0$ for $h \neq 0$. For example,

$$\begin{aligned} H_0(y) &= 1, & H_1(y) &= y, \\ H_2(y) &= y^2 - 1, & H_3(y) &= y^3 - 3y, \\ H_4(y) &= y^4 - 6y^2 + 3, & H_5(y) &= y^5 - 10y^3 + 15y. \end{aligned}$$

Observe that the joint Gaussian distribution of (x, y) with correlation ρ and unit standard deviation may be written in terms of the Hermite polynomials $\{H_k\}$ as follows:

$$p(x, y) = \frac{1}{2\pi} \exp\left(-\frac{x^2 + y^2}{2}\right) \sum_{k=0}^{\infty} \frac{\rho^k}{k!} H_k(x) H_k(y).$$

Thus,

$$
\begin{aligned}
\langle H_i(y_t), H_j(y_s)\rangle &= \int_{-\infty}^{\infty}\int_{-\infty}^{\infty} H_i(y_t)H_j(y_s)p(y_t, y_s)\, dy_t\, dy_s \\
&= \int_{-\infty}^{\infty}\int_{-\infty}^{\infty} H_i(x)H_j(y)\frac{1}{2\pi} \\
&\quad \times \sum_{k=0}^{\infty} \frac{\rho(t-s)^k}{k!} H_k(x)H_k(y)\exp\left(-\frac{x^2+y^2}{2}\right) dx\, dy \\
&= \sum_{k=0}^{\infty} \frac{\rho(t-s)^k}{k!} \langle H_i(y_t), H_k(y_t)\rangle \langle H_j(y_s), H_k(y_s)\rangle \\
&= \sum_{k=0}^{\infty} \frac{\rho(t-s)^k}{k!} i!\delta(i-k)j!\delta(j-k) \\
&= j!\rho(t-s)^j\delta(i-j),
\end{aligned}
$$

where $\rho(t-s)$ is the correlation between y_t and y_s.

Therefore, from (7.1)

$$
\langle f(y_t), f(y_s)\rangle = \sum_{j=0}^{\infty} \frac{\alpha_j^2 \rho(t-s)^j}{j!}. \tag{7.2}
$$

■ EXAMPLE 7.1

Let $\{y_t : t \in \mathbb{Z}\}$ be a Gaussian process with $E[y_t] = 0$ and $\mathrm{Var}[y_t) = 1$. For the transformation $f(y_t) = y_t^2$, the coefficients of the Hermite expansion are $\alpha_0 = 1$, $\alpha_2 = 1$, and $\alpha_j = 0$ for all $j \neq 0, 2$. Thus, we have $\langle f(y_t), f(y_s)\rangle = 1 + \rho_y^2(h)/2$. But, $E(y_t^2) = 1$ so that

$$
\mathrm{Cov}[f(y_t), f(y_s)] = \langle f(y_t), f(y_s)\rangle - 1 = \rho_y^2(h)/2,
$$

and then the autocorrelation function of $f(y_t) = y_t^2$ is

$$
\rho_{y^2}(h) = \rho_y^2(h). \tag{7.3}
$$

From this example we observe that since $|\rho_y| \leq 1$, $\rho_{y^2}(h)$ is smaller or equal than $\rho_y(h)$. Consequently, the autocorrelation of the squares is smaller than the autocorrelation of the original series. Actually, for a Gaussian process this reduction of the dependence is true for any transformation, as stated in the following lemma.

Lemma 7.1. *Let $\{y_t : t \in \mathbb{Z}\}$ be a Gaussian process and let $\mathcal{F}$ be the class of all measurable transformations such that $E[f(y_t)] = 0$ and $E[f(y_t)^2] = 1$. Then,*

$$
\sup_{f\in\mathcal{F}} E[f(y_t)f(y_s)] = |\rho_y(t-s)|,
$$

where $\rho_y(t-s)$ is the correlation between y_t and y_s.

Proof. To simplify the notation, define $\rho = \rho_y(t-s)$. Observe that $E[f(y_t)f(y_s)] = \langle f(y_t), f(y_s)\rangle$. Therefore,

$$E[f(y_t)f(y_s)] = \sum_{j=0} \frac{\alpha_j^2 \rho^j}{j!}.$$

Since $f \in \mathcal{F}$, we have

$$E[f(y_t)] = \sum_{j=0}^{\infty} \frac{\alpha_j}{j!} E[H_j(y_t)] = \alpha_0 = 0, \tag{7.4}$$

and

$$E[f(y_t)^2] = \sum_{j=0}^{\infty} \frac{\alpha_j^2 \rho^j}{j!} = 1. \tag{7.5}$$

Since $|\rho| \leq 1$, $|\rho^j| \leq |\rho|$ for all $j \geq 1$. Consequently, from (7.4) and (7.5) we conclude that

$$|\, E[f(y_t)f(y_s)]| \leq |\rho| \sum_{j=1}^{\infty} \frac{\alpha_j^2}{j!} = |\rho|.$$

Since the equality is attained for $f(y_t) = [y_t - E(y_t)]/\sqrt{\mathrm{Var}(y_t)}$, the lemma is proved. □

As a consequence of our previous discussion, in order to account for situations where the squares exhibit more dependence than the series itself, we must abandon Gaussianity. In the next section we examine this issue in detail.

7.2 AUTOCORRELATION OF SQUARES

Consider the regular linear process $\{y_t\}$ with Wold expansion

$$y_t = \psi(B)\varepsilon_t, \tag{7.6}$$

where $\psi(B) = \sum_{i=0}^{\infty} \psi_i B^i$, $\psi_0 = 1$, and $\sum_{i=0}^{\infty} \psi_i^2 < \infty$. The *input noise* sequence $\{\varepsilon_t\}$ is assumed to be white noise. However, the behavior of the squared series $\{y_t^2\}$ may be substantially different whether the sequence $\{\varepsilon_t\}$ is independent (*strict white noise*) or dependent (e.g., martingale difference), as will be discussed later. On the other hand, the autocorrelation structure of the squares will depend on whether the filter $\psi(B)$ has short or long memory. For simplicity, in this study we call *short-memory filter* to the case $\psi_i = O(v^i)$ for some $|v| < 1$ and *long-memory filter* if

$\psi_i = \mathcal{O}(i^{d-1})$ for $d \in (0, \frac{1}{2})$; see, for example, the definition of a strongly dependent process given by (3.4).

Furthermore, we assume that $\{\varepsilon_t\}$ are random variables with zero mean, finite kurtosis, $\eta = E(\varepsilon_t^4)/[E(\varepsilon_t^2)]^2$, uncorrelated but not necessarily independent with

$$E(\varepsilon_u \varepsilon_v) = \begin{cases} \sigma^2 & \text{if } u = v, \\ 0 & \text{otherwise,} \end{cases} \tag{7.7}$$

$$E(\varepsilon_s \varepsilon_t \varepsilon_u \varepsilon_v) = \begin{cases} [1 + (\eta - 1)\rho_{\varepsilon^2}(s - v)]\sigma^4 & s = t, u = v \text{ or } s = u, t = v, \\ [1 + (\eta - 1)\rho_{\varepsilon^2}(s - t)]\sigma^4 & s = v, t = u, \\ 0 & \text{otherwise.} \end{cases} \tag{7.8}$$

The next theorem establishes an explicit expression for the autocorrelation function of $\{y_t^2\}$ for linear and nonlinear processes satisfying both (7.7) and (7.8). This formula plays a key role in the analysis of the asymptotic behavior of the autocorrelation function of the square observations; see Palma and Zevallos (2004).

Theorem 7.1. *For the process defined by* (7.6) *with errors satisfying* (7.7) *and* (7.8) *with finite kurtosis* η, *the autocorrelation function of the squared process is given by*

$$\rho_{y^2}(h) = \frac{2}{\kappa - 1}\rho_y^2(h) + \frac{\kappa - 3}{\kappa - 1}\alpha(h) + \frac{\eta - 1}{\kappa - 1}[\tau(h) + 2\Delta(h) - 3\Delta(0)\alpha(h)], \tag{7.9}$$

where the autocorrelation function of the process $\{y_t\}$ *is defined as*

$$\rho_y(h) = \left(\sum_{i=0}^{\infty} \psi_i^2\right)^{-1} \sum_{i=0}^{\infty} \psi_i \psi_{i+|h|},$$

and

$$\alpha(h) = \left(\sum_{i=0}^{\infty} \psi_i^4\right)^{-1} \sum_{i=0}^{\infty} \psi_i^2 \psi_{i+|h|}^2,$$

$$\Delta(h) = \left(\sum_{i=0}^{\infty} \psi_i^2\right)^{-2} \sum_{i=0}^{\infty}\sum_{j=0}^{\infty} \psi_i \psi_j \psi_{i+|h|} \psi_{j+|h|} \rho_{\varepsilon^2}(i - j),$$

$$\tau(h) = \left(\sum_{i=0}^{\infty} \psi_i^2\right)^{-2} \sum_{i=0}^{\infty}\sum_{j=0}^{\infty} \psi_i^2 \psi_j^2 \rho_{\varepsilon^2}(|h| + |i - j|),$$

$$\kappa = 3 - 2\eta \left(\sum_{i=0}^{\infty} \psi_i^2\right)^{-2} \sum_{i=0}^{\infty} \psi_i^4 + 3(\eta - 1)\Delta(0),$$

where κ *the is kurtosis of* y_t *and* ρ_{ε^2} *is the autocorrelation function of* $\{\varepsilon_t^2\}$.

Observe that if the sequence $\{\varepsilon_t\}$ is a strict white noise, then

$$\tau(h) = \Delta(h) = \Delta(0)\alpha(h) = \left(\sum_{i=0}^{\infty} \psi_i^2\right)^{-2} \sum_{i=0}^{\infty} \psi_i^2 \psi_{i+h}^2,$$

and therefore $\tau(h) + 2\Delta(h) - 3\Delta(0)\alpha(h) = 0$. Replacing this in expression (7.9) yields the following result for a linear process; see Taylor (1986).

Corollary 7.2. *(Linear Process) Assume that $\{\varepsilon_t\}$ are independent identically distributed random variables with zero mean and finite kurtosis η. Then,*

$$\rho_{y^2}(h) = \frac{2}{\kappa - 1}\rho_y^2(h) + \frac{\kappa - 3}{\kappa - 1}\alpha(h), \tag{7.10}$$

where κ is the kurtosis of y_t given by

$$\kappa = (\eta - 3)\left(\sum_{i=0}^{\infty} \psi_i^2\right)^{-2} \sum_{i=0}^{\infty} \psi_i^4 + 3. \tag{7.11}$$

Furthermore, if y_t is Gaussian, then $\eta = 3$ and $\kappa = 3$. Therefore, $\rho_{y^2} = \rho_y^2$, which coincides with formula (7.3).

7.3 ASYMPTOTIC BEHAVIOR

In this section we examine the asymptotic behavior of the autocorrelation function of squares for two important situations,

$$\rho_{\varepsilon^2}(m) = \mathcal{O}(a^{|m|}), \qquad 0 < a < 1, \tag{7.12}$$

$$\rho_{\varepsilon^2}(m) = \mathcal{O}(|m|^{2\delta-1}), \qquad 0 < \delta < \frac{1}{2}. \tag{7.13}$$

Note that under condition (7.12), the square of the input sequence $\{\varepsilon_t\}$ has short memory. On the other hand, under condition (7.13) this sequence has long memory. In what follows, *short-memory input* means that ε_t^2 has short memory. Besides, *long-memory input* indicates that ε_t^2 is a long-memory process.

Table 7.1 to Table 7.3 summarize the asymptotic behavior of $\rho_y(h)$ and $\rho_{y^2}(h)$ as the lag h increases to infinity for different combinations of filters ψ and input noise $\{\varepsilon_t\}$ structures. Here, SM, IM, and LM mean short, intermediate, and long memory, respectively.

From Table 7.1, the square of a linear process with independent input noise has an autocorrelation structure similar to the square of the autocorrelation of the original

Table 7.1 Asymptotic Behavior of the Autocorrelation Function of the Squares of the Process $y_t = \sum_{i=0}^{\infty} \psi_i \varepsilon_{t-i}$ where ε_t Has Finite Kurtosis η, $|a| < 1$, $0 < \delta < \frac{1}{2}$, $|v| < 1$, $d \in (0, \frac{1}{2})$, and $0 < d + \delta < \frac{1}{2}$

ψ_i	$\mathcal{O}(v^i)$	$\mathcal{O}(i^{d-1})$	$\mathcal{O}(v^i)$	$\mathcal{O}(i^{d-1})$	$\mathcal{O}(v^i)$	$\mathcal{O}(i^{d-1})$
$\rho_{\varepsilon^2}(h)$	$= 0$		$\mathcal{O}(a^{\lvert h\rvert})$		$\mathcal{O}(h^{2\delta-1})$	
$\rho_y(h)$	$v^{\lvert h\rvert}$	h^{2d-1}	$v^{\lvert h\rvert}$	h^{2d-1}	$v^{\lvert h\rvert}$	h^{2d-1}
$\rho_{y^2}(h)$	$v^{2\lvert h\rvert}$	h^{4d-2}	$\max\{v^2, a\}^{\lvert h\rvert}$	h^{4d-2}	$h^{2\delta-1}$	$h^{2(d+\delta)-1}$

Table 7.2 Asymptotic Behavior of the Autocorrelation Function of the Squares of a Regular Linear Process

	Filter ψ					
	SM	LM	SM	LM	SM	LM
	ε^2 Independent		ε^2 SM		ε^2 LM	
	Memory					
y	Short	Long	Short	Long	Short	Long
	$0 < d < \frac{1}{4}$					
y^2	Short	Intermediate	Short	Intermediate	Long	Long
	$\frac{1}{4} < d < \frac{1}{2}$					
y^2	Short	Long	Short	Long	Long	Long

series, no matter the distribution of the process or the memory introduced by the Wold expansion. However, when the input noise is not independent, the autocorrelation of the squares may depend upon both the memory structure of ε_t^2 and the memory structure of ψ. From Table 7.2 and assuming that $\frac{1}{4} < d < \frac{1}{2}$, if y_t has long memory, then y_t^2 has long memory too. On the other hand, if y_t has short memory, then y_t^2 has also short memory, excepting the case where ε_t^2 is itself a long-memory process. Finally, Table 7.3 displays the memory structure of y_t and y_t^2 for several well-known time series models.

Table 7.3 Analysis of the Memory Structure of y_t and y_t^2 for Several Time Series Models[a]

		y^2		
		SM	IM	LM
y	SM	ARMA GARCH EGARCH ARMA-GARCH SV		FIEGARCH LMSV
			$0 < d < \frac{1}{4}$	$\frac{1}{4} < d < \frac{1}{2}$
	LM		ARFIMA ARFIMA-GARCH ARFIMA-EGARCH ARFIMA-FIEGARCH	ARFIMA ARFIMA-GARCH ARFIMA-EGARCH ARFIMA-FIEGARCH

[a] SM, IM, and LM mean short, intermediate, and long memory, respectively.

EXAMPLE 7.2

Consider the following AR(1)-ARCH(1) process described by the equations

$$\begin{aligned} y_t &= \phi y_{t-1} + \varepsilon_t, \\ \varepsilon_t &= \epsilon_t \sigma_t, \\ \sigma_t^2 &= \alpha_0 + \beta \varepsilon_{t-1}^2, \end{aligned}$$

where ϵ_t is sequence of independent and identically distributed random variables with distribution $N(0,1)$. As shown in Problem 7.6, the autocorrelation function of $\{y_t^2\}$ is given by

$$\rho_{y^2}(h) = \phi^{2|h|}\left[1 + \frac{\eta - 1}{\kappa - 1}\Delta(0)[\beta^{|h|} - 1]\right].$$

In this case, $\rho_{y^2}(h) = \mathcal{O}(\phi^{2|h|})$ and therefore the squared process $\{y_t^2\}$ has short memory.

7.4 ILLUSTRATIONS

Figure 7.1 shows the sample ACF of a series of 1000 observations from a Gaussian ARFIMA$(0, d, 0)$ process with $d = 0.4$ and the sample ACF of the squares. Since

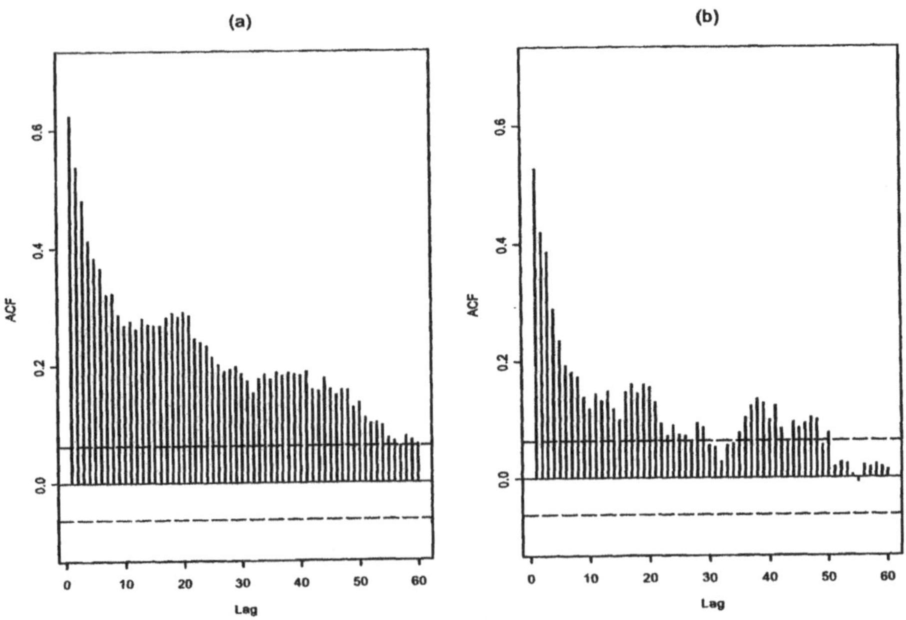

Figure 7.1 Simulated fractional noise process FN(d), 1000 observations with $d = 0.4$. (a) ACF of the series and (b) ACF of the squared series.

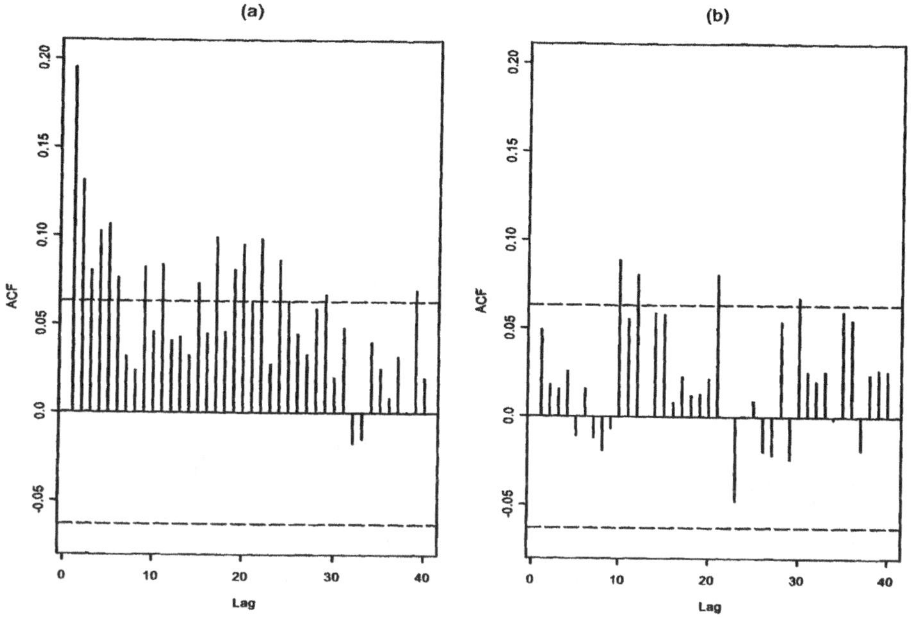

Figure 7.2 Simulated fractional noise process FN(d), 1000 observations with $d = 0.2$. (a) ACF of the series and (b) ACF of the squared series.

this is a Gaussian process, from formula (7.3) we expect that $\widehat{\rho}_{y^2} \sim \widehat{\rho}_y^2$. Additionally, given that $\rho_y(h) \sim Ch^{2d-1}$, the ACF of the squared series should behave like $\rho_{y^2}(h) \sim C^2 h^{2\widetilde{d}-1}$, where $\widetilde{d} = 2d - \frac{1}{2}$. In this case, $\widetilde{d} = 0.3$. Thus, the sample ACF of y_t^2 should decay a bit more rapidly than the ACF of y_t as it seems to be the case when comparing panels (a) and (b). A similar behavior occurs when $d = 0.2$; see Figure 7.2, where $\widetilde{d} = -0.1$.

Figure 7.3 displays the sample ACF from 1000 simulated observations of the ARCH(1) process:

$$\begin{aligned} y_t &= \epsilon_t \sigma_t, \\ \sigma_t^2 &= 0.1 + 0.8 y_{t-1}^2, \end{aligned}$$

where ϵ_t is assumed to be a sequence of independent and identically distributed $N(0, 1)$ random variables. Note that in panel (a), as expected, the sample ACF of the series y_t shows no significant correlations. On the contrary, the sample ACF of the squared series shown in panel (b) exhibits a substantial level of autocorrelation, which decays at an exponential rate. A similar behavior of the autocorrelation is displayed by Figure 7.4, which depicts the sample ACF of 1000 observations from the following GARCH(1, 1) model:

$$\begin{aligned} y_t &= \epsilon_t \sigma_t, \\ \sigma_t^2 &= 0.1 + 0.7 y_{t-1}^2 + 0.2 \sigma_t^2, \end{aligned}$$

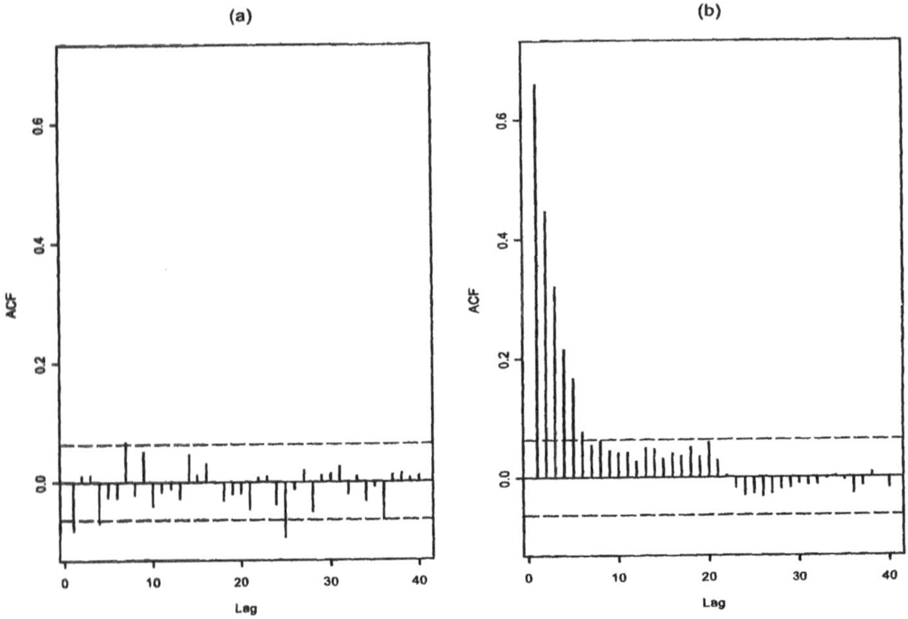

Figure 7.3 Simulated ARCH(1) process: 1000 observations with $\alpha_0 = 0.1$ and $\alpha_1 = 0.8$. (a) ACF of the series and (b) ACF of the squared series.

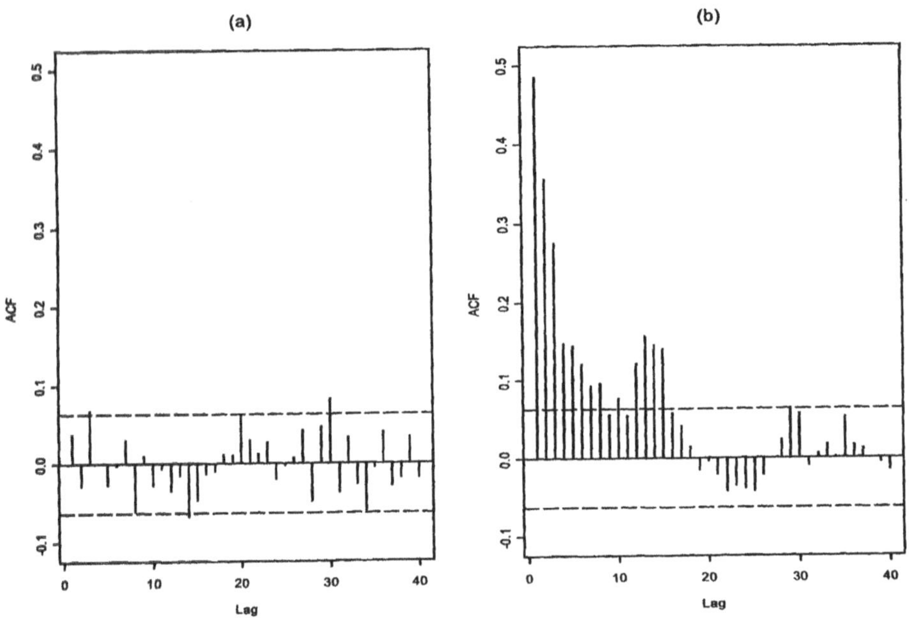

Figure 7.4 Simulated GARCH(1,1) process: 1000 observations with $\alpha_0 = 0.1$, $\alpha_1 = 0.7$, and $\beta_1 = 0.2$. (a) ACF of the series and (b) ACF of the squared series.

where $\{\epsilon_t\}$ is a sequence of independent and identically distributed $N(0,1)$ random variables.

Figure 7.5 exhibits a trajectory of 1000 observations from the ARFIMA$(0, d, 0)$-GARCH$(1, 1)$ process:

$$
\begin{aligned}
y_t &= \sum_{j=0}^{\infty} \psi_j \varepsilon_{t-j}, \\
\varepsilon_t &= \epsilon_t \sigma_t, \\
\sigma_t^2 &= 0.1 + 0.7\varepsilon_{t-1}^2 + 0.2\sigma_t^2,
\end{aligned}
$$

where $d = 0.4$,

$$\psi_j = \frac{\Gamma(0.4 + j)}{\Gamma(1 + j)\Gamma(0.4)},$$

and ϵ_t is an independent and identically distributed Gaussian sequence with zero mean and unit variance. Panel (a) shows the series while panel (b) shows the squares. On the other hand, Figure 7.6 shows the sample ACF of this series; see panel (a) and the sample ACF of the squares; see panel (b). Note that in this case both panels seem to exhibit long-memory behavior, as suggested by Table 7.3 because $d \in (\frac{1}{4}, \frac{1}{2})$.

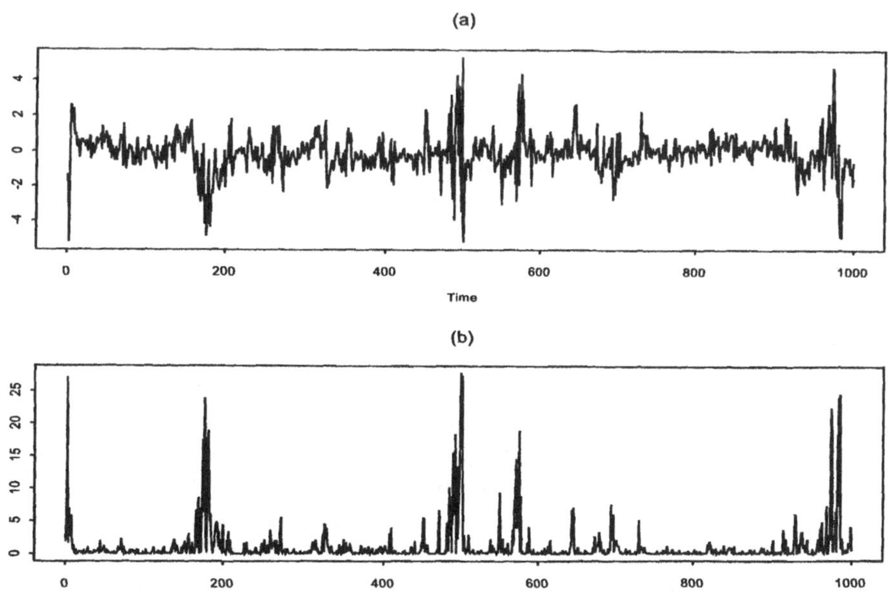

Figure 7.5 Simulated ARFIMA$(0, d, 0)$-GARCH$(1, 1)$ process: 1000 observations with $d = 0.4$, $\alpha_0 = 0.1$, $\alpha_1 = 0.7$, and $\beta_1 = 0.2$. (a) Series and (b) squared series.

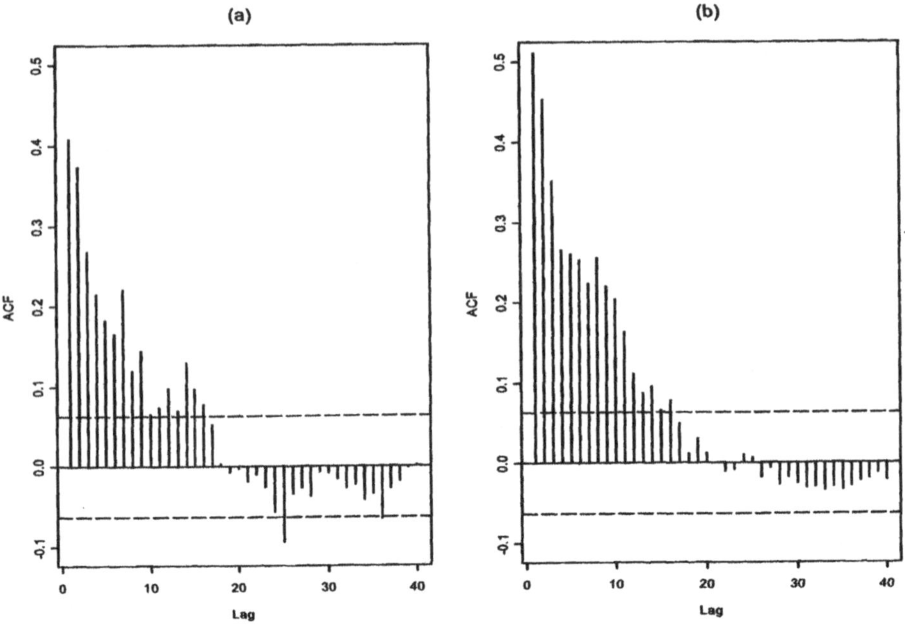

Figure 7.6 Simulated ARFIMA$(0, d, 0)$-GARCH$(1, 1)$ process: 1000 observations with $d = 0.4, \alpha_0 = 0.1, \alpha_1 = 0.7$, and $\beta_1 = 0.2$. (a) ACF of the series and (b) ACF of the squared series.

7.5 BIBLIOGRAPHIC NOTES

The analysis of transformations of a process, including the squares, has been studied by several authors; see, for example, Rosenblatt (1961), Rozanov (1967), Robinson (2001), Dittmann and Granger (2002), and Palma and Zevallos (2004), among others. The analysis of the autocovariance structure of heteroskedastic series has been advanced by exact expressions for the autocorrelations of the squares obtained by Karanasos (1999) and He and Teräsvirta (1999a,b) for the GARCH model, by Karanasos (2001) and He, Teräsvirta, and Malmsten (2002) for the EGARCH process, by Demos (2002) for a model that nests both the EGARCH and stochastic volatility specifications, and by Karanasos and Kim (2003) for FIGARCH and LMGARCH models. On the other hand, asymptotic expressions for the autocovariance function of squares and other nonlinear transformations of a class of stochastic volatility models have been established by Robinson (2001). Furthermore, model identification by analyzing the autocorrelation function of squares is addressed by Bollerslev (1988) for GARCH processes and by Karanasos (2001) for EGARCH processes. Diagnostic checking of nonlinear models with conditional heteroskedasticity through the squares of residuals is discussed by Li and Mak (1994), Granger and Andersen (1978), Maravall (1983), and McLeod and Li (1983) for the univariate case and Ling and Li (1997a) for the multivariate case, among others.

Section 7.1 is mostly based on Section IV.10 of Rozanov (1967). In particular Lemma 7.1 appears in that book. The results from Section 7.2, including Theorem 7.2 and Table 7.1 to Table 7.3, are based on Palma and Zevallos (2004) while Corollary 7.2 was first obtained by Taylor (1986).

Problems

7.1 Verify that the Hermite polynomials $\{H_j(y)\}$ satisfy the following equations:

$$\begin{aligned} H_j'(y) &= jH_{j-1}(y), \\ H_{j+1}(y) &= yH_j(y) - jH_{j-1}(y). \end{aligned}$$

7.2 Let y_t be a stationary process with mean μ and unit variance.

a) Show that

$$E[H_j(y_t)] = \mu^j,$$

for $j = 0, 1, 2, \ldots$

b) Prove that for a transformation (7.1) with $E\,|f(y_t)| < \infty$ we have that

$$E[f(y_t)] = \sum_{j=0}^{\infty} \frac{\alpha_j \mu^j}{j!}.$$

7.3 Show that

$$e^x = \sqrt{e} \sum_{j=0}^{\infty} \frac{H_j(x)}{j!}.$$

7.4 Suppose that $\{y_t : t \in \mathbb{Z}\}$ is a zero mean and unit variance Gaussian stationary process with autocorrelation function ρ and consider the nonlinear transformation $x_t = y_t^3$. By means of expression (7.2), verify that

a)

$$\langle x_t, x_{t+k} \rangle = 9\rho(k) + \frac{\rho(k)^3}{6}.$$

b) If ρ_x is the autocorrelation function of the process $\{x_t\}$, then

$$\rho_x(k) = \frac{(54 + \rho(k)^2)\rho(k)}{55}.$$

7.5 Let $\{y_t : t \in \mathbb{Z}\}$ be a zero mean and unit variance Gaussian stationary process with autocorrelation function ρ and consider the nonlinear transformation $x_t = y_t^4$. Show that

a)

$$\langle x_t, x_{t+k}\rangle = 9 + 18\rho(k)^2 + \frac{\rho(k)^4}{24}.$$

b) If ρ_x is the autocorrelation function of the process $\{x_t\}$, then

$$\rho_x(k) = \frac{(432 + \rho(k)^2)\rho(k)^2}{433}.$$

7.6 [AR(1)-ARCH(1) Process] Consider the AR(1) process

$$y_t = \phi y_{t-1} + \varepsilon_t,$$

where the noise ε_t follows the ARCH(1) model

$$\begin{aligned}\varepsilon_t &= \epsilon_t \sigma_t,\\ \sigma_t^2 &= \alpha_0 + \beta\varepsilon_{t-1}^2,\end{aligned}$$

where $\{\epsilon_t\}$ is a sequence of independent and identically distributed random variables with distribution $N(0, 1)$.

a) Verify that the autocorrelation function of ε_t^2 is

$$\rho_{\varepsilon^2}(h) = \beta^{|h|}.$$

b) Show that

$$\begin{aligned}\alpha(h) &= \phi^{2|h|},\\ \Delta(h) &= \frac{1+\phi^2\beta}{(1+\phi^2)(1-\phi^2\beta)}\phi^{2|h|},\\ \tau(h) &= \frac{1+\phi^2\beta}{(1+\phi^2)(1-\phi^2\beta)}(\phi^2\beta)^{|h|},\\ \kappa &= \frac{3\phi^2(1-2\beta-\phi^2\beta)+\eta(1+5\phi^2\beta)}{(1+\phi^2)(1-\phi^2\beta)},\\ \eta &= \frac{3(1-\beta^2)(1-\beta)^2}{\alpha_0(1-3\beta^2)},\\ \alpha(h) &= \rho_y^2(h),\\ \Delta(h) &= \Delta(0)\rho_y^2(h),\\ \tau(h) &= \Delta(0)\rho_y^2(h)\beta^{|h|}.\end{aligned}$$

c) Show that

$$\rho_{y^2}(h) = \phi^{2|h|}\left\{1 + \frac{\eta-1}{\kappa-1}\Delta(0)[\beta^{|h|}-1]\right\}.$$

7.7 Let $\delta \in (1,2)$, $n \geq 1$. Show that

$$\sum_{j=1}^{\infty} [j(j+n)]^{-\delta} = \mathcal{O}(n^{-\delta}).$$

7.8 Show that for $a < 1$,

$$\lim_{m\to\infty} m^{2\beta} \sum_{i=1}^{m-1} a^i (m-i)^{-2\beta} = \frac{a}{1-a}.$$

7.9 Let $\psi_i \sim i^{-\beta}$ with $\beta \in (\frac{1}{2}, 1)$ and suppose that $\rho_\varepsilon(m) = \mathcal{O}(m^{2\delta-1})$. Verify that

a) $\sum_{i=0}^{\infty} \psi_i \psi_{i+n} = \mathcal{O}(n^{1-2\beta})$.
b) $\sum_{i=0}^{\infty} \psi_i^2 \psi_{i+n}^2 = \mathcal{O}(n^{-2\beta})$.
c) If ρ_{ε^2} satisfies (7.12), then $\sum_{i=0}^{\infty} \psi_i^2 \rho_{\varepsilon^2}(m-i) = \mathcal{O}(m^{-2\beta})$ for $m \geq 1$.
d) If ρ_{ε^2} follows (7.13) with $\delta > 0$ and $\beta - \delta > \frac{1}{2}$, then

$$\sum_{i=0}^{\infty} \psi_i^2 \rho_{\varepsilon^2}(m-i) = \mathcal{O}(m^{1+2\delta-2\beta}),$$

for $m \geq 1$.

7.10 Let $\psi_i \sim v^i$ with $|v| < 1$. Show that if ρ_{ε^2} satisfies (7.13), then

$$\sum_{i=0}^{\infty} \psi_i^2 \rho_{\varepsilon^2}(m-i) = \mathcal{O}(m^{2\delta-1}),$$

for $m \geq 1$.

7.11 Verify that $\Delta(h)$ and $\rho_y^2(h)$ have the same asymptotic order as $n \to \infty$.

7.12 Assume that $\epsilon_t \sim t_\nu$ with $\nu > 4$. Given that for even n

$$E(\epsilon_t^n) = \frac{\Gamma\left(\frac{n+1}{2}\right)\Gamma\left(\frac{\nu-n}{2}\right)}{\sqrt{\pi}\,\Gamma\left(\frac{\nu}{2}\right)}\nu^{n/2},$$

prove the following results:

a) The kurtosis of ϵ_t is

$$\eta_\nu = 3\left(\frac{\nu-2}{\nu-4}\right),$$

and $\eta_\nu > 3$ for any fixed ν.

b) The kurtosis of the process y_t defined by the Wold expansion (7.6) is

$$\kappa_\nu = \frac{6}{\nu - 4}\left(\sum_{j=0}^{\infty} \psi_j^2\right)^{-2}\left(\sum_{j=0}^{\infty} \psi_j^4\right) + 3,$$

and $\kappa_\nu \to 3$ as $\nu \to \infty$.

Hint: Recall that $x\Gamma(x) = \Gamma(x+1)$ and $\Gamma(\frac{1}{2}) = \sqrt{\pi}$.

CHAPTER 8

BAYESIAN METHODS

This chapter addresses some applications of the Bayesian methodology to the analysis of long-memory time series data. Section 8.1 describes a general Bayesian framework for the analysis of ARFIMA processes while Section 8.2 is devoted to the analysis of the Markov chain Monte Carlo (MCMC) methodology, an important computational tool for obtaining samples from a posterior distribution. In particular, we describe applications of the Metropolis-Hastings algorithm and the Gibbs sampler in the context of long-memory processes. Mathematical support for the convergence of the MCMC methodology is provided in Theorem 8.1.

The implementation of these computational procedures are illustrated with an example of Bayesian estimation of a stationary Gaussian process. Specific issues such as selection of initial values and proposal distributions are also discussed. Techniques for monitoring the convergence of MCMC procedures are studied in Section 8.3 while Section 8.4 presents an application of the MCMC methodology to a simulated ARFIMA$(1, d, 1)$ process. Furthermore, Section 8.5 illustrates these Bayesian techniques with a statistical analysis of the well-known Nile River data. Further readings are suggested in Section 8.6 and a list of exercises can be found at the end of the chapter.

Long-Memory Time Series. By Wilfredo Palma

8.1 BAYESIAN MODELING

Consider the time series data $\boldsymbol{y} = (y_1, \ldots, y_n)'$ and a statistical model described by the parameter $\boldsymbol{\theta}$. Let $f(y|\boldsymbol{\theta})$ be the likelihood function of the model and $\pi(\boldsymbol{\theta})$ a prior distribution for the parameter. According to the Bayes theorem, the posterior distribution of $\boldsymbol{\theta}$ given the data $\boldsymbol{y}$ is proportional to

$$\pi(\boldsymbol{\theta}|\boldsymbol{y}) \propto f(\boldsymbol{y}|\boldsymbol{\theta})\pi(\boldsymbol{\theta}).$$

More specifically, suppose that the time series follows an ARFIMA(p, d, q) model described by

$$\phi(B)(y_t - \mu) = \theta(B)(1 - B)^{-d}\varepsilon_t,$$

where the polynomials $\phi(B) = 1 + \phi_1 B + \cdots + \phi_p B^p$ and $\theta(B) = 1 + \theta_1 B + \cdots + \theta_p B^q$ do not have common roots and $\{\varepsilon_t\}$ is a white noise sequence with zero mean and variance σ^2. Define $C_d = \{d : y_t \text{ is stationary and invertible}\}$, $C_\phi = \{\phi_1, \ldots, \phi_p : y_t \text{ is stationary}\}$, and $C_\theta = \{\theta_1, \ldots, \theta_q : y_t \text{ is invertible}\}$.

For this model, the parameter vector may be written as

$$\boldsymbol{\theta} = (d, \phi_1, \ldots, \phi_p, \theta_1, \ldots, \theta_q, \mu, \sigma^2),$$

and the parameter space can be expressed as

$$\boldsymbol{\Theta} = C_d \times C_\phi \times C_\theta \times \mathbb{R} \times (0, \infty).$$

Sometimes, in order to simplify the specification of a prior distribution over the parameter space $\boldsymbol{\Theta}$, one may consider assigning prior distributions individually to subsets of parameters. For instance, we may assume uniform priors for $d, \phi_1, \ldots, \phi_p$, and $\theta_1 \ldots, \theta_q$, that is, $\pi(d) = U(C_d)$, $\pi(\phi_1, \ldots, \phi_p) = U(C_\phi)$, and $\pi(\theta_1, \ldots, \theta_q) = U(C_\theta)$. Besides, we may assume an improper prior μ, $\pi(\mu) \propto 1$ and a prior $\pi(\sigma^2)$ for σ^2. With this specification, the prior distribution of $\boldsymbol{\theta}$ is simply

$$\pi(\boldsymbol{\theta}) \propto \pi(\sigma^2),$$

and the posterior distribution of $\boldsymbol{\theta}$ is given by

$$\pi(\boldsymbol{\theta}|\boldsymbol{y}) \propto f(\boldsymbol{y}|\boldsymbol{\theta})\pi(\sigma^2). \tag{8.1}$$

Apart from the calculation of this posterior distribution, we are usually interested in finding Bayes estimators for $\boldsymbol{\theta}$. For example, we may consider finding the value of $\boldsymbol{\theta}$ such that the posterior loss be minimal. That is, if $L(\boldsymbol{\theta}, \boldsymbol{y})$ is the loss function, then

$$\widehat{\boldsymbol{\theta}} = \text{argmin} \int L(\boldsymbol{\theta}, \boldsymbol{y})\pi(\boldsymbol{\theta}|\boldsymbol{y})\, dy.$$

As a particular case, under the quadratic loss $L(\boldsymbol{\theta}, \boldsymbol{y}) = \|\boldsymbol{\theta} - \boldsymbol{y}\|^2$ we have that the estimate of $\boldsymbol{\theta}$ is the posterior mean

$$\widehat{\boldsymbol{\theta}} = E[\boldsymbol{\theta}|\boldsymbol{y}].$$

Obtaining any of these quantities requires integration. However, in many practical situations the calculation of these integrals may be extremely difficult. To circumvent this problem, several methodologies have been proposed in the Bayesian literature, including conjugate prior distributions, numerical integration, Monte Carlo simulations, Laplace analytical approximation, and Markov chain Monte Carlo (MCMC) procedures. The analysis of all these methods is beyond the scope of this text; here we will focus on MCMC techniques. Discussions about other techniques are referenced in Section 8.6.

8.2 MARKOV CHAIN MONTE CARLO METHODS

A MCMC algorithm produces a sample of a distribution of interest by a method that combines Monte Carlo techniques and Markov chains. Consider, for example, that we want to obtain a sample of the posterior distribution $\pi(\boldsymbol{\theta}|\boldsymbol{y})$. Two well-known procedures for this purpose are the Metropolis-Hastings algorithm and the Gibbs sampler.

8.2.1 Metropolis-Hastings Algorithm

Following the Metropolis-Hastings algorithm, we start with an initial value for $\boldsymbol{\theta}$, $\boldsymbol{\theta}^{(0)}$, say. Suppose that at the stage m we have obtained the value $\boldsymbol{\theta}^{(m)}$. We update this value to $\boldsymbol{\theta}^{(m+1)}$ according to the following procedure:

1. Generate the random variable ξ from the proposal distribution $q(\xi|\boldsymbol{\theta}^{(m)})$,

$$\xi \sim q(\xi|\boldsymbol{\theta}^{(m)}).$$

2. Define

$$\alpha = \min\left\{ \frac{\pi(\xi|\boldsymbol{y})q(\boldsymbol{\theta}^{(m)}|\xi)}{\pi(\boldsymbol{\theta}^{(m)}|\boldsymbol{y})q(\xi|\boldsymbol{\theta}^{(m)})}, 1 \right\}.$$

3. Generate $\omega \sim \text{Ber}(\alpha)$.

4. Obtain

$$\boldsymbol{\theta}^{(m+1)} = \begin{cases} \xi & \text{if } \omega = 1, \\ \boldsymbol{\theta}^{(m)} & \text{if } \omega = 0. \end{cases}$$

The convergence of this procedure is guaranteed by the following result; see Theorem 6.2.3 of Robert and Casella (2004).

Theorem 8.1. *Assume that the support of the proposal distribution q contains the support of the posterior distribution π. Then $\pi(\boldsymbol{\theta}^{(m)}|\boldsymbol{y})$ converges to the unique stationary distribution of the Markov chain $\pi(\boldsymbol{\theta}|\boldsymbol{y})$ as $m \to \infty$.*

8.2.2 Gibbs Sampler

Another well-known iterative method is the Gibbs sampler. Suppose that the random variable $\boldsymbol{\theta}$ can be decomposed as $\boldsymbol{\theta} = (\theta_1, \ldots, \theta_r)$ and we are able to simulate from the conditional densities

$$\theta_j|\theta_1, \ldots, \theta_{j-1}, \theta_{j+1}, \ldots, \theta_r \sim f_j(\theta_j|\theta_1, \ldots, \theta_{j-1}, \theta_{j+1}, \ldots, \theta_r),$$

for $j = 1, \ldots, r$. In order to sample from the joint density of $(\theta_1, \ldots, \theta_r)$ we proceed according to the following algorithm:

0. Given the sample $(\theta_1^{(m)}, \ldots, \theta_r^{(m)})$, generate

1. $\theta_1^{(m+1)} \sim f_1(\theta_1|\theta_2^{(m)}, \theta_3^{(m)}, \ldots, \theta_r^{(m)})$,

2. $\theta_2^{(m+1)} \sim f_2(\theta_2|\theta_1^{(m)}, \theta_3^{(m)}, \ldots, \theta_r^{(m)})$,

$\vdots$

r. $\theta_r^{(m+1)} \sim f_r(\theta_r|\theta_1^{(m)}, \theta_2^{(m)}, \ldots, \theta_{r-1}^{(m)})$.

The acceptance rate in this algorithm is always one, that is, all simulated values are accepted. A nice property of the Gibbs sampler is that all the simulations may be univariate. On the other hand, this algorithm *requires* that we can actually simulate samples from every conditional density f_j for $j = 1, \ldots, r$. By choosing adequately these densities, it can be shown that the Gibbs sampler is a particular case of the Metropolis-Hastings algorithm; see, for example, Theorem 7.1.16 of Robert and Casella (2004).

■ EXAMPLE 8.1

If the process $\{y_t\}$ is Gaussian, then the likelihood function is given by

$$\begin{aligned} f(\boldsymbol{y}|\boldsymbol{\theta}) &= (2\pi\sigma^2)^{-n/2}|\Gamma(\boldsymbol{\theta})|^{-1/2} \\ &\quad \times \exp\left\{-\frac{(\boldsymbol{y}-\mathbf{1}\mu)'\,\Gamma(\boldsymbol{\theta})^{-1}(\boldsymbol{y}-\mathbf{1}\mu)}{2\sigma^2}\right\}, \end{aligned} \tag{8.2}$$

where $\Gamma(\boldsymbol{\theta}) = \mathrm{Var}(\boldsymbol{y})$. Hence, the posterior distribution of $\boldsymbol{\theta}$ given by (8.1) is

$$\begin{aligned} \pi(\boldsymbol{\theta}|\boldsymbol{y}) &\propto (2\pi)^{-n/2}\sigma^{-n}|\Gamma(\boldsymbol{\theta})|^{-1/2} \\ &\quad \times \exp\left\{-\frac{(\boldsymbol{y}-\mathbf{1}_\mu)'\Gamma(\boldsymbol{\theta})^{-1}(\boldsymbol{y}-\mathbf{1}_\mu)}{2\sigma^2}\right\}\pi(\sigma^2). \end{aligned}$$

In this case, the MCMC method can be implemented as follows. To simplify the notation, we write the parameter $\boldsymbol{\theta}$ as $(d, \phi, \theta, \sigma^2, \mu)$ where $\phi = (\phi_1, \ldots, \phi_p)$ and $\theta = (\theta_1, \ldots, \theta_q)$.

Consider an initial sample for (d, ϕ, θ), for example,

$$\left(d^{(0)}, \phi^{(0)}, \theta^{(0)}\right) \sim N_r\left[\left(\widehat{d}, \widehat{\phi}, \widehat{\theta}\right), \Sigma\right],$$

where N_r is a multivariate Gaussian distribution with $r = p+q+3$ and $\left(\widehat{d}, \widehat{\phi}, \widehat{\theta}\right)$ is the MLE of (d, ϕ, θ) and Σ may be obtained from the Fisher information matrix, that is,

$$\Sigma = \left[H\left(\widehat{d}, \widehat{\phi}, \widehat{\theta}\right)\right]^{-1},$$

where H is the Hessian matrix of the log-likelihood function of the data derived from (8.2).

Given the value $\left(d^{(m)}, \phi^{(m)}, \theta^{(m)}\right)$, we generate ξ from the proposal distribution $q\left(\xi | d^{(m)}, \phi^{(m)}, \theta^{(m)}\right)$ according to

$$\xi \sim N_r\left[\left(d^{(m)}, \phi^{(m)}, \theta^{(m)}\right), \Sigma\right],$$

and restricting the random variable ξ to the space $C_d \times C_\phi \times C_\theta$ to ensure the stationarity and the invertibility of the ARFIMA process.

Now, we calculate α as

$$\alpha = \min\left\{\frac{\pi[\xi|\boldsymbol{y}]}{\pi[d^{(m)}, \phi^{(m)}, \theta^{(m)}|\boldsymbol{y}]}, 1\right\},$$

since in this case $q(\boldsymbol{\theta}|\xi) = q(\xi|\boldsymbol{\theta})$.

Then we proceed to steps 3 and 4 of the MCMC method described above. Once (d, ϕ, θ) has been updated, we update μ and σ^2.

For updating μ, one may start with an initial drawing from a normal distribution

$$\mu \sim N(\widehat{\mu}, \widehat{\sigma}_\mu^2),$$

where $\widehat{\mu}$ is an estimate of the location (e.g., the sample mean) and

$$\widehat{\sigma}_\mu^2 = \frac{\sigma^2}{\pi} \frac{|\theta(1)|^2}{|\phi(1)|^2} \frac{\Gamma(1-2d)}{d(2d+1)} \sin(\pi d) n^{2d-1},$$

cf., equation (3.12), where the values of d, $\phi(B)$, $\theta(B)$, and σ^2 are replaced by their respective MLEs. Notice, however, that we may use an *overdispersed distribution* (e.g., Student distribution) as discussed in Subsection 8.2.3.

Given the sample $\mu^{(m)}$, this term may be updated to $\mu^{(m+1)}$ by generating a random variable

$$\xi \sim N(\mu^{(m)}, \widehat{\sigma}^2_{\mu^{(m)}}),$$

and then calculating

$$\alpha = \min\left\{\frac{\pi[d^{(m)}, \phi^{(m)}, \theta^{(m)}, \xi, \widehat{\sigma}^2|\boldsymbol{y}]}{\pi[d^{(m)}, \phi^{(m)}, \theta^{(m)}, \mu^{(m)}, \widehat{\sigma}^2|\boldsymbol{y}]}, 1\right\}.$$

Finally, for updating σ^2 one may draw samples from an inverse gamma distribution $\mathrm{IG}(\alpha, \beta)$ where the coefficients α and β can be chosen in many different ways. A simple approach is to consider that for an ARFIMA model with r parameters the MLE of σ^2, $\widehat{\sigma}^2$, satisfies approximately

$$\widehat{\sigma}^2 \sim \frac{\sigma^2}{n-r}\chi^2_{n-r},$$

for large n. Therefore, $E[\widehat{\sigma}^2] = \sigma^2$ and $\mathrm{Var}[\widehat{\sigma}^2] = 2\sigma^4/(n-r)$. Hence, by matching these moments with the coefficients α and β we have

$$\alpha = \frac{n-r+4}{2}, \qquad \beta = \frac{\widehat{\sigma}^2(n-r+2)}{2}.$$

Naturally, there are many other choices for drawing samples for σ^2, including gamma distributions and *overdispersed distributions* as discussed next.

8.2.3 Overdispersed Distributions

A frequently used approach to draw the initial samples for the MCMC procedure is employing *overdispersed distributions.* Suppose that the random variable x has distribution $x \sim F$. An overdispersed initial distribution for x may be defined by the random variable y given by

$$y = \frac{x}{g(\chi^2_\eta/\eta)},$$

where g is a positive function and χ^2_η is an *independent* chi-square random variable with η degrees of freedom.

For instance, if $x_1 \sim N(0, \sigma^2)$, then with $g(x) = \sqrt{x}$ we have

$$y_1 = x_1\frac{\sqrt{\eta}}{\sqrt{\chi^2_\eta}} \sim \sigma t_\eta,$$

where t_η is a univariate Student distribution with η degrees of freedom. On the other hand, if $x_2 \sim \mathrm{gamma}(\alpha, \beta)$, then $2\beta x_2 \sim \chi^2_{2\alpha}$. Thus, with $g(x) = x$ we have that

$$y_2 = \eta\frac{x_2}{\chi^2_\eta} \sim \frac{\alpha}{\beta}\mathrm{Fisher}(2\alpha, \eta);$$

see Problem 8.6.

The key idea behind overdispersed distributions is that the variance of y be greater than the variance of x. This is true for both examples above since

$$\mathrm{Var}(y_1) = \sigma^2 \frac{\eta}{\eta - 2} > \sigma^2 = \mathrm{Var}(x_2),$$

for $\eta > 2$ and

$$\mathrm{Var}(y_2) = \frac{\alpha}{\beta^2} \left(\frac{\eta}{\eta - 2} \right)^2 \frac{2\alpha + \eta - 2}{\eta - 4} > \frac{\alpha}{\beta^2} = \mathrm{Var}(x_2),$$

for $\eta > 4$. On the other hand, note that $\mathrm{Var}(y_i) \to \mathrm{Var}(x_i)$ as $\eta \to \infty$ for $i = 1, 2$.

8.3 MONITORING CONVERGENCE

A crucial aspect of the MCMC procedure is to devise tools to verify how close is the distribution of $\boldsymbol{\theta}^{(m)}$ to the stationary distribution of the Markov chain provided by Theorem 8.1.

There are several ways to do this; here we will focus on the methodology proposed by Gelman and Rubin (1992). To simplify the notation, we assume a univariate parameter $\boldsymbol{\theta}$ and that the target posterior distribution $\pi(\boldsymbol{\theta}|\boldsymbol{y})$ is normal, that is,

$$\pi(\boldsymbol{\theta}|\boldsymbol{y}) \sim N(\mu, \sigma^2).$$

Consider approximating the posterior distribution at step m, $\pi(\boldsymbol{\theta}^{(m)}|\boldsymbol{y})$, by a Student distribution centered at $\widehat{\mu}$ with squared scale $\widehat{v}$ and ν degrees of freedom:

$$\pi(\boldsymbol{\theta}^{(m)}|\boldsymbol{y}) \sim t_\nu(\widehat{\mu}, \widehat{v}).$$

If ν is small, then the t_ν distribution is highly diffuse. On the other hand, if ν is large, then the t_ν distribution becomes very similar to the normal distribution. Thus, a measure of how close is $\pi(\boldsymbol{\theta}^{(m)}|\boldsymbol{y})$ from $\pi(\boldsymbol{\theta}|\boldsymbol{y})$ could be given by the ratio of the variances between both distributions. In other words, if the Fisher information calculated from the distribution $\pi(\boldsymbol{\theta}^{(m)}|\boldsymbol{y})$ is similar to the Fisher information from $\pi(\boldsymbol{\theta}|\boldsymbol{y})$, then we could say that no relevant information is lost in the approximation.

However, neither of these two distributions is known, so we will develop a framework to estimate the ratio of variances by carrying out several parallel Markov chains.

Let $m \geq 2$ be the number of independent runs of the Markov chain, each of length $n_0 + n$. The first n_0 iterations are discarded to allow for a burn-in period and we keep the n remaining iterations.

Let x_{ij} be the jth value of the ith sequence with $i = 1, \ldots, m$ and $j = 1, \ldots, n$. Let $\bar{x}_{i\cdot}$ be the mean of the n values of the ith sequence and let b be the sample variance between these means:

$$b = \frac{1}{m - 1} \sum_{i=1}^{m} (\bar{x}_{i\cdot} - \bar{x}_{\cdot\cdot})^2,$$

where $\bar{x}_{..}$ is the average of the means over all the sequences. Let s_i^2 be the within-sequence variance of the ith run:

$$s_i^2 = \frac{1}{n-1}\sum_{j=1}^{n}(\bar{x}_{ij} - \bar{x}_{i\cdot})^2,$$

and let s^2 be the average of these variances:

$$s^2 = \frac{1}{m}\sum_{i=1}^{m} s_i^2.$$

Consider now the following *analysis of variance* (ANOVA) decomposition:

$$\frac{1}{n}\sum_{j=1}^{n}(x_{ij} - \mu)^2 = \frac{1}{n}\sum_{j=1}^{n}(x_{ij} - \bar{x}_{i\cdot})^2 + (\bar{x}_{i\cdot} - \mu)^2.$$

Then, by taking expectations in both sides we have

$$\sigma^2 = E\left(\frac{n-1}{n}s_i^2\right) + \mathrm{Var}(\bar{x}_{i\cdot}) = E\left(\frac{n-1}{n}s^2\right) + E(b).$$

Thus, an unbiased estimator of the variance σ^2 is

$$\widehat{\sigma}^2 = \frac{n-1}{n}s^2 + b.$$

If the posterior distribution $\pi_{mn}(\boldsymbol{\theta}|\boldsymbol{y})$ is approximated by a Student distribution, then we have

$$\pi_{mn}(\boldsymbol{\theta}|\boldsymbol{y}) \approx t_\nu(\boldsymbol{\theta}|\widehat{\mu}, \widehat{v}),$$

where $\widehat{\mu}$ is the location and

$$\widehat{v} = \widehat{\sigma}^2 + b/m,$$

is the squared scale. This term involves two sources of variability, an estimate of the posterior variance of the target normal distribution, $\widehat{\sigma}^2$, plus an estimate of the uncertainty about the mean μ, $\widehat{\mathrm{Var}}(\widehat{\mu}) = b/m$.

Observe that

$$\frac{\widehat{v}}{v} \sim \frac{\chi_\nu^2}{\nu},$$

so that a plausible estimate of the degree of freedom ν is

$$\widehat{\nu} = 2\frac{\widehat{v}^2}{\widehat{\mathrm{Var}}(\widehat{v})},$$

where the variance of $\widehat{v}$ is estimated by

$$\begin{aligned}\widehat{\mathrm{Var}}(\widehat{v}) &= \left(\frac{n-1}{n}\right)^2 \frac{1}{m}\widehat{\mathrm{Var}}(s_i^2) + \left(\frac{m+1}{m}\right)^2 \left(\frac{2}{m-1}\right) b^2 \\ &+2\left(\frac{m+1}{m}\right)\left(\frac{n-1}{n}\right)\frac{1}{m}\left[\widehat{\mathrm{Cov}}(s_i^2,\bar{x}_{i\cdot}^2) - 2\bar{x}\,\widehat{\mathrm{Cov}}(s_i^2,\bar{x}_{i\cdot})\right].\end{aligned}$$

Under the t_ν distribution, we have that for $\nu \geq 2$:

$$\mathrm{Var}(\boldsymbol{\theta}|\boldsymbol{y})_{\mathrm{Student}} = \frac{v\nu}{\nu-2},$$

and under the normal distribution

$$\mathrm{Var}(\boldsymbol{\theta}|\boldsymbol{y})_{\mathrm{Normal}} = \sigma^2.$$

Thus, the Fisher information ratio about $\widehat{\boldsymbol{\theta}}$ is given by

$$R = \frac{v}{\sigma^2}\frac{\nu}{\nu-2}.$$

This ratio indicates how much the posterior distribution $\pi_{mn}(\boldsymbol{\theta}|\boldsymbol{y})$ has shrunk as n becomes large. Thus, a measure of the scale reduction is given by

$$\widehat{R}^{1/2} = \sqrt{\frac{\widehat{v}}{s^2}\frac{\nu}{\nu-2}}.$$

8.4 A SIMULATED EXAMPLE

In order to illustrate how the MCMC methodology works in the context of long-memory processes consider the following ARFIMA$(1,d,1)$ model:

$$(1-\phi B)(y_t-\mu) = (1-\theta B)(1-B)^{-d}\varepsilon_t,$$

for $t = 1,\ldots,n$ with $\varepsilon_t \sim n(0,\sigma^2)$. In this case, the parameter of interest is $\boldsymbol{\theta} = (d,\phi,\theta,\mu,\sigma^2)$. Consider a simulated process with $\boldsymbol{\theta} = (0.3, 0.2, 0.6, 0.0, 1.0)$ and $n = 1000$. In this case, 10 parallel Markov chains of 2000 iterations each were run and the first 1000 iterations were discarded.

Following Example 8.1, the initial samples for (d,ϕ,θ) are drawn from the multivariate normal distribution

$$\left(d^{(0)},\phi^{(0)},\theta^{(0)}\right) \sim N_r\left[\left(\widehat{d},\widehat{\phi},\widehat{\theta}\right),\Sigma\right],$$

with $r = 3$, $(\widehat{d},\widehat{\phi},\widehat{\theta}) = (0.2618, 0.2055, 0.4996)$ and

$$\Sigma = \begin{bmatrix} 0.00933 & 0.00305 & 0.01230 \\ 0.00305 & 0.00905 & 0.01068 \\ 0.01230 & 0.01068 & 0.02246 \end{bmatrix}.$$

The initial drawing distribution of μ is

$$\mu \sim N(\widehat{\mu}, \widehat{\sigma}_\mu^2),$$

with $\widehat{\mu} = -0.2627$ and

$$\widehat{\sigma}_\mu^2 = \frac{\widehat{\sigma}^2}{\pi} \frac{(1-\widehat{\theta})^2}{(1-\widehat{\phi})^2} \frac{\Gamma(1-2\widehat{d})}{\widehat{d}(2\widehat{d}+1)} \sin(\pi \widehat{d}) n^{2\widehat{d}-1},$$

where $\widehat{\sigma}^2 = 1.0499$. To complete an iteration of the Metropolis-Hastings algorithm, the parameter σ^2 is drawn from an inverse gamma distribution:

$$\sigma^2 \sim \text{IG}(\alpha, \beta),$$

where the coefficients α, β are given by

$$\alpha = \frac{n-1}{2}, \qquad \beta = \frac{\widehat{\sigma}^2(n-3)}{2}.$$

Additionally, we use an initial overdispersed distribution obtained by dividing the samples of σ^2 by an independent random deviate χ_η^2/η with $\eta = 4$.

The results from this algorithm are displayed in Figures 8.1–8.3 and Table 8.1. Figures 8.1(a)–8.1(c) display kernel estimates of the posterior density of the parameters d, ϕ, and θ, respectively.

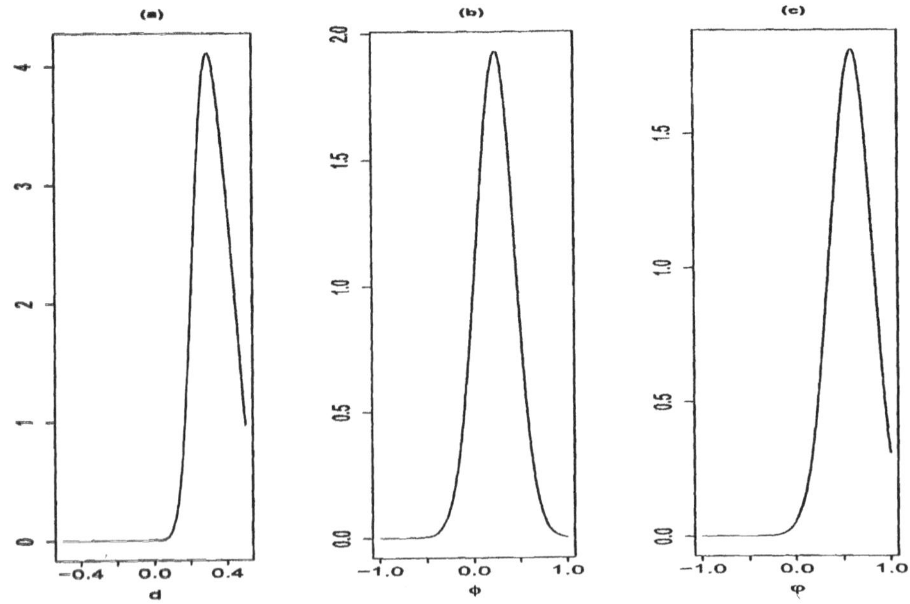

Figure 8.1 MCMC simulation example: Posterior densities (a) d, (b) ϕ, and (c) θ.

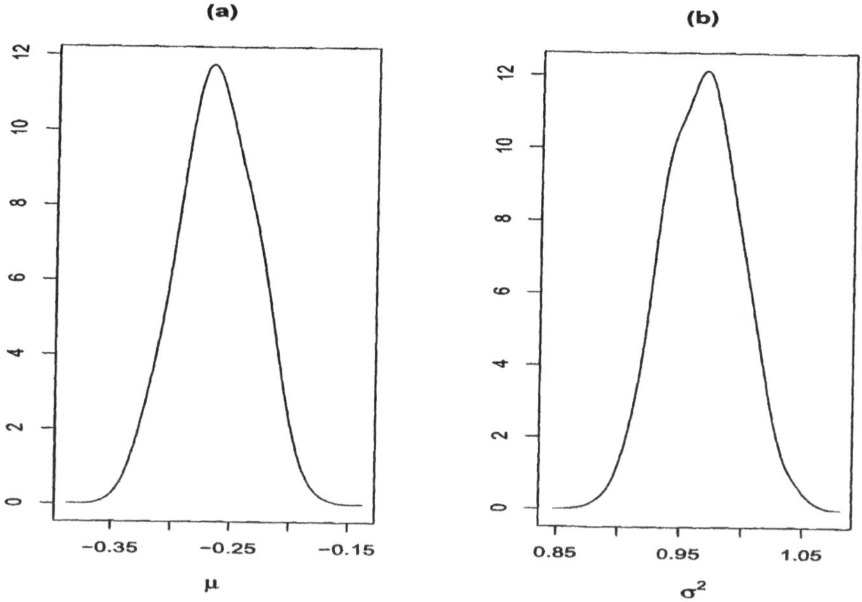

Figure 8.2 MCMC simulation example: Posterior densities (a) μ and (b) σ^2.

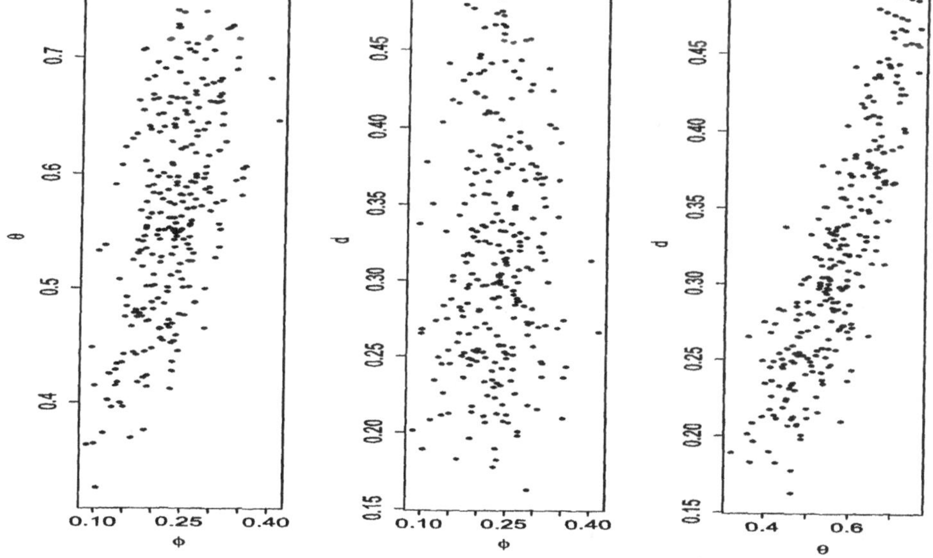

Figure 8.3 MCMC simulation example: Scatterplots for d, ϕ, and θ.

Table 8.1 MCMC Simulation Example: Bayesian Estimation and Convergence Monitoring

	d	ϕ	θ	μ	σ^2
			Estimation		
Posterior mean	0.3327	0.2511	0.5983	-0.2624	0.9675
Posterior SD	0.0768	0.0586	0.0885	0.0320	0.0307
			Convergence monitoring		
$\widehat{R}^{1/2}$	1.0077	1.0065	1.0063	1.0103	1.0061

Figure 8.1(a) shows the estimated posterior density of the long-memory parameter d. Observe that this density appears truncated on the right due to the upper limit for the long-memory parameter, $d < \frac{1}{2}$. Figure 8.1(b) displays the posterior density of the autoregressive parameter ϕ while Figure 8.1(c) shows the posterior density of the moving-average parameter θ. Furthermore, kernel estimates of the posterior densities of μ and σ^2 are displayed in Figure 8.2(a) and Figure 8.2(b), respectively. Additionally, scatterplots for parameters d, ϕ, and θ are displayed in Figure 8.3.

The posterior mean and standard deviation of each parameter are presented in Table 8.1. Furthermore, the convergence of the MCMC runs was monitored by the scale reduction factor $\widehat{R}^{1/2}$, shown in Table 8.1. From this table observe that all the scale reduction factors in this example are close to 1.

8.5 DATA APPLICATION

Consider the hydrologic time series plotted in Figure 8.4. These data represent the annual minimum water level of the Nile River at the Rhoda gauge from A.D. 622 to A.D. 1281; see Toussoun (1925).

Notice that from the sample ACF, plotted in Figure 8.5, this series seems to exhibit long-range dependence. This behavior is corroborated by the variance plot (cf., Subsection 4.5.3) shown in Figure 8.6.

From these observations and previous studies (see Subsection 8.6 for further details) the following ARFIMA$(0, d, 0)$ model is fitted to these data:

$$y_t - \mu = (1 - B)^{-d}\varepsilon_t,$$

for $t = 1, \ldots, n$ with $\varepsilon_t \sim n(0, \sigma^2)$. In this case, the parameter of interest is $\boldsymbol{\theta} = (d, \mu, \sigma^2)$.

As in the previous simulated example, 10 parallel Markov chains were run with 2000 iterations each. The first 1000 iterations were discarded from the analysis.

The starting distribution for the long-memory parameter d is

$$d \sim N(\widehat{d}, \widehat{\sigma}^2_d),$$

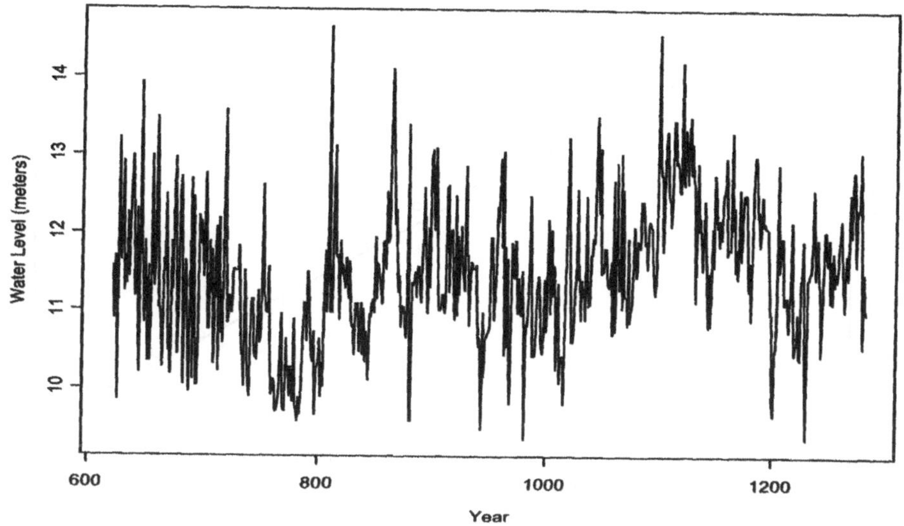

Figure 8.4 MCMC data application: Minimum water levels of the Nile River at the Rhoda Gauge from A.D. 622 to A.D. 1281.

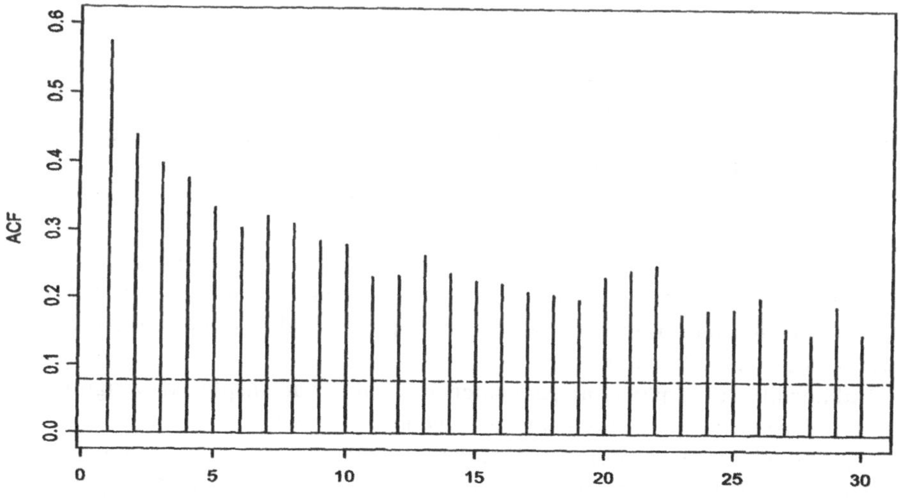

Figure 8.5 MCMC data application: Sample autocorrelation function (ACF) of the Nile River series.

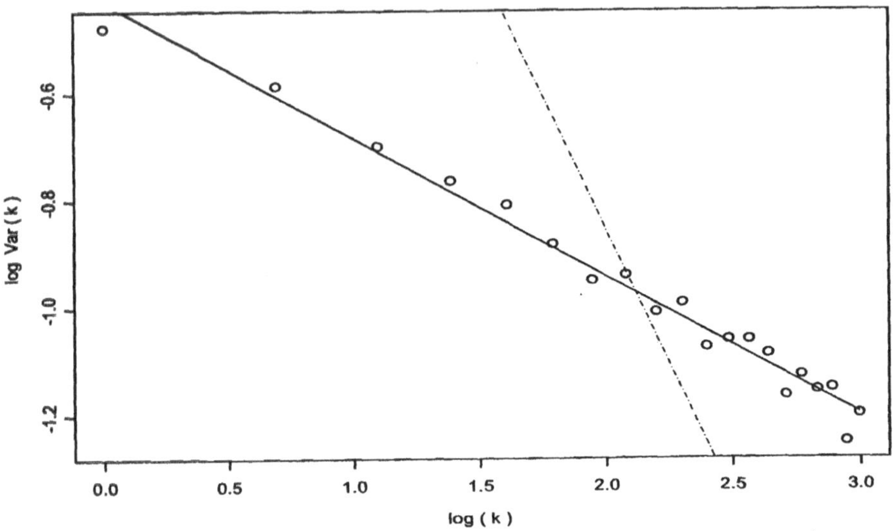

Figure 8.6 MCMC data application: Variance plot of the Nile River series.

where $\widehat{d} = 0.4223$ and $\widehat{\sigma}_d = 0.0317$. Furthermore, the noise variance is $\widehat{\sigma}^2 = 0.4559$. Observe that in order to ensure the stationarity and invertibility of the ARFIMA model, we restrict the long-memory parameter to belong to $\mathcal{C}_d$.

The initial drawing distribution for the mean of the process is given by

$$\mu \sim N(\widehat{\mu}, \widehat{\sigma}_\mu^2),$$

with $\widehat{\mu} = 11.4852$ and $\widehat{\sigma}_\mu^2 = 0.3690$.

On the other hand, the starting distribution for σ^2 is an overdispersed distribution obtained from an inverse gamma IG(331.5, 150.7) random variable divided by an independent random deviate from $\chi_4^2/4$.

The results from these iterative procedures are shown in Figure 8.7, Figure 8.8, and Table 8.2.

Kernel estimates of the posterior densities of parameters d, μ, and σ^2 are displayed in panel (a) to panel (c) of Figure 8.7, respectively. Similarly to the simulated example of Subsection 8.4, the posterior distribution of d appears slightly truncated on the right because of the restriction $d < \frac{1}{2}$.

Additionally, scatterplots fore parameters d, μ, and σ^2 are displayed in Figure 8.8. No clear dependence patterns can be observed in these plots.

The posterior mean and standard deviation of the parameter d, μ, and σ^2 are presented in Table 8.2. This table also displays an analysis of the convergence of the MCMC runs, monitored by the scale reduction factor $\widehat{R}^{1/2}$. Notice that these three values are close to 1 indicating a good level of approximation to the target distributions.

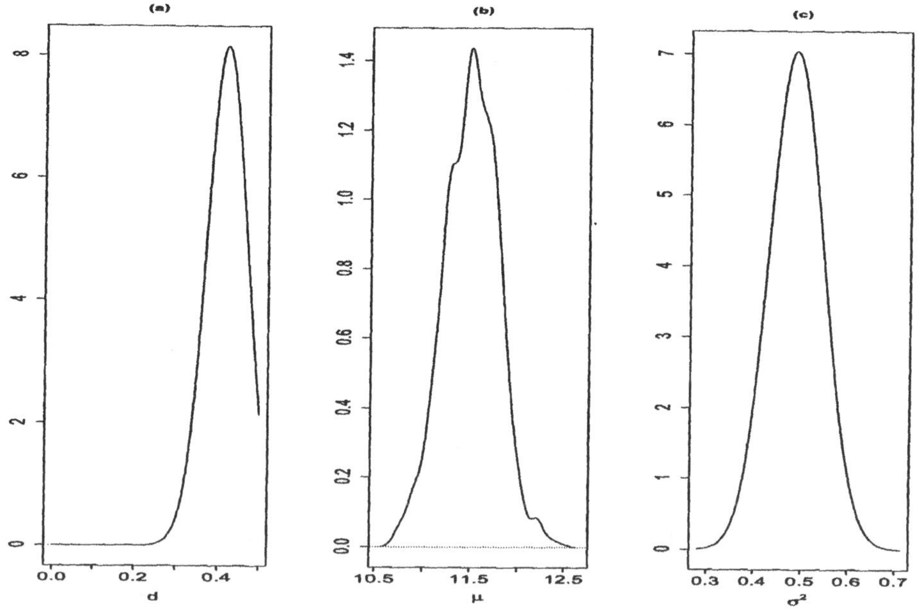

Figure 8.7 MCMC Nile River: Posterior density of (a) d, (b) μ, and (c) σ^2.

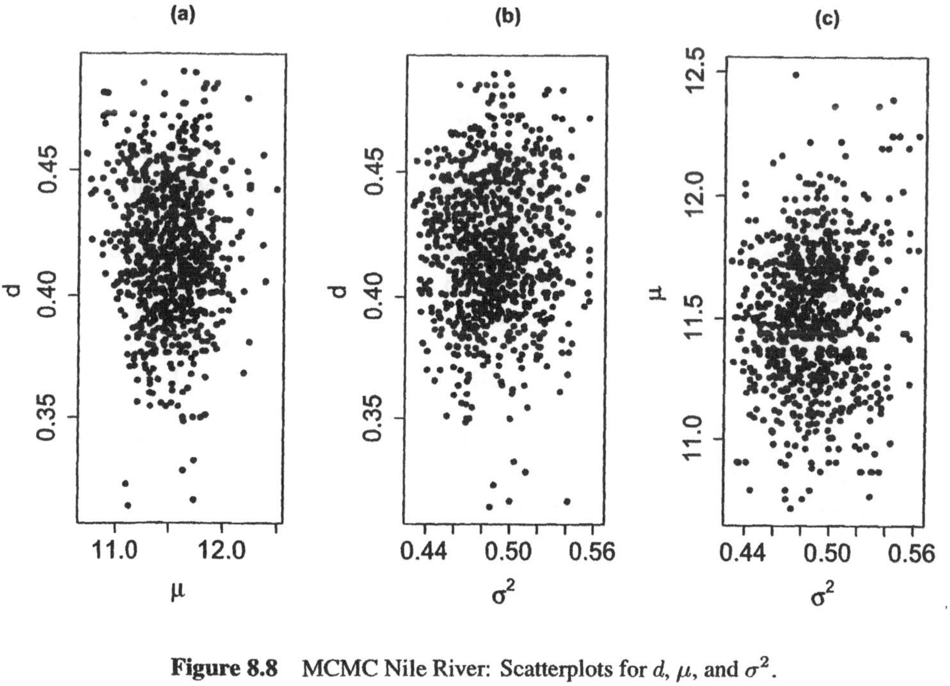

Figure 8.8 MCMC Nile River: Scatterplots for d, μ, and σ^2.

Table 8.2 MCMC Application to the Nile River Data: Bayesian Estimation and Convergence Monitoring

	d	μ	σ^2
		Estimation	
Posterior mean	0.4195	11.5045	0.4884
Posterior SD	0.02826	0.2799	0.0264
		Convergence monitoring	
$\widehat{R}^{1/2}$	1.0021	1.0060	1.0060

8.6 BIBLIOGRAPHIC NOTES

There is an extensive literature on Bayesian methods for time series analysis. Most of the procedures discussed in this chapter are based on the article by Pai and Ravishanker (1998). Extensions of these techniques to multivariate long-memory time series can be found in Ravishanker and Ray (1997).

The books by Robert (2001) and Robert and Casella (2004) are excellent references on Bayesian methodologies. In particular, Robert (2001, Chapter 9) and Robert and Casella (2004, Chapters 6 and 7) describe several computational techniques including MCMC algorithms and the Gibbs sampler. Other good general references on Bayesian methods are the books by Box and Tiao (1992) and Press (2003).

There are several versions of the MCMC algorithm. For example, the one discussed here is based on the works by Metropolis, Rosenbluth, Teller, and Teller (1953) and Hastings (1970). A comprehensive revision of Markov chain methods is provided by Tierney (1994). Section 8.3 on monitoring convergence of the MCMC runs is based on the methodology proposed by Gelman and Rubin (1992).

Some theoretical aspects such as the convergence of the MCMC method by virtue of the ergodic theorem can be found in Meyn and Tweedie (1993).

Apart from the MCMC algorithms there is a plethora of methods for carrying out Bayesian calculations. Among them we have the use of conjugate priors, numerical integration [e.g., Naylor and Smith (1982)], importance sampling, Laplace analytical approximation [e.g., Tierney and Kadane (1986), Tierney, Kass, and Kadane (1989)], Monte Carlo methods [e.g., Stewart (1979), Geweke (1989), Ripley (1987)]. The book by Givens and Hoeting (2005) offers an excellent treatment of several numerical algorithms, including an analysis of Markov chain Monte Carlo procedures in Chapter 7 and Chapter 8.

Problems

8.1 Let $\mathbf{y}_n = (y_1, \ldots, y_n)$ be a sequence of an ARFIMA(p, d, q) process with innovations $n(0, \sigma^2)$ and let $\boldsymbol{\theta}$ be a vector containing the ARFIMA parameters.

a) Show that

$$f(y_{n+1}, \boldsymbol{y}_n|\boldsymbol{\theta}) = f(y_{n+1}|\boldsymbol{y}_n, \boldsymbol{\theta})f(\boldsymbol{y}_n|\boldsymbol{\theta}).$$

b) Verify that

$$y_{n+1}|\boldsymbol{y}_n, \boldsymbol{\theta} \sim n(\mu_{n+1}, \nu_{n+1}^2\sigma^2),$$

with

$$\begin{aligned}\mu_{n+1} &= \sum_{j=1}^{n} \phi_{nj} y_{n+1-j}, \\ \nu_{n+1}^2 &= \frac{\gamma(0)}{\sigma^2 \prod_{j=1}^{n}(1-\phi_{jj}^2)},\end{aligned}$$

where ϕ_{ij} are the partial linear regression coefficients.

8.2 Show that $E(b) = \text{Var}(\bar{x}_{i\cdot})$.

8.3 Prove that

$$\begin{aligned}\text{Var}(\widehat{v}) &= \left(\frac{n-1}{n}\right)^2 \text{Var}(s^2) + \left(\frac{m+1}{m}\right)^2 \text{Var}(b) \\ &+ \left(\frac{m-1}{m}\right)\left(\frac{n-1}{n}\right) \text{Cov}(s^2, b).\end{aligned}$$

8.4 Verify that

$$\text{Cov}(s^2, b) = \frac{1}{m}[\text{Cov}(s_i^2, \bar{x}_{i\cdot}^2) - 2\,E(\bar{x}_{j\cdot})\,\text{Cov}(s_i^2, \bar{x}_{i\cdot})]$$

8.5 Show that

$$\widehat{R} = \left(\frac{n-1}{n} + \frac{m+1}{m}\frac{b}{s^2}\right)\frac{\nu}{\nu-2}.$$

8.6 Suppose that $\sigma^2 \sim \text{Gamma}(\alpha, \beta)$. An *overdispersed distribution* is obtained as

$$\mathcal{F} = \frac{\eta\sigma^2}{\chi_\eta^2},$$

where η is a parameter indicating the *degrees of freedom* of the χ^2 distribution. Verify that $\mathcal{F} \propto \text{Fisher}(a, b)$ and find the parameters a and b.

8.7 Assume that the process y_t is Gaussian such that

$$\boldsymbol{y} = (y_1, \ldots, y_n)' \sim N(\boldsymbol{\theta}, \Sigma).$$

a) Suppose that the prior distributions of $\boldsymbol{\theta}$ and Σ are

$$\pi(\boldsymbol{\theta}) \propto 1,$$

and

$$\pi(\Sigma) \propto |\Sigma|^{-(n+1)/2}.$$

Calculate the posterior distribution $\pi(\boldsymbol{\theta}, \Sigma|\boldsymbol{y})$.

b) Verify that the likelihood function may be written as

$$L(\boldsymbol{\theta}, \Sigma|\boldsymbol{y}) \propto |\Sigma|^{-(n/2)} e^{-(1/2)\,\mathrm{tr}[\Sigma^{-1} S(\boldsymbol{y})]},$$

where $S(\boldsymbol{y}) = [(y_i - \theta_i)(y_j - \theta_j)]_{i,j=1,\ldots,n}$.

c) Suppose that $\Sigma^{-1} \sim \mathrm{Wishart}_n(B^{-1}, n)$, that is,

$$\pi(\Sigma^{-1}) \propto |\Sigma|^{-(n-2)/2} e^{-(1/2)\,\mathrm{tr}[\Sigma^{-1} B]}.$$

d) Prove that the posterior distribution of $\boldsymbol{\theta}$ and Σ given $\boldsymbol{y}$ may be written as

$$\pi(\boldsymbol{\theta}, \Sigma|\boldsymbol{y}) \propto |\Sigma|^{-n+1} e^{-(1/2)\,\mathrm{tr}[B+S(\boldsymbol{y})]},$$

and therefore

$$\boldsymbol{\theta}, \Sigma|\boldsymbol{y} \sim \mathrm{Wishart}_n(B + S(\boldsymbol{y}), 2n).$$

e) Given that

$$\int |Z|^{(1/2)q-1} e^{-(1/2)\,\mathrm{tr}\, ZC} dZ = |C|^{-(1/2)(q+m-1)} 2^{(1/2)m(q+m-1)} \times \Gamma_m\left(\frac{q+m-1}{2}\right),$$

where $\Gamma_p(b)$ is the generalized gamma function

$$\Gamma_p(b) = \left[\Gamma\left(\tfrac{1}{2}\right)\right]^{(1/2)p(p-1)} \prod_{\alpha=1}^{p} \Gamma\left(b + \frac{\alpha - p}{2}\right),$$

with $b > (p-1)/2$, show that $\pi(\boldsymbol{\theta}|\boldsymbol{y}) \propto |S(\boldsymbol{y})|^{-n/2}$.

8.8 [cf., Hastings (1970)] Let $Q = (q_{ij})$ be a transition matrix of an arbitrary Markov chain on the states $0, 1, \ldots, S$ and let α_{ij} be given by

$$\alpha_{ij} = \frac{s_{ij}}{1 + \dfrac{\pi_i q_{ij}}{\pi_j q_{ji}}},$$

where $\pi_i > 0$, $i = 0, \ldots, S$, $\sum \pi_i = 1$ and s_{ij} is a symmetric function of i and j chosen such that $0 \leq \alpha_{ij} \leq 1$ for all i, j.

Consider the Markov chain on the states $0, 1, \ldots, S$ with transition matrix P given by

$$p_{ij} = q_{ij}\alpha_{ij},$$

for $i \neq j$ and

$$p_{ii} = 1 - \sum_{i \neq j} p_{ij}.$$

a) Show that the matrix P is in fact a transition matrix.

b) Prove that P satisfies the *reversibility* condition that

$$\pi_i p_{ij} = \pi_j p_{ji},$$

for all i and j.

c) Verify that $\pi = (\pi_0, \ldots, \pi_S)$ is the unique stationary distribution of P, that is,

$$\pi = \pi P.$$

8.9 Consider the following two choices of the function s_{ij} on the method discussed in Problem 8.8:

$$s_{ij}^M = \begin{cases} 1 + \dfrac{\pi_i q_{ij}}{\pi_j q_{ji}} & \text{if } \dfrac{\pi_i q_{ij}}{\pi_j q_{ji}} \leq 1, \\[2ex] 1 + \dfrac{\pi_j q_{ji}}{\pi_i q_{ij}} & \text{if } \dfrac{\pi_i q_{ij}}{\pi_j q_{ji}} > 1. \end{cases}$$

proposed by Metropolis, Rosenbluth, Teller, and Teller (1953) and

$$s_{ij}^B = 1,$$

used by Barker (1965).

a) Show that both choices satisfy the condition $1 \leq \alpha_{ij} \leq 1$.

b) Suppose that the matrix Q is symmetric and consider the sampling scheme

$$x_{t+1} = \begin{cases} j & \text{with probability } \alpha_{ij}, \\ i & \text{with probability } 1 - \alpha_{ij}. \end{cases}$$

Verify that if $\pi_i = \pi_j$, then $P(x_{t+1} = j) = 1$ for the Metropolis algorithm and $P(x_{t+1} = j) = 1$ for Barker's method.

c) According to part (b), which method is preferable?

d) Consider the choice

$$s_{ij} = g\left[\min\left\{\frac{\pi_i q_{ij}}{\pi_j q_{ji}}, \frac{\pi_j q_{ji}}{\pi_i q_{ij}}\right\}\right],$$

where the function $g(x)$ is symmetric and satisfies $0 \le g(x) \le 1 + x$. Verify that for this choice, the condition $0 \le \alpha_{ij} \le 1$ holds.

e) Consider the particular choice $g(x) = 1 + 2(x/2)^\gamma$ for a constant $\gamma \ge 1$. Prove that s_{ij}^M is obtained with $\gamma = 1$ and s_{ij}^B is obtained with $\gamma = \infty$.

8.10 Consider the following Poisson sampling with

$$\pi_i = e^{-\lambda}\frac{\lambda^i}{i!},$$

for $i = 0, 1, \ldots,$ and

$$q_{00} = q_{01} = \tfrac{1}{2},$$

and

$$q_{ij} = \tfrac{1}{2}, \text{ for } j = i-1, i+1, i \neq 0.$$

a) Show that in this case

$$x_{t+1} = \begin{cases} i+1 & \text{with probability} \begin{cases} \dfrac{\lambda}{i+1} & \text{if } \lambda \le i+1, \\ 1 & \text{if } \lambda > i+1, \end{cases} \\ \\ i-1 & \text{with probability} \begin{cases} \dfrac{i}{\lambda} & \text{if } \lambda \le i, \\ 1 & \text{if } \lambda > i. \end{cases} \end{cases}$$

b) What is the disadvantage of this method when λ is large?

CHAPTER 9

PREDICTION

Some aspects of the prediction theory of stationary long-memory processes are examined in this chapter. The prediction of future values of regular stationary time series follows well-known procedures which are summarized in Section 9.1 and Section 9.2. These sections address the formulation of one-step and multistep ahead predictors based on finite and infinite past. Additionally, those sections contain some recently established asymptotic expressions for the partial autocorrelations and the mean-squared prediction errors. The prediction of future volatility is a crucial aspect in the context of heteroskedastic time series. Consequently, Section 9.3 discusses techniques for forecasting volatility for some of the models described in Chapter 6. An illustrative application of the long-memory forecasting techniques is discussed in Section 9.4. Alternative methods for carrying out prediction procedures for long-memory processes are examined in Section 9.5 where the attention is focused on the so-called *rational approximation* techniques. In this approach, a long-memory model is approximated by a short-memory ARMA process. An attractive feature of this methodology is that one-step and multistep ahead forecasts are more easily calculated, since they are based on the approximating ARMA model. Bibliographic notes are given in Section 9.6 and a list of problems is proposed at the end of this chapter.

Long-Memory Time Series. By Wilfredo Palma

9.1 ONE-STEP AHEAD PREDICTORS

Let $\{y_t\}$ be a regular invertible linear process with Wold representation

$$y_t = \sum_{j=0}^{\infty} \psi_j \varepsilon_{t-j}, \tag{9.1}$$

and AR(∞) expansion

$$y_t - \sum_{j=1}^{\infty} \pi_j y_{t-j} = \varepsilon_t, \tag{9.2}$$

where $\mathrm{Var}(\varepsilon_t) = \sigma^2$.

9.1.1 Infinite Past

Let $\mathcal{F}_t = \overline{\mathrm{sp}}\{y_t, y_{t-1}, \dots\}$ be the Hilbert space generated by the infinite past at time $t+1$. The best linear one-step predictor of y_{t+1} given its past $\mathcal{F}_t$ is given by the orthogonal projection in $\mathcal{L}_2$,

$$\widehat{y}_{t+1} = E[y_{t+1}|\mathcal{F}_t] = \sum_{j=1}^{\infty} \pi_j y_{t+1-j} = \sum_{j=1}^{\infty} \psi_j \varepsilon_{t+1-j},$$

with prediction error variance $E[y_{t+1} - \widehat{y}_{t+1}]^2 = \sigma^2$.

9.1.2 Finite Past

In practice, we only have a finite stretch of data $\{y_1, \dots, y_t\}$. Under this circumstance, if $\mathcal{P}_t = \overline{\mathrm{sp}}\{y_t, y_{t-1}, \dots, y_1\}$ denotes the Hilbert space generated by the observations $\{y_1, \dots, y_t\}$, then the best linear predictor of y_{t+1} based on its finite past is given by

$$\widetilde{y}_{t+1} = E[y_{t+1}|\mathcal{P}_t] = \phi_{t1} y_t + \cdots + \phi_{tt} y_1,$$

where $\phi_t = (\phi_{1t}, \dots, \phi_{tt})'$ is the unique solution of the linear equation

$$\Gamma_t \phi_t = \gamma_t,$$

with $\Gamma_t = [\gamma(i-j)]_{i,j=1,\dots,t}$ and $\gamma_t = [\gamma(1), \dots, \gamma(t)]'$. The calculation of these coefficients can be carried out using the Durbin-Levinson algorithm described in Chapter 4. In particular, the prediction error variance of the one-step finite sample predictor $\widetilde{y}_{t+1}$,

$$\nu_t = E[y_{t+1} - \widetilde{y}_{t+1}]^2,$$

can be calculated by means of the recursive equations

$$\nu_t = \nu_{t-1}(1 - \phi_{tt}^2),$$

for $t \geq 1$, where ϕ_{tt} are the *partial autocorrelation coefficients* and $\nu_0 = \gamma(0)$. Thus, ν_t may be written as

$$\nu_t = \gamma(0) \prod_{j=1}^{t} (1 - \phi_{jj}^2).$$

EXAMPLE 9.1

For a fractional noise FN(d), the partial autocorrelation function is given by

$$\phi_{tt} = \frac{d}{t-d}, \tag{9.3}$$

for $t \geq 1$. Hence, the prediction error variance is given by

$$\begin{aligned} \nu_t &= \gamma(0) \prod_{j=1}^{t} \left[1 - \left(\frac{d}{j-d} \right)^2 \right] \\ &= \gamma(0) \frac{\prod_{j=1}^{t} j \prod_{j=1}^{t} (j-2d)}{\left[\prod_{j=1}^{t} (j-d) \right]^2}. \end{aligned}$$

But, for any real α we have that

$$\prod_{j=1}^{t} (j - \alpha) = \frac{\Gamma(t+1-\alpha)}{\Gamma(1-\alpha)}.$$

Hence,

$$\nu_t = \gamma(0) \frac{\Gamma(t+1)\Gamma(t+1-2d)\Gamma(1-d)^2}{[\Gamma(t+1-d)]^2 \, \Gamma(1-2d)}.$$

Therefore, since from (3.21) the autocovariance function at zero lag is

$$\gamma(0) = \sigma^2 \frac{\Gamma(1-2d)}{[\Gamma(1-d)]^2},$$

we obtain the formula

$$\nu_t = \sigma^2 \frac{\Gamma(t+1)\Gamma(t+1-2d)}{[\Gamma(t+1-d)]^2},$$

for $t \geq 1$. Now, an application of expression (3.31) yields

$$\lim_{t \to \infty} \nu_t = \sigma^2.$$

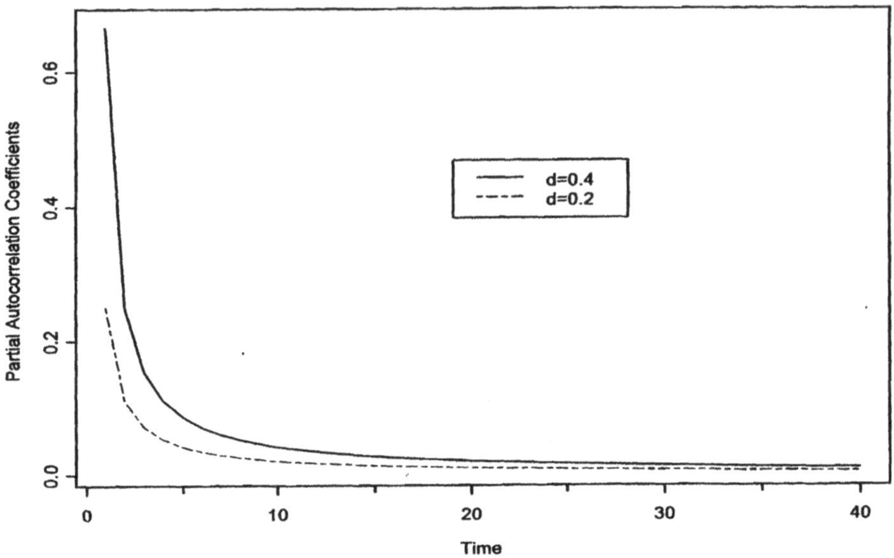

Figure 9.1 Evolution of the partial autocorrelation coefficients ϕ_{tt} for $t = 1, \ldots, 40$ for a fractional noise FN(d) process with $d = 0.2$ and $d = 0.4$.

Figure 9.1 displays the evolution of the coefficients (9.3) for $d = 0.2$, $d = 0.4$, and $t = 1, \ldots, 40$.

Notice that for the fractional noise processes, $\phi_{tt} = \mathcal{O}(1/t)$ for large t. Despite the difficulty of finding explicit expressions for the partial autocorrelations for a general class of long-memory models, this rate of convergence to zero can be extended to any ARFIMA(p, d, q) process, as stated in the following result that corresponds to Theorem 1.1 of Inoue (2002).

Theorem 9.1. *Let $\{y_t\}$ be an ARFIMA(p, d, q) process with $0 < d < \frac{1}{2}$. Then, the partial autocorrelations ϕ_{tt} satisfy*

$$|\phi_{tt}| \sim \frac{d}{t}, \tag{9.4}$$

as $t \to \infty$.

Observe that the rate on t given by expression (9.4) does not depend on the value of the long-memory parameter d.

Figure 9.2 displays the evolution of the mean-squared prediction error ν_t for a fractional noise FN(d) with $d = 0.2$, $d = 0.4$, $\sigma^2 = 1$, and $t = 1, \ldots, 40$.

As t increases, the effect of the remote past fades out and $\widetilde{y}_{t+1}$ becomes similar to $\widehat{y}_{t+1}$. In turn, the prediction error variance of the finite sample predictor ν_t becomes

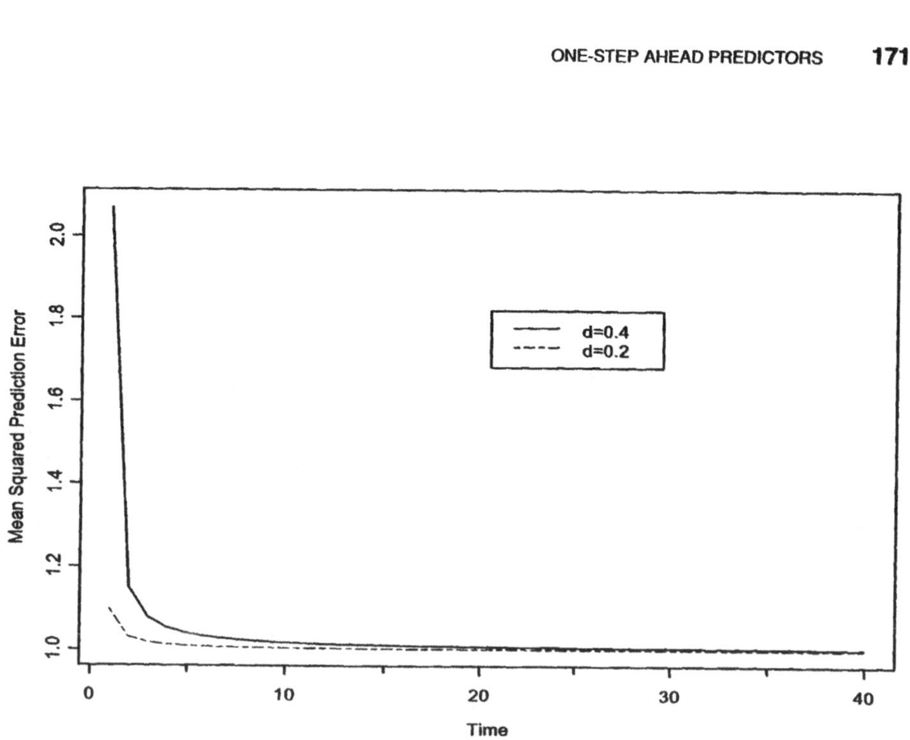

Figure 9.2 Evolution of the mean-squared prediction error ν_t for $t = 1, \dots, 40$ for a fractional noise FN(d) process with $d = 0.2$, $d = 0.4$, and $\sigma^2 = 1$.

similar to the prediction error variance of the infinite past predictor, σ^2. This is formally established in the following result; see Lemma 7.1 of Pourahmadi (2001).

Theorem 9.2. *If $\{y_t\}$ is a stationary process, then $\|\widehat{y}_{t+1} - \widetilde{y}_{t+1}\| \to 0$ and $\nu_t \to \sigma^2$ as $t \to \infty$.*

Let $\delta_t = \|\widehat{y}_{t+1} - \widetilde{y}_{t+1}\|^2$ be the squared distance between the optimal predictor based on the full past and the best predictor based on the finite past. Then, we may write

$$\begin{aligned}
\|\widehat{y}_{t+1} - \widetilde{y}_{t+1}\|^2 &= \|(\widehat{y}_{t+1} - y_{t+1}) + (y_{t+1} - \widetilde{y}_{t+1})\|^2 \\
&= \|\widehat{y}_{t+1} - y_{t+1}\|^2 + \|y_{t+1} - \widetilde{y}_{t+1}\|^2 \\
&\quad -2\langle y_{t+1} - \widehat{y}_{t+1}, y_{t+1} - \widetilde{y}_{t+1}\rangle \\
&= \sigma^2 + \nu_t - 2\langle \varepsilon_{t+1}, y_{t+1} - \widetilde{y}_{t+1}\rangle.
\end{aligned}$$

Since $\langle \varepsilon_{t+1}, y_{t+1}\rangle = \sigma^2$ and $\langle \varepsilon_{t+1}, \widetilde{y}_{t+1}\rangle = 0$, we have

$$\delta_t = \nu_t - \sigma^2.$$

A precise rate at which ν_t converges to σ^2 as t increases is specified in the next result which corresponds to Theorem 4.3 of Inoue (2002).

Theorem 9.3. *Let $\{y\}$ be an ARFIMA(p, d, q) process with unit variance noise and $0 < d < \frac{1}{2}$. Then,*

$$\delta_t \sim \frac{d^2}{t}, \tag{9.5}$$

as $t \to \infty$.

9.1.3 An Approximate Predictor

Since the Durbin-Levinson algorithm for calculating the coefficients ϕ_{tj} is order $\mathcal{O}(n^2)$, for very large sample sizes it could be desirable a faster algorithm to obtain finite sample forecasts. One way to do this is approximating the regression coefficients ϕ_{tj} by π_j, based on the following lemma; see Theorem 7.14 of Pourahmadi (2001).

Lemma 9.1. *If $\{y_t\}$ is a stationary process, then $\phi_{tj} \to \pi_j$ as $t \to \infty$.*

With this approximation, we introduce the finite sample predictor

$$\check{y}_{t+1} = \sum_{j=1}^{t} \pi_j y_{t+1-j},$$

with prediction error variance

$$\text{Var}(y_{t+1} - \check{y}_{t+1}) = \sigma^2 + r_t,$$

where $r_t = \text{Var}(\sum_{j=t+1}^{\infty} \pi_j y_{t+1-j})$ or equivalently

$$r_t = \text{Var}(\widehat{y}_{t+1} - \check{y}_{t+1}).$$

As expected, the prediction error variance of the approximate forecast is larger than the prediction error variance of $\widetilde{y}_{t+1}$. However, as t increases, these two predictors become similar, as stated in the following result.

Lemma 9.2. *If $\{y_t\}$ is a stationary process with AR(∞) representation satisfying $\sum_{j=1}^{\infty} |\pi_j| < \infty$, then $r_t \to 0$ as $t \to \infty$.*

Proof. Notice that

$$r_t = \text{Var}\left(\sum_{j=t+1}^{\infty} \pi_j y_{t+1-j}\right) \leq \text{Var}(y_0)\left(\sum_{j=t+1}^{\infty} |\pi_j|\right)^2. \tag{9.6}$$

Thus, since the autoregressive coefficients $\{\pi_j\}$ are absolutely summable, the result is proved. □

For short-memory processes with autoregressive coefficients $\{\pi_j\}$ satisfying

$$|\pi_j| \sim c|\phi|^j,$$

for large j and positive constant c we have from (9.6) that

$$r_t \leq c_1 |\phi|^{2t},$$

where $c_1 = \mathrm{Var}(y_0)[c|\phi|/(1 - |\phi|)]^2$. Therefore, r_t converges to zero at an exponential rate.

However, for long-memory processes this rate is slower. As established in the following theorem due to Bondon and Palma (2006), the decay rate of r_t to zero is $\mathcal{O}(1/t)$. Specifically we have the following theorem.

Theorem 9.4. *If $\{y_t\}$ is a stationary and invertible process with AR(∞) and MA(∞) satisfying*

$$\pi_j \sim \frac{j^{-d-1}}{\ell(j)\Gamma(-d)},$$
$$\psi_j \sim \frac{j^{d-1}\ell(j)}{\Gamma(d)},$$

as $j \to \infty$, with $0 < d < \frac{1}{2}$ and $\ell(\cdot)$ is a slowly varying function, then

$$r_t \sim \frac{d \tan(\pi d)}{\pi t}, \tag{9.7}$$

as $t \to \infty$.

Comparing expressions (9.5) and (9.7), we observe that for small values of d both terms δ_t and r_t behave similarly since

$$\frac{d \tan(\pi d)}{\pi t} \sim \frac{d^2}{t},$$

as $d \to 0$. On the contrary, when d approaches $\frac{1}{2}$, δ_t is bounded but r_t increases to infinity since $\tan(\pi/2) = \infty$.

9.2 MULTISTEP AHEAD PREDICTORS

9.2.1 Infinite Past

Let $\widehat{y}_t(h)$ be the best linear predictor of y_{t+h} based on the infinite past $\mathcal{F}_t$ for $h \geq 1$, which may be written as

$$\widehat{y}_t(h) = E[y_{t+h}|\mathcal{F}_t] = \sum_{j=0}^{\infty} \pi_j(h) y_{t-j} = \sum_{j=0}^{\infty} \psi_{j+h}\varepsilon_{t-j} = \sum_{j=h}^{\infty} \psi_j \varepsilon_{t+h-j}, \tag{9.8}$$

where the coefficients $\pi_j(h)$ for $j \geq 1$ are given by

$$\pi_j(h) = \sum_{i=0}^{h-1} \psi_i \pi_{j+h-i}.$$

The prediction error variance of $\widehat{y}_t(h)$, $\sigma^2(h) = E[y_{t+h} - \widehat{y}_t(h)]^2$, is

$$\sigma^2(h) = \sigma^2 \sum_{j=0}^{h-1} \psi_j^2. \tag{9.9}$$

9.2.2 Finite Past

The best linear predictor of y_{t+h} based on the finite past $\mathcal{P}_t$ is

$$\widetilde{y}_t(h) = \phi_{t1}(h)y_t + \cdots + \phi_{tt}(h)y_1,$$

where $\phi_t(h) = [\phi_{t1}(h), \ldots, \phi_{tt}(h)]'$ satisfies

$$\Gamma_t \phi_t(h) = \gamma_t(h),$$

with $\gamma_t(h) = [\gamma(h), \ldots, \gamma(t+h-1)]'$. Besides, the mean-squared prediction error is defined by

$$\sigma_t^2(h) = \|y_{t+h} - \widetilde{y}_t(h)\|^2.$$

Analogously to the one-step prediction case, we can use the approximate finite sample h-step ahead forecasts given by

$$\check{y}_t(h) = \sum_{j=0}^{t} \pi_j(h) y_{t-j},$$

with prediction error variance

$$\mathrm{Var}[y_{t+h} - \check{y}_t(h)] = \sigma^2 + r_t(h),$$

where $r_t(h) = \mathrm{Var}[\sum_{j=t+1}^{\infty} \pi_j(h) y_{t-j}]$. For fixed h, this term behaves similarly to r_t, excepting a constant, as described in the following theorem due to Bondon and Palma (2006).

Theorem 9.5. *Let $\{y_t\}$ be a stationary and invertible process with AR(∞) and MA(∞) satisfying*

$$\begin{aligned} \pi_j &\sim \frac{j^{-d-1}}{\ell(j)\Gamma(-d)}, \\ \psi_j &\sim \frac{j^{d-1}\ell(j)}{\Gamma(d)}, \end{aligned}$$

as $j \to \infty$, with $0 < d < \frac{1}{2}$ and $\ell(\cdot)$ is a slowly varying function. If $\sum_{j=0}^{h-1} \psi_j \neq 0$, then

$$r_t(h) \sim \left(\sum_{j=0}^{h-1} \psi_j \right)^2 \frac{d \tan(\pi d)}{\pi t},$$

as $t \to \infty$.

9.3 HETEROSKEDASTIC MODELS

Consider a general heteroskedastic process $\{y_t : t \in \mathbb{Z}\}$ specified by the equations

$$y_t = \sum_{j=0}^{\infty} \psi_j \varepsilon_{t-j}, \tag{9.10}$$

$$\varepsilon_t = \epsilon_t \sigma_t, \tag{9.11}$$

$$\sigma_t^2 = f(\epsilon_{t-1}, \epsilon_{t-2}, \ldots, \eta_t, \eta_{t-1}, \ldots), \tag{9.12}$$

where f is a measurable transformation of the σ-algebra generated by $\{\epsilon_t\}$ and $\{\eta_t\}$, which are sequences of independent and identically distributed random variables with zero mean and unit variance and $\{\epsilon_t\}$ is independent of $\{\eta_t\}$.

Observe that this specification includes the ARFIMA-GARCH, the ARCH-type and the LMSV processes, among others, as shown in the following examples.

EXAMPLE 9.2

The conditional variance of the ARFIMA-GARCH$(1,1)$ model may be written as in (6.4),

$$f(\epsilon_{t-1}, \epsilon_{t-2}, \ldots, \eta_t, \eta_{t-1}, \ldots) = \alpha_0 \left[1 + \sum_{k=0}^{\infty} \prod_{j=1}^{k+1} (\alpha_1 \epsilon_{t-j}^2 + \beta_1) \right],$$

and $\eta_t = 0$ for all t.

EXAMPLE 9.3

The ARCH-type process is also included in specification (9.10)–(9.12) with the Volterra expansion

$$f(\epsilon_{t-1}, \epsilon_{t-2}, \ldots) = \alpha_0 \sum_{k=0}^{\infty} \sum_{j_1, \ldots, j_k = 1}^{\infty} \alpha_{j_1} \alpha_{j_2} \cdots \alpha_{j_k} \epsilon_{t-j_1}^2 \epsilon_{t-j_2}^2 \cdots \epsilon_{t-j_k}^2,$$

and $\eta_t = 0$ for all t, cf., expression (6.9).

EXAMPLE 9.4

For the LMSV model introduced in Section 6.4, the conditional variance is given by

$$f(\eta_t, \eta_{t-1}, \ldots) = \sigma^2 \exp\{\phi(B)^{-1}(1-B)^{-d}\theta(B)\eta_t\}.$$

Notice that in this case, the conditional variance does not depend directly on the sequence $\{\epsilon_t\}$ but it depends on random perturbations $\{\eta_t\}$ such as those appearing in (6.16).

EXAMPLE 9.5

The conditional variance for the FIEGARCH model may be written as

$$f(\epsilon_{t-1}, \epsilon_{t-2}, \dots) \\ = \exp\left\{\phi(B)^{-1}(1-B)^{-d}[\theta(B)(|\epsilon_{t-1}| - E\,|\epsilon_{t-1}|) + \lambda(B)\epsilon_{t-1}]\right\};$$

see Problem 6.9.

The following lemma is fundamental for the formulation of prediction of the volatility techniques for the heteroskedastic models described by (9.10)–(9.12).

Lemma 9.3. *For all $t \in \mathbb{Z}$ we have*

$$E[\varepsilon_{t+h}^2|\mathcal{F}_t] = \begin{cases} E[\sigma_{t+h}^2|\mathcal{F}_t] & h \geq 1, \\ \varepsilon_{t+h}^2 & h < 1. \end{cases}$$

Proof. First, notice that for $h < 1$ the result is trivial. For $h \geq 1$ observe that by the definition of ε_{t+h} we may write $E[\varepsilon_{t+h}^2|\mathcal{F}_t] = E[\epsilon_{t+h}^2\sigma_{t+h}^2|\mathcal{F}_t]$. But, since $\sigma_{t+h}^2 = f(\epsilon_{t+h-1}, \epsilon_{t+h-2}, \dots, \eta_{t+h}, \eta_{t+h-1}, \dots)$, σ_{t+h}^2 and ϵ_{t+h} are independent. Furthermore, ϵ_{t+h} is independent of $\mathcal{F}_t$ for any $h \geq 1$. Thus,

$$E[\varepsilon_{t+h}^2|\mathcal{F}_t] = E[\epsilon_{t+h}^2]\,E[\sigma_{t+h}^2|\mathcal{F}_t].$$

Finally, by noting that $E[\epsilon_{t+h}^2] = 1$, the result is obtained. □

9.3.1 Prediction of Volatility

Observe that from (9.8) the multistep ahead prediction error for $h \geq 1$ is given by

$$e_t(h) = y_{t+h} - \widehat{y}_t(h) = \varepsilon_{t+h} + \psi_1\varepsilon_{t+h-1} + \cdots + \psi_{h-1}\varepsilon_{t+1}.$$

Therefore, the mean-squared prediction error is

$$E[e_t^2(h)|\mathcal{F}_t] = E[\varepsilon_{t+h}^2|\mathcal{F}_t] + \psi_1^2\,E[\varepsilon_{t+h-1}^2|\mathcal{F}_t] + \cdots + \psi_{h-1}^2\,E[\varepsilon_{t+1}^2|\mathcal{F}_t].$$

Thus, by Lemma 9.3, we may write

$$E[e_t^2(h)|\mathcal{F}_t] = E[\sigma_{t+h}^2|\mathcal{F}_t] + \psi_1^2\,E[\sigma_{t+h-1}^2|\mathcal{F}_t] + \cdots + \psi_{h-1}^2\,E[\sigma_{t+1}^2|\mathcal{F}_t].$$

Let $\sigma_t^2(h) = E[\sigma_{t+h}^2|\mathcal{F}_t]$, then the h-step prediction error conditional variance is

$$E[e_t^2(h)|\mathcal{F}_t] = \sum_{j=0}^{h-1}\psi_j^2\sigma_t^2(h-j).$$

The calculation of the conditional variances $\sigma_t^2(j)$ depends on the specification of the heterokedastic model. Some specific examples are discussed next.

EXAMPLE 9.6

For the ARFIMA-GARCH$(1,1)$ we have

$$\sigma_{t+h}^2 = \alpha_0 + \alpha_1 \varepsilon_{t+h-1}^2 + \beta_1 \sigma_{t+h-1}^2.$$

Therefore, an application of Lemma 9.3 yields $\sigma_t^2(1) = \sigma_{t+1}^2$ and for $h \geq 2$,

$$\sigma_t^2(h) = \alpha_0 + (\alpha_1 + \beta_1)\sigma_t^2(h-1).$$

Solving this recursive equation we find the following solution for $h \geq 1$:

$$\sigma_t^2(h) = \alpha_0 \frac{1 - (\alpha_1 + \beta_1)^{h-1}}{1 - (\alpha_1 + \beta_1)} + (\alpha_1 + \beta_1)^{h-1} \sigma_{t+1}^2.$$

Since $0 \leq \alpha_1 + \beta_1 < 1$, we have that

$$\lim_{h \to \infty} \sigma_t^2(h) = \frac{\alpha_0}{1 - (\alpha_1 + \beta_1)},$$

where the term on the left hand of this equation corresponds to the variance of $\{e_t\}$.

EXAMPLE 9.7

For the general ARFIMA-GARCH(r,s) it is not hard to check that for $h > \max\{r,s\}$ we have

$$\sigma_t^2(h) = \alpha_0 + \sum_{j=1}^{r} \alpha_j \sigma_t^2(h-j) + \sum_{j=1}^{s} \beta_j \sigma_t^2(h-j),$$

and since $0 < \sum_{j=1}^{r} \alpha_j + \sum_{j=1}^{s} \beta_j < 1$,

$$\lim_{h \to \infty} \sigma_t^2(h) = \frac{\alpha_0}{1 - (\sum_{j=1}^{r} \alpha_j + \sum_{j=1}^{s} \beta_j)} = \text{Var}(\varepsilon_t).$$

EXAMPLE 9.8

For the ARCH(∞) model, we have

$$\sigma_{t+h}^2 = \alpha_0 + \sum_{j=1}^{\infty} \alpha_j y_{t+h-j}^2 = \alpha_0 + \sum_{j=1}^{h-1} \alpha_j y_{t+h-j}^2 + \sum_{j=h}^{\infty} \alpha_j y_{t+h-j}^2.$$

Thus,

$$E[\sigma_{t+h}^2 | \mathcal{F}_t] = \alpha_0 + \sum_{j=1}^{h-1} \alpha_j \, E[y_{t+h-j}^2 | \mathcal{F}_t] + \sum_{j=h}^{\infty} \alpha_j y_{t+h-j}^2,$$

and then,

$$\sigma_t^2(h) = \alpha_0 + \sum_{j=1}^{h-1} \alpha_j \sigma_t^2(h-j) + \sum_{j=h}^{\infty} \alpha_j y_{t+h-j}^2.$$

Now, if $0 < \sum_{j=1}^{\infty} \alpha_j < 1$, then

$$\lim_{h\to\infty} \sigma_t^2(h) = \frac{\alpha_0}{1 - \sum_{j=1}^{\infty} \alpha_j} = \operatorname{Var}(y_t).$$

9.4 ILLUSTRATION

Figure 9.3 displays a simulated Gaussian ARFIMA$(0, d, 0)$ process with zero mean and unit variance white noise, long-memory parameter $d = 0.40$ and sample size $n = 1000$. The last 100 observations (dotted line) will be predicted using the estimated model.

The MLE of d calculated from the first 900 observations is $\widehat{d} = 0.3917$ with standard deviation $\widehat{\sigma}_d = 0.02712$. Besides, the estimated standard deviation of the white noise is $\widehat{\sigma} = 0.9721$. Figure 9.4 shows the observations from $t = 800$ to $t =$

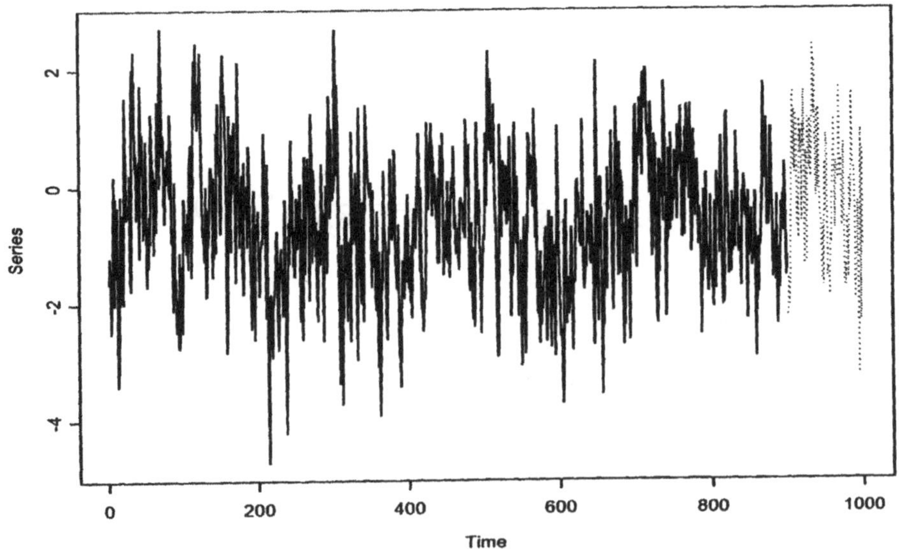

Figure 9.3 Simulated fractional noise process FN(d), with $d = 0.40$ and unit variance white noise. The dotted line indicates the last 100 observations that will be predicted using the estimated model.

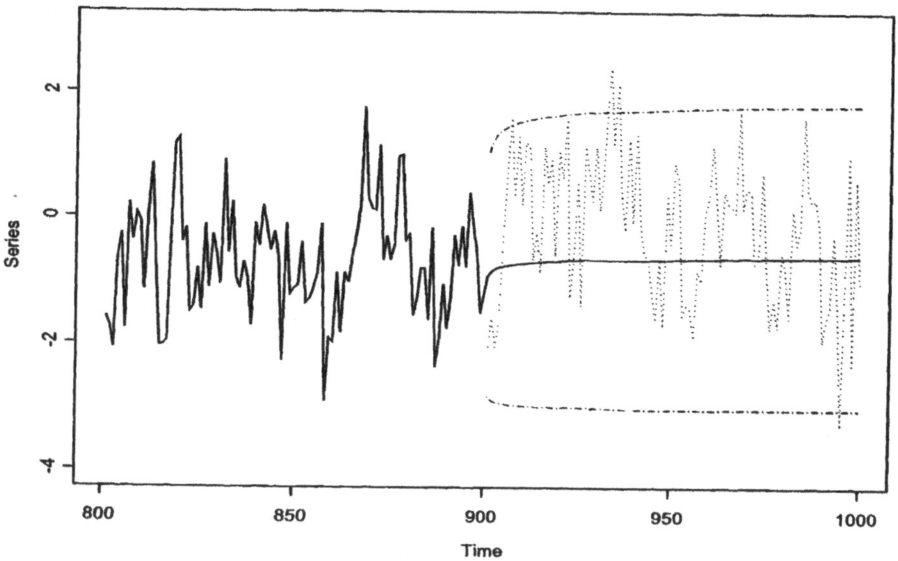

Figure 9.4 Simulated fractional noise process FN(d): Multistep forecasts of the last 100 values and 95% prediction bands.

1000, along with $\widehat{y}_{900}(h)$ forecasts with $h = 1, 2, \ldots, 100$ and 95% prediction bands. These multistep ahead predictors are based on the fitted model and on observations $t = 1, 2, \ldots, 900$. From Figure 9.4, notice that most of the future observations fall inside the prediction bands.

Figure 9.5 displays the theoretical and the empirical evolution of the prediction error standard deviation, from $t = 901$ to $t = 1000$. The *theoretical* prediction error standard deviation is based on the multistep prediction error variance formula (9.9) which yields

$$\widehat{\sigma}(h) = \widehat{\sigma}\sqrt{\sum_{j=0}^{h-1} \psi_j^2(\widehat{d})}, \tag{9.13}$$

for $h = 1, 2, \ldots, 100$. Equation (9.13) provides only an approximation of the theoretical prediction error standard deviation at step h since we are dealing with a finite past of 900 observations. On the other hand, the empirical prediction error standard deviations are based on the Kalman filter output from a truncated state space representation with $m = 50$. Notice from this graph that the sample prediction error standard deviations are very close to their theoretical counterparts.

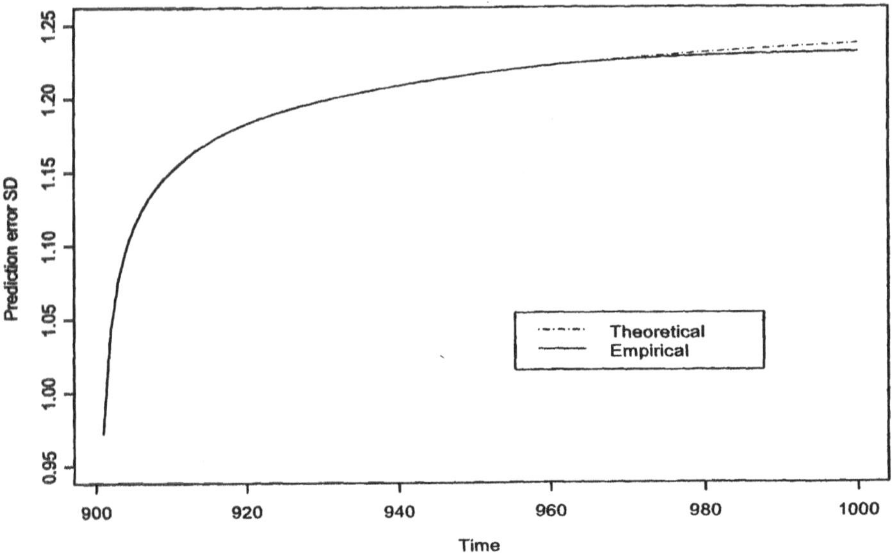

Figure 9.5 Simulated fractional noise process FN(d): Theoretical and empirical prediction error standard deviations of the last 100 observations.

9.5 RATIONAL APPROXIMATIONS

Under the Fisherian paradigm, one should use the same model for both fitting the data and predicting the value of future observations. However, in some situations one may be interested in obtaining quick and good predictions through a simpler model. This is the case, for example, when the time series analyzed have several thousands observations. Time series models with rational spectrum such as the ARMA family are easier to handle computationally than long-memory processes. Thus, in this section we discuss the issue approximating an ARFIMA process by an ARMA model. Recall, however, that the main motivation for this approximation is finding good predictions through simple numerical procedures. It is not intended for making inference about the parameters of the model.

Consider the regular linear model with Wold expansion

$$y_t = \sum_{j=0}^{\infty} \psi_j \varepsilon_{t-j},$$

and the approximating ARMA$(1,1)$ process

$$x_t = \sum_{j=0}^{\infty} a_j \varepsilon_{t-j}, \tag{9.14}$$

where a_j are the coefficients of the infinity moving-average expansion

$$a(B) = \sum_{j=0}^{\infty} a_j B^j = \frac{1 - \theta B}{1 - \phi B}.$$

Thus, $a_0 = 1$ and $a_j = (\phi - \theta)\phi^{j-1}$ for $j \geq 1$.

For $h \geq 1$, let $\widehat{x}_t(h)$ be the h-step ahead forecast based on the infinite past $\mathcal{F}_t$ given by model (9.14),

$$\widehat{x}_t(h) = \sum_{j=h}^{\infty} a_j \varepsilon_{t+h-j}.$$

Since we are looking for an ARMA$(1,1)$ model that produce good h-step ahead forecasts for the value y_{t+h} in terms of mean-squared prediction error (MSPE), we must minimize

$$\begin{aligned} G_h(\phi,\theta) &= \operatorname{Var}\left[y_{t+h} - \widehat{x}_t(h)\right] \\ &= \operatorname{Var}\left[\sum_{j=0}^{\infty} \psi_j \varepsilon_{t+h-j} - \sum_{j=h}^{\infty} a_j \varepsilon_{t+h-j}\right] \\ &= \operatorname{Var}\left[\sum_{j=0}^{h-1} \psi_j \varepsilon_{t+h-j} + \sum_{j=h}^{\infty} (\psi_j - a_j)\varepsilon_{t+h-j}\right] \\ &= \operatorname{Var}\left[\sum_{j=0}^{h-1} \psi_j \varepsilon_{t+h-j}\right] + \operatorname{Var}\left[\sum_{j=h}^{\infty} (\psi_j - a_j)\varepsilon_{t+h-j}\right] \\ &= \sigma^2 \sum_{j=0}^{h-1} \psi_j^2 + \sigma^2 \sum_{j=h}^{\infty} (\psi_j - a_j)^2, \end{aligned}$$

with respect to ϕ and θ. Consequently, if $J_h(\phi,\theta) = \sum_{j=h}^{\infty} (\psi_j - a_j)^2$, then minimizing $G_h(\phi,\theta)$ is equivalent to solving the problem

$$\min_{(\phi,\theta)\in\Theta} J_h(\phi,\theta). \tag{9.15}$$

By making $\nabla J_h(\phi,\theta) = 0$, we obtain the equations

$$\sum_{j=h}^{\infty} (\psi_j - a_j) a_j = 0, \tag{9.16}$$

$$\sum_{j=h}^{\infty} j(\psi_j - a_j) a_j = 0. \tag{9.17}$$

Generally speaking, equations (9.16) and (9.17) are nonlinear and their solutions can be found by numerical procedures.

EXAMPLE 9.9

For a fractional noise model FN(d) and $h = 1$ we have that

$$\begin{aligned}
\sum_{j=1}^{\infty} \psi_j a_j &= \frac{\phi-\theta}{\phi}[(1-\phi)^{-d}-1], \\
\sum_{j=1}^{\infty} a_j^2 &= \frac{(\phi-\theta)^2}{1-\phi^2}, \\
\sum_{j=1}^{\infty} j\psi_j a_j &= \frac{d(\phi-\theta)}{(1-\phi)^{d+1}}, \\
\sum_{j=1}^{\infty} j a_j^2 &= \left(\frac{\phi-\theta}{1-\phi^2}\right)^2.
\end{aligned}$$

Thus, equations (9.16) and (9.17) in this case are reduced to

$$\begin{aligned}
(1-\phi)^{-d} &= \frac{1-\phi\theta}{1-\phi^2}, \\
d(1-\phi)^{1-d} &= \frac{\phi-\theta}{(1+\phi)^2}.
\end{aligned}$$

For $h = 2$, the corresponding equations are

$$\begin{aligned}
(1-\phi)^{-d} - 1 - d\phi &= \frac{(\phi-\theta)\phi^3}{1-\phi^2}, \\
\frac{d}{(1-\phi)^{d+1}}[1-(1-\phi)^{d+1}] &= \frac{(\phi-\theta)\phi^2(2-\phi^2)}{(1-\phi^2)^2}.
\end{aligned}$$

For a general ARFIMA(p, d, q) model, the nonlinear equations (9.16) and (9.17) may be solved numerically. Alternatively, we may solve computationally the minimization problem (9.15).

9.5.1 Illustration

Figure 9.6 and Figure 9.7 display the optimal ARMA(1, 1) parameters $[\phi_h(d), \theta_h(d)]$ for horizons $h = 1$ and $h = 2$, respectively. These parameters are plotted as functions of d for a fractional noise process FN(d), where the long-memory parameter d ranges from -1 to $\frac{1}{2}$. The optimal ARMA(1, 1) parameters ϕ and θ were calculated by solving problem (9.15) with

$$J_1(\phi,\theta) = \frac{\Gamma(1-2d)}{\Gamma(1-d)^2} - 1 + \frac{(\phi-\theta)^2}{1-\phi^2} - \frac{2(\phi-\theta)}{\phi}[(1-\phi)^{-d}-1],$$

and

$$J_2(\phi,\theta) = J_1(\phi,\theta) - [d-(\phi-\theta)]^2,$$

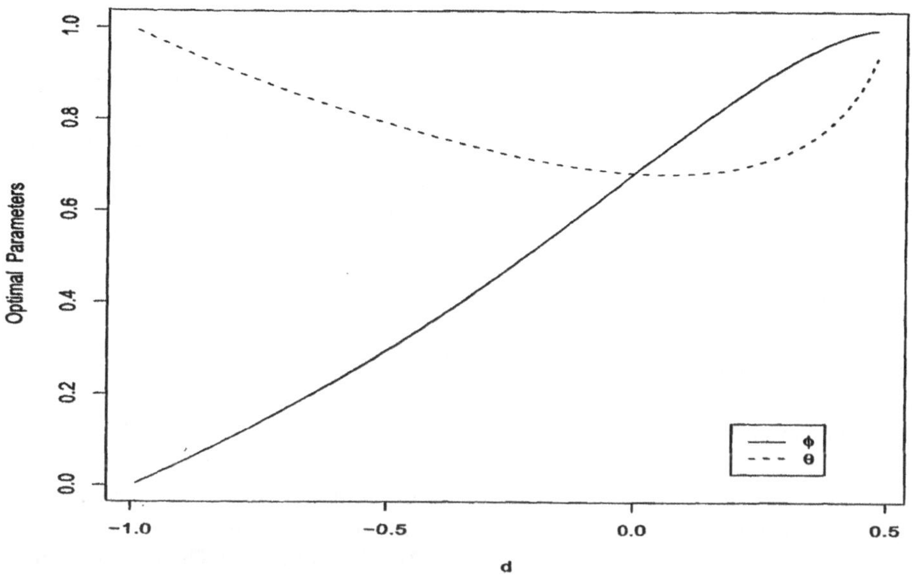

Figure 9.6 Fractional noise process FN(d): Solutions ϕ (heavy line) and θ (broken line) for the approximating ARMA(1,1) model, for different values of d (horizontal axis) for $h = 1$.

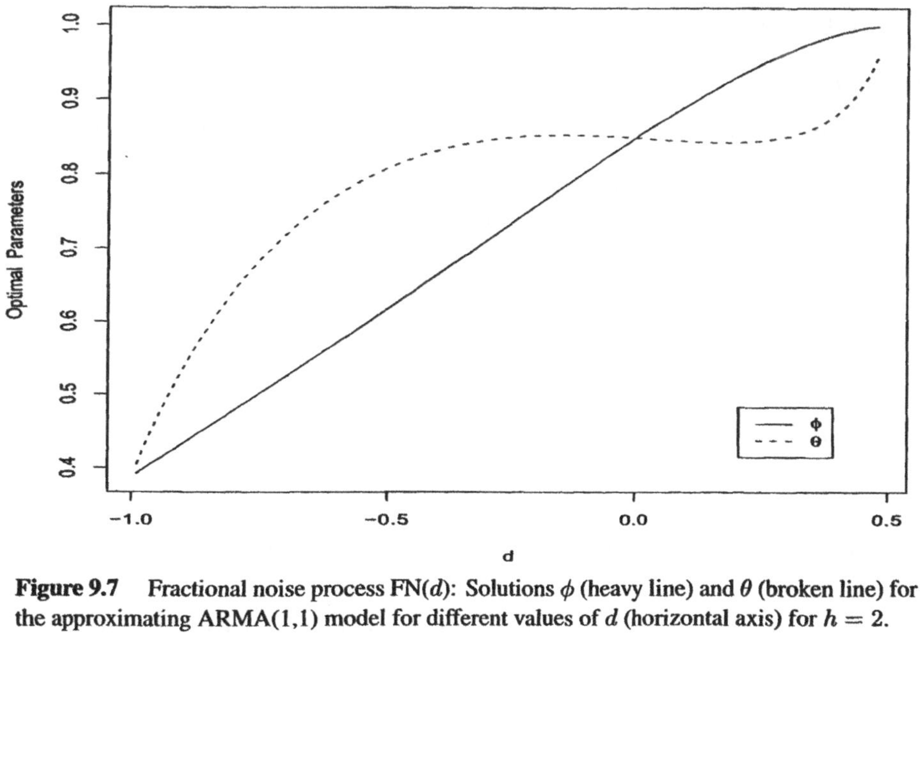

Figure 9.7 Fractional noise process FN(d): Solutions ϕ (heavy line) and θ (broken line) for the approximating ARMA(1,1) model for different values of d (horizontal axis) for $h = 2$.

where the parameter space is $\Theta = (-1, 1) \times (-1, 1)$.

Observe from Figures 9.6 and 9.7 that the parameters of the optimal approximating ARMA(1, 1) model, $[\phi_h(d), \theta_h(d)]$, tends to 1 as d approaches $\frac{1}{2}$, for $h = 1, 2$. Besides, notice that $\phi_h(0) = \theta_h(0)$ for the two horizons analyzed.

The optimal autoregressive parameter $\phi_h(d)$ increases with d in both figures. On the contrary, the behavior of the optimal parameter $\theta_h(d)$ is not as monotone as that of $\phi_h(d)$. In Figure 9.6, $\theta_h(d)$ decreases up to a point and then it increases until it reaches the value $\phi_h(\frac{1}{2}) = 1$. On the other hand, in Figure 9.7 the moving-average parameter $\theta_h(d)$ displays a more complex structure.

9.6 BIBLIOGRAPHIC NOTES

The literature about prediction of stationary processes is extensive and spans several decades since the pioneering works on linear predictors by Kolmogorov, Wiener, and Wold, among others. The monographs by Rozanov (1967), Hannan (1970), and Pourahmadi (2001) offer excellent overviews of the theoretical problems involved in the prediction of linear processes. For a review of prediction methods in the context of long-memory processes; see, for example, Bhansali and Kokoszka (2003).

Theorems 9.1 and 9.3 are due to Inoue (2002), while proofs of Theorem 9.2 and Lemma 9.1 can be found in Pourahmadi (2001). Theorems 9.4 and 9.5 are due to Bondon and Palma (2006).

The approximation of long-memory processes by ARMA models has been discussed by Tiao and Tsay (1994) and Basak, Chan, and Palma (2001), among others, while optimal prediction with long-range-dependent models has been analyzed, for instance, by Ray (1993b).

Problems

9.1 Show explicitly Lemma 9.1 for the fractional noise process FN(d) where

$$\phi_{tj} = -\binom{n}{j} \frac{\Gamma(j-d)\Gamma(t-d-j+1)}{\Gamma(-d)\Gamma(t-d+1)},$$

and

$$\pi_j = \frac{\Gamma(j-d)}{\Gamma(j+1)\Gamma(-d)}.$$

9.2 Consider the AR(p) model

$$y_t + \phi_1 y_{t-1} - \cdots - \phi_p y_{t-p} = \varepsilon_t,$$

where $\{\varepsilon_t\}$ is a white noise sequence with variance σ^2. Verify that ν_t converges to σ^2 after p steps.

9.3 Find an expression for $\sigma_t^2(h)$, $h \geq 1$, for the ARFIMA-GARCH(r, s) model.

9.4 Show that for the ARFIMA-GARCH$(1,1)$ model, the conditional variance at time $t+1$ may be written as

$$\sigma_{t+1}^2 = \frac{\alpha_0}{1-\beta-1} + \alpha_1 \sum_{j=0}^{\infty} \beta_1^j \varepsilon_{t-j}^2.$$

9.5 Find explicit solutions to equations (9.16) and (9.17) for $h=1$ for a fractional noise process with long-memory parameter d.

9.6 Show that for $d \in (-\frac{1}{2}, \frac{1}{2})$,

$$(1-\phi)^{-d} = \sum_{j=0}^{\infty} \psi_j \phi^j,$$

where

$$\psi_j = \frac{\Gamma(j+d)}{\Gamma(j+1)\Gamma(d)}.$$

9.7 Prove that for $d \in (-\frac{1}{2}, \frac{1}{2})$,

$$\sum_{j=1}^{\infty} j\psi_j \phi^{j-1} = \frac{\partial}{\partial \phi}(1-\phi)^{-d} = \frac{d}{(1-\phi)^{1+d}}.$$

9.8 Verify the following result for the random variables x, y, z. If x is independent of y and z, then

$$E(xy|z) = E(x)\,E(y|z).$$

9.9 Show explicitly Theorem 9.3 for the fractional noise process, FN(d).

CHAPTER 10

REGRESSION

In the previous chapters we have studied methods for dealing with long-memory time series but we have not discussed the problem of relating those time series to other covariates or trends. Nevertheless, in many practical applications, the behavior of a time series may be related to the behavior of other data. A widely used approach to model these relationships is the linear regression analysis. Consequently, in this chapter we explore several aspects of the statistical analysis of linear regression models with long-range-dependent disturbances.

Basic definitions about the model under study are given in Section 10.1. We then proceed to the analysis of some large sample properties of the least squares estimators (LSE) and the best linear unbiased estimators (BLUE). Aspects such as strong consistency, the asymptotic variance of the estimators, normality, and efficiency are discussed in Section 10.2 for the LSE and in Section 10.3 for the BLUE. Sections 10.4–10.6 present some important examples, including the estimation of the mean of a long-memory process, the polynomial, and the harmonic regression. A real life data application to illustrate these regression techniques is presented in Section 10.7 while some references and further readings are discussed in the Section 10.8. This chapter concludes with a section containing proposed problems.

Long-Memory Time Series. By Wilfredo Palma

10.1 LINEAR REGRESSION MODEL

Consider the linear regression model

$$y_t = x_t \beta + \varepsilon_t, \tag{10.1}$$

for $t = 1, 2, \ldots$, where $x_t = (x_{t1}, \ldots, x_{tp})$ is a sequence of regressors, $\beta \in \mathbb{R}^p$ is a vector of parameters, and $\{\varepsilon_t\}$ is a long-range-dependent stationary process with spectral density

$$f(\lambda) = |1 - e^{i\lambda}|^{-2d} f_0(\lambda), \tag{10.2}$$

where $f_0(\lambda)$ is a symmetric, positive, piecewise continuous function for $\lambda \in (-\pi, \pi]$ and $0 < d < \frac{1}{2}$.

The least squares estimator (LSE) of β is given by

$$\widehat{\beta}_n = (\boldsymbol{x}_n' \boldsymbol{x}_n)^{-1} \boldsymbol{x}_n' \boldsymbol{y}_n, \tag{10.3}$$

where $\boldsymbol{x}_n$ is the $n \times p$ matrix of regressors, $[\boldsymbol{x}_n]_{ij} = x_{ij}$, $i = 1, \ldots, n$, $j = 1, \ldots, p$, and $\boldsymbol{y}_n = (y_1, \ldots, y_n)'$. The variance of the LSE is

$$\operatorname{Var}(\widehat{\beta}_n) = (\boldsymbol{x}_n' \boldsymbol{x}_n)^{-1} \boldsymbol{x}_n' \Gamma \boldsymbol{x}_n (\boldsymbol{x}_n' \boldsymbol{x}_n)^{-1}, \tag{10.4}$$

where Γ is the variance-covariance matrix of $\{y_t\}$ with elements

$$\Gamma_{ij} = \int_{-\pi}^{\pi} f(\lambda) e^{i\lambda(i-j)} \, d\lambda. \tag{10.5}$$

On the other hand, the BLUE of β is

$$\widetilde{\beta}_n = (\boldsymbol{x}_n' \Gamma^{-1} \boldsymbol{x}_n)^{-1} \boldsymbol{x}_n' \Gamma^{-1} \boldsymbol{y}_n, \tag{10.6}$$

which has variance equal to

$$\operatorname{Var}(\widetilde{\beta}_n) = (\boldsymbol{x}_n' \Gamma^{-1} \boldsymbol{x}_n)^{-1}. \tag{10.7}$$

10.1.1 Grenander Conditions

In order to analyze the large sample properties of the LSE and the BLUE, we introduce the following so-called Grenander conditions on the regressors. Let $\boldsymbol{x}_n(j)$ be the jth column of the design matrix, $\boldsymbol{x}_n$.

(1) $\|\boldsymbol{x}_n(j)\| \to \infty$ as $n \to \infty$ for $j = 1, \ldots, p$.

(2) $\lim_{n\to\infty} \|\boldsymbol{x}_{n+1}(j)\| / \|\boldsymbol{x}_n(j)\| = 1$ for $j = 1, \ldots, p$.

(3) Let $\boldsymbol{x}_{n,h}(j) = (x_{h+1,j}, x_{h+2,j}, \ldots, x_{n,j}, 0 \ldots, 0)'$ for $h \geq 0$ and $\boldsymbol{x}_{n,h}(j) = (0, \ldots, 0, x_{1,j}, x_{2,j}, \ldots, x_{n+h,j})'$ for $h < 0$. Then, there exists a $p \times p$ finite matrix $R(h)$ such that

$$\frac{\langle \boldsymbol{x}_{n,h}(i), \boldsymbol{x}_n(j) \rangle}{\|\boldsymbol{x}_{n,h}(i)\| \|\boldsymbol{x}_n(j)\|} \to R_{ij}(h), \tag{10.8}$$

as $n \to \infty$ with $\langle x, y \rangle = x\bar{y}$ for complex numbers $x, y \in \mathbb{C}$ where $\bar{y}$ is the complex conjugate of y.

(4) The matrix $R(0)$ is nonsingular.

Notice first that under conditions (1)–(4), the matrix $R(h)$ may be written as

$$R(h) = \int_{-\pi}^{\pi} e^{ih\lambda}\, dM(\lambda), \tag{10.9}$$

where $M(\lambda)$ is a Hermitian matrix function with positive semidefinite increments; see Yajima (1991).

The asymptotic properties of the LSE estimates and the BLUE depend upon the behavior of $M_{jj}(\lambda)$ around frequency zero. Consequently, assume that for $j = 1, \dots, s$, $M_{jj}(\lambda)$ suffers a jump at the origin, that is,

$$M_{jj}(0+) > M_{jj}(0), \qquad j = 1, \dots, s,$$

and for $j = s+1, \dots, p$, $M_{jj}(\lambda)$ is right continuous at $\lambda = 0$, that is,

$$M_{jj}(0+) = M_{jj}(0), \qquad j = s+1, \dots, p,$$

where $M_{jj}(0+) = \lim_{\lambda \to 0+} M_{jj}(\lambda)$.

Before stating the asymptotic behavior of these estimators, the following definitions and technical results are needed. Define the *characteristic function* of the design matrix by

$$M_{ij}^n(\lambda) = \int_{-\pi}^{\lambda} m_{ij}^n(\omega)\, d\omega,$$

where

$$m_{ij}^n(\lambda) = \frac{\langle \sum_{t=1}^n x_{ti} e^{-it\lambda}, \sum_{t=1}^n x_{tj} e^{-it\lambda} \rangle}{2\pi \|\boldsymbol{x}_n(i)\| \|\boldsymbol{x}_n(j)\|}.$$

Also, we define the function $\delta(\cdot)$ such that $\delta(x) = 1$ for $x = 0$ and $\delta(x) = 0$ for $x \neq 0$.

Lemma 10.1. *Under assumption* (3), *$M^n(\lambda)$ converges weakly to $M(\lambda)$, that is,*

$$\int_{-\pi}^{\pi} g(\lambda)\, dM^n(\lambda) \to \int_{-\pi}^{\pi} g(\lambda)\, dM(\lambda),$$

as $n \to \infty$ for any continuous function $g(\lambda)$ with $\lambda \in [-\pi, \pi]$.

Proof. Since g is continuous on $[-\pi, \pi]$ for any $\varepsilon > 0$ there is a trigonometric polynomial $\phi_k(\lambda) = \sum_{\ell=-k}^{k} c_\ell e^{i\lambda\ell}$ such that $|g(\lambda) - \phi_k(\lambda)| < \varepsilon$ for all $\lambda \in [-\pi, \pi]$.

For this polynomial we have

$$\begin{aligned}
\int_{-\pi}^{\pi} \phi_k(\lambda) m_{ij}^n(\lambda)\, d\lambda &= \frac{1}{2\pi} \sum_{\ell=-k}^{k} c_\ell \int_{-\pi}^{\pi} e^{i\lambda\ell} m_{ij}^n(\lambda)\, d\lambda \\
&= \frac{1}{2\pi} \sum_{\ell=-k}^{k} \sum_{s=1}^{n} \sum_{t=1}^{n} c_\ell \frac{\langle x_{s,i}, x_{t,j}\rangle}{\|x_n(i)\| \|x_n(j)\|} \int_{-\pi}^{\pi} e^{i\lambda(t-s+\ell)} d\lambda \\
&= \sum_{\ell=-k}^{k} \sum_{s=1}^{n} \sum_{t=1}^{n} c_\ell \frac{\langle x_{s,i}, x_{t,j}\rangle}{\|x_n(i)\| \|x_n(j)\|} \delta(t-s+\ell) \\
&= \sum_{\ell=-k}^{k} \sum_{t=\max\{1,1-\ell\}}^{\min\{n,n-\ell\}} c_\ell \frac{\langle x_{t+\ell,i}, x_{t,j}\rangle}{\|x_n(i)\| \|x_n(j)\|}.
\end{aligned}$$

Hence, by condition (3) and fixed k we conclude that

$$\lim_{n\to\infty} \int_{-\pi}^{\pi} \phi_k(\lambda) m_{ij}^n(\lambda)\, d\lambda = \sum_{\ell=-k}^{k} R_{ij}(\ell) c_\ell = \int_{-\pi}^{\pi} \phi_k(\lambda)\, dM_{ij}(\lambda). \quad (10.10)$$

Thus,

$$\begin{aligned}
\left| \int_{-\pi}^{\pi} g(\lambda)\, dM_{ij}^n(\lambda) \right. &- \left. \int_{-\pi}^{\pi} g(\lambda)\, dM_{ij}(\lambda) \right| \\
&\leq \left| \int_{-\pi}^{\pi} \phi_k(\lambda)\, dM_{ij}^n(\lambda) - \int_{-\pi}^{\pi} \phi_k(\lambda)\, dM_{ij}(\lambda) \right| \\
&+ \left| \int_{-\pi}^{\pi} h(\lambda)\, dM_{ij}^n(\lambda) \right| + \left| \int_{-\pi}^{\pi} h(\lambda)\, dM_{ij}(\lambda) \right|,
\end{aligned}$$

where $h(\lambda) = g(\lambda) - \phi_k(\lambda)$. By (10.10), the first term on the right-hand side is bounded by ε as $n \to \infty$. Besides, by decomposing $h(\lambda) = h^+(\lambda) - h^-(\lambda)$, where $h^+, h^- \geq 0$, the second term on the right-hand side satisfies; see Problem 10.4,

$$\begin{aligned}
\left| \int_{-\pi}^{\pi} h(\lambda)\, dM_{ij}^n(\lambda) \right| &\leq \left| \int_{-\pi}^{\pi} h^+(\lambda)\, dM_{ij}^n(\lambda) \right| + \left| \int_{-\pi}^{\pi} h^-(\lambda)\, dM_{ij}^n(\lambda) \right| \\
&\leq \int_{-\pi}^{\pi} h^+(\lambda)\, dM_{ii}^n(\lambda) + \int_{-\pi}^{\pi} h^+(\lambda)\, dM_{jj}^n(\lambda) \\
&+ \int_{-\pi}^{\pi} h^-(\lambda)\, dM_{ii}^n(\lambda) + \int_{-\pi}^{\pi} h^-(\lambda)\, dM_{jj}^n(\lambda).
\end{aligned}$$

But, $h^+(\lambda) \leq \varepsilon$ and $h^-(\lambda) \leq \varepsilon$. Therefore,

$$\left| \int_{-\pi}^{\pi} h(\lambda)\, dM_{ij}^n(\lambda) \right| \leq 4\varepsilon,$$

since $\int_{-\pi}^{\pi} dM_{jj}^n(\lambda) = 1$. Analogously we can bound the third term on the right-hand side by 4ε and the result is proved. □

The following very useful result is a consequence of Lemma 10.1.

Corollary 10.1. *Let the function $f_n(\cdot)$ be defined for $n \geq 1$ as*

$$f_n(\lambda) = \frac{1}{2\pi n} \left| \sum_{t=1}^{n} e^{i\lambda t} \right|^2.$$

Then, the sequence of functions $\{f_n(\lambda)\}$ converges weakly to the Dirac operator $\delta_D(\lambda)$ as $n \to \infty$. That is, for any continuous function $g(\lambda)$, $\lambda \in [-\pi, \pi]$ we have that

$$\int_{-\pi}^{\pi} f_n(\lambda) g(\lambda)\, d\lambda \to \int_{-\pi}^{\pi} g(\lambda) \delta_D(\lambda)\, d\lambda = g(0),$$

as $n \to \infty$.

Proof. By taking $x_n(1) = (1, \ldots, 1)'$ in Lemma 10.1 and observing that $\|x_n(1)\| = \sqrt{n}$ the result is obtained. □

The following is a very simple but useful result.

Lemma 10.2. *Let the sequence of functions $\{f_n\}$ be defined by*

$$f_n(\lambda) = \frac{1}{n} \sum_{t=1}^{n} e^{i\lambda t}.$$

Then, $f_n(\lambda) \to \delta(\lambda)$ as $n \to \infty$.

Proof. For $\lambda = 0$ we have that $f_n(0) = 1$ for all n. On the other, for $\lambda \neq 0$ we have

$$\left| \frac{1}{n} \sum_{t=1}^{n} e^{i\lambda t} \right| = \frac{1}{n} \left| \frac{e^{i\lambda n} - 1}{e^{i\lambda} - 1} \right| \leq \frac{2}{n} \left| \frac{1}{e^{i\lambda} - 1} \right| \to 0,$$

as $n \to \infty$. □

10.2 PROPERTIES OF THE LSE

The large sample behavior of the LSE is studied in this section. We begin this analysis by establishing the strong consistency of the LSE in Theorem 10.2 and Theorem 10.3. Then, in Theorem 10.4 we turn our attention to the problem of obtaining expressions for its asymptotic variance. Sufficient conditions to achieve normality are discussed in Theorem 10.5. In what follows, the regressors are assumed to be nonstochastic.

10.2.1 Consistency

Let $\sigma_n = \lambda_{\min}(x_n' x_n)$ be the smallest eigenvalue of $x_n' x_n$ and let β be the true parameter. The strong consistency of the LSE is established in the next result due to Yajima (1988).

Theorem 10.2. *Consider the linear model* (10.1) *where the sequence of disturbances* $\{\varepsilon_t\}$ *is a stationary process with spectral density satisfying* (10.2) *and* f_0 *is a bounded function. If the following two conditions hold*

(a) $n^{-2d}\sigma_n \to \infty$ *as* $n \to \infty$,

(b) $\sum_{n=p+2}^{\infty} \sigma_{n-1}^{-1} n^{2d-1} \log^2 n < \infty$,

then $\widehat{\beta}_n \to \beta$ *almost surely as* $n \to \infty$.

Finding the eigenvalues of the matrix $x_n' x_n$ may not be a simple task in some situations. Fortunately, there is a set of conditions that guarantees the strong consistency of the LSE which are simpler to verify. The following result corresponds to Corollary 2.1 of Yajima (1988).

Theorem 10.3. *Consider the linear model* (10.1) *where the sequence of disturbances* $\{\varepsilon_t\}$ *is a stationary process with spectral density satisfying* (10.2) *and* f_0 *is a bounded function. Suppose that the Grenander conditions* (3, *with* $h = 0$) *and* (4) *hold. If*

$$0 < \liminf_{n\to\infty} \frac{\|x_n(j)\|^2}{n^\delta},$$

for some $\delta > 2d$ *and* $j = 1, \ldots, p$, *then* $\widehat{\beta}_n \to \beta$ *almost surely as* $n \to \infty$.

Proof. Let $D_n = \operatorname{diag}(\|x_n(1)\|, \ldots, \|x_n(p)\|)$ and $G_n = D_n^{-1}(x_n' x_n)D_n^{-1}$. Then for any $\alpha \in \mathbb{R}^n$ with $\|\alpha\| = 1$ we have

$$\alpha'(x_n' x_n)\alpha \geq \lambda_{\min}(G_n)\alpha' D_n^2 \alpha \geq \lambda_{\min}(G_n) \min_{1\leq j\leq p} \|x_n(j)\|^2.$$

Hence,

$$\sigma_n \geq \lambda_{\min}(G_n) \min_{1\leq j\leq p} \|x_n(j)\|^2.$$

Therefore,

$$0 < \lambda_{\min}[R(0)] \liminf_{n\to\infty} \min_{1\leq j\leq p} \frac{\|x_n(j)\|^2}{n^\delta} \leq \liminf_{n\to\infty} \frac{\sigma_n}{n^\delta}.$$

Thus, the sequence $\{\sigma_n\}$ satisfies conditions (a) and (b) of Theorem 10.2. Consequently, according to that theorem $\widehat{\beta}_n$ is strongly consistent. □

10.2.2 Asymptotic Variance

The asymptotic variance of the LSE is analyzed in this section. Consider now the following assumption. For any $\delta > 0$, there exists a positive constant c such that

$$\int_{-c}^{c} f(\lambda)\, dM_{jj}^{n}(\lambda) < \delta, \tag{10.11}$$

for every n and $j = s+1, \ldots, p$. With this additional condition the following result due to Yajima (1991) is obtained.

Theorem 10.4. *Assume that the Grenander conditions (1) to (4) are satisfied and that* (10.11) *holds. Define the* $p \times p$ *diagonal matrix*

$$D_n = \operatorname{diag}(\|\boldsymbol{x}_n(1)\|n^d, \ldots, \|\boldsymbol{x}_n(s)\|n^d, \|\boldsymbol{x}_n(s+1)\|, \ldots, \|\boldsymbol{x}_n(p)\|).$$

Then,

(*a*) *For* $s = 0$,

$$D_n \operatorname{Var}(\widehat{\boldsymbol{\beta}}_n) D_n \to 2\pi R(0)^{-1} \int_{-\pi}^{\pi} f(\lambda)\, dM(\lambda) R(0)^{-1}, \tag{10.12}$$

as $n \to \infty$ *if and only if condition* (10.11) *holds.*

(*b*) *For* $s > 0$,

$$D_n^{-1}(\boldsymbol{x}_n'\boldsymbol{x}_n) \operatorname{Var}(\widehat{\boldsymbol{\beta}}_n)(\boldsymbol{x}_n'\boldsymbol{x}_n) D_n^{-1} \to 2\pi A, \tag{10.13}$$

as $n \to \infty$ *where*

$$A = \begin{pmatrix} B & 0 \\ 0 & C \end{pmatrix}, \tag{10.14}$$

and the elements of the $s \times s$ *matrix* B *are given by*

$$b_{ij} = f_0(0) \lim_{n\to\infty} n^{-2d} \int_{-\pi}^{\pi} |1 - e^{i\lambda}|^{-2d}\, dM_{ij}^{n}(\lambda),$$

for $i, j = 1, \ldots, s$ *and the elements of the* $(p-s) \times (p-s)$ *matrix* C *are given by*

$$c_{ij} = \int_{-\pi}^{\pi} f(\lambda)\, dM_{i+s,j+s}(\lambda).$$

10.2.3 Asymptotic Normality

Assume that the regression errors $\{\varepsilon_t\}$ corresponds to a strictly stationary process with spectral density satisfying (10.2) and Wold decomposition:

$$\varepsilon_t = \sum_{j=0}^{\infty} \psi_j \nu_{t-j}. \tag{10.15}$$

Let $c_j(t_1, \ldots, t_j)$ be the $(j+1)$th order cumulant of $(\nu_t, \nu_{t+t_1}, \ldots, \nu_{t+t_j})$; see Section 1.1.10, and let $f_j(\lambda_1, \ldots, \lambda_j)$ be the $(j+1)$th order cumulant spectral density of $\{\nu_t\}$. Thus,

$$c_j(t_1, \ldots, t_j) = \int_{\Pi^j} \exp\left(i \sum_{k=1}^{j} \lambda_k t_k\right) f_j(\lambda_1, \ldots, \lambda_j) d\lambda_1 \cdots \lambda_j,$$

where $\Pi = [-\pi, \pi]$. In order to establish the asymptotic normality of the LSE, the following boundedness condition on the cumulants will be imposed:

$$\sum_{t_1, \ldots, t_j = -\infty}^{\infty} |c_j(t_1, \ldots, t_j)| < \infty. \tag{10.16}$$

The asymptotic normality of the LSE is established in the following theorem due to Yajima (1991).

Theorem 10.5. *Assume that the cumulant c_j satisfies condition* (10.16) *and that the conditions of Theorem* 10.4 *(b) hold. If*

$$\max_{|\lambda| \leq \delta} m_{jj}^n(\lambda) = o(n^{-1/2-d}), \tag{10.17}$$

for $j = s+1, \ldots, p$ for some $\delta > 0$ and

$$\int_{-\pi}^{\pi} \sqrt{m_{jj}^n(\lambda)}\, d\lambda = o(n^{-1/4-d/2}), \tag{10.18}$$

for $j = 1, \ldots, p$, then

$$D_n^{-1}(\boldsymbol{x}_n' \boldsymbol{x}_n)(\widehat{\boldsymbol{\beta}}_n - \boldsymbol{\beta}) \to N_p(0, A),$$

as $n \to \infty$ where A is the variance-covariance matrix defined by (10.14).

10.3 PROPERTIES OF THE BLUE

In order to derive a formula for the asymptotic variance of the BLUE, we shall introduce three additional conditions on the behavior of the regressors.

(5) For some $\delta > 1 - 2d$,

$$\max_{1 \leq t \leq n} \frac{\boldsymbol{x}_{tj}^2}{\|\boldsymbol{x}_n(j)\|^2} = o(n^{-\delta}),$$

for $j = 1, \ldots, p$.

(6) $0 < M_{jj}(0+) - M_{jj}(0) < 1$ for $j = 1, \ldots, s$.

(7) $0 < \int_{-\pi}^{\pi} f(\lambda)^{-1} dM(\lambda) < \infty$.

The following theorem that describes the limiting behavior of the variance of the BLUE was proved by Yajima (1991).

Theorem 10.6. *Let* $\widetilde{D}_n = \text{diag}(\|\boldsymbol{x}_n(1)\|, \ldots, \|\boldsymbol{x}_n(p)\|)$ *and assume that conditions* (1)–(7) *hold. Then,*

$$\widetilde{D}_n \operatorname{Var}(\widetilde{\boldsymbol{\beta}}_n)\widetilde{D}_n \to 2\pi \left[\int_{-\pi}^{\pi} f(\lambda)^{-1} dM(\lambda)\right]^{-1}, \tag{10.19}$$

as $n \to \infty$.

10.3.1 Efficiency of the LSE Relative to the BLUE

When $s > 0$, that is, $M(\lambda)$ displays a jump at the origin for some regressors, the LSE is not asymptotically efficient compared to the BLUE.

On the other hand, if $s = 0$ [that is, $M(\lambda)$ does not jump at zero frequency] then, under some conditions on the regressors, it can be established that the LSE is asymptotically efficient. This is stated in the next theorem due to Yajima (1991).

Theorem 10.7. *Assuming that* $s = 0$ *and conditions* (1)–(5) *and* (10.11) *hold, the LSE is efficient compared to the BLUE if and only if M increases at no more than p frequencies and the sum of the ranks of increases is p.*

■ **EXAMPLE 10.1**

As an illustration of these theorems, consider the following trigonometric regression:

$$y_t = \beta x_t + \varepsilon_t$$

where $x_t = \sum_{j=1}^{q} e^{i\lambda_j t}$, $\lambda_j \neq 0$ for $j = 1, \ldots, q$ are *known* frequencies and the error sequence has spectral density (10.2).

In this case,

$$\frac{\|\boldsymbol{x}_n\|^2}{n} = \frac{1}{n}\sum_{t=1}^{n}\left(\sum_{j=1}^{q} e^{i\lambda_j t}\right)\left(\sum_{k=1}^{q} e^{-i\lambda_k t}\right) = \sum_{j=1}^{q}\sum_{k=1}^{q}\left(\frac{1}{n}\sum_{t=1}^{n} e^{i(\lambda_j - \lambda_k)t}\right).$$

Consequently, by Lemma 10.2 we conclude that

$$\lim_{n\to\infty} \frac{\|\boldsymbol{x}_n\|^2}{n} = \sum_{j=1}^{q}\sum_{k=1}^{q} \delta(\lambda_j - \lambda_k) = q.$$

Furthermore, $dM(\lambda) = (1/q)\sum_{j=1}^{q} \delta_D(\lambda - \lambda_j)\, d\lambda$, or equivalently,

$$M(\lambda) = \frac{1}{q}\int_{-\pi}^{\lambda} \sum_{j=1}^{q} \delta_D(\omega - \lambda_j)\, d\omega.$$

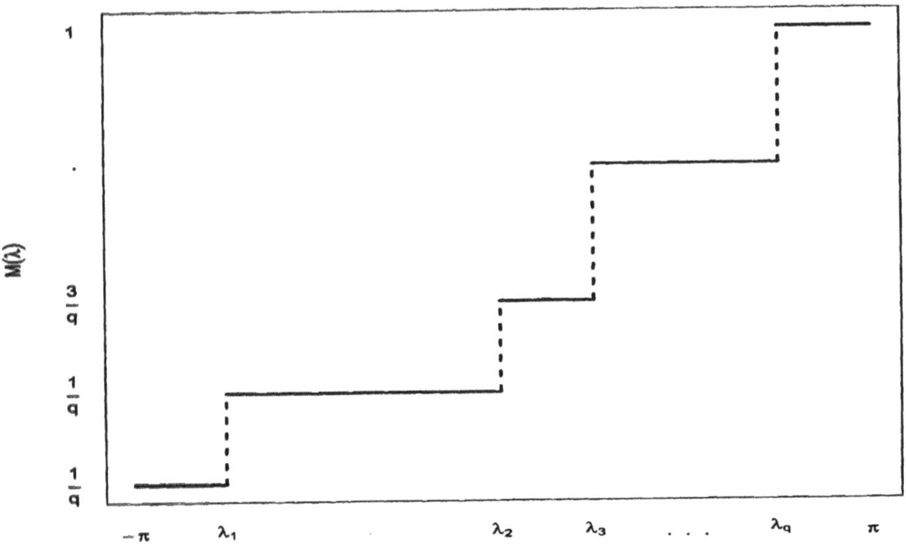

Figure 10.1 An example of trigonometric regression: $M(\lambda)$.

This function is displayed in Figure 10.1. Notice that $M(\lambda)$ exhibits q jumps located at frequencies $\lambda_1, \ldots, \lambda_q$. But, it does not jump at the origin. Therefore, $s = 0$ and $R(0) = 1$.

By Theorem 10.4(a) the asymptotic variance of the LSE satisfies

$$\begin{aligned}\lim_{n\to\infty} \|\boldsymbol{x}_n\|^2 \operatorname{Var}(\widehat{\boldsymbol{\beta}}_n) &= \frac{2\pi}{q}\int_{-\pi}^{\pi} f(\lambda)\sum_{j=1}^{q}\delta_D(\lambda-\lambda_j)\,d\lambda \\ &= \frac{2\pi}{q}[f(\lambda_1)+\cdots+f(\lambda_q)].\end{aligned}$$

On the other hand, the asymptotic variance of the BLUE is by Theorem 10.6:

$$\begin{aligned}\lim_{n\to\infty} \|\boldsymbol{x}_n\|^2 \operatorname{Var}(\widetilde{\boldsymbol{\beta}}_n) &= 2\pi q\left[\int_{-\pi}^{\pi} f(\lambda)^{-1}\sum_{j=1}^{q}\delta_D(\lambda-\lambda_j)\,d\lambda\right]^{-1} \\ &= 2\pi q\left[f(\lambda_1)^{-1}+\cdots+f(\lambda_q)^{-1}\right]^{-1}.\end{aligned}$$

The relative efficiency of the LSE compared to the BLUE is defined by

$$r(d) = \lim_{n\to\infty}\frac{\det \operatorname{Var}(\widetilde{\boldsymbol{\beta}}_n)}{\det \operatorname{Var}(\widehat{\boldsymbol{\beta}}_n)}. \tag{10.20}$$

In this case we have

$$r(q) = \frac{q^2}{[f(\lambda_1)^{-1} + \cdots + f(\lambda_q)^{-1}][f(\lambda_1) + \cdots + f(\lambda_q)]}.$$

Thus, for $q = 1$, the LSE is asymptotically efficient and by Jensen's inequality, for $q \geq 2$ the LSE is not efficient.

Recalling that in this example $p = 1$, the above conclusions agree with Theorem 10.7, since for $q = 1$, $M(\lambda)$ has only one jump at frequency λ_1 and consequently the LSE is efficient. On the other hand, for $q \geq 2$, $M(\lambda)$ has more than one jump and therefore the LSE is not efficient.

Consider, for instance, the frequencies $\lambda_j = \pi j/q$ for $j = 1, \ldots, q$. In this case we have

$$\lim_{q \to \infty} r(q) = \frac{\pi^2}{\int_0^\pi f(\lambda)\, d\lambda \int_0^\pi f(\lambda)^{-1}\, d\lambda}.$$

Figure 10.2 shows the behavior of $r(q)$ for a fractional noise model for different values of the long-memory parameter d and q.

For this model, we have that

$$\lim_{q \to \infty} r(q) = \frac{\Gamma(1-d)^2\Gamma(1+d)^2}{\Gamma(1-2d)\Gamma(1+2d)}.$$

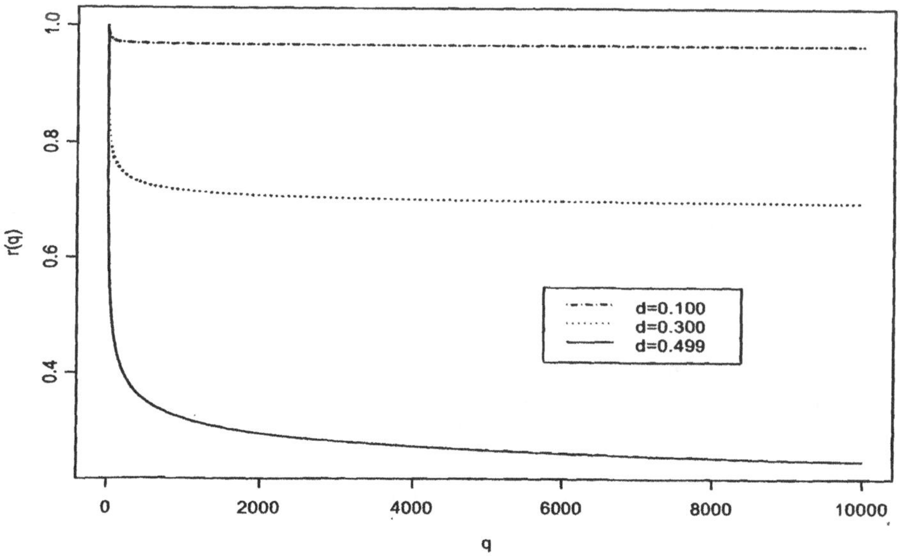

Figure 10.2 An example of trigonometric regression: $M(\lambda)$.

As expected, for $d = 0$ the asymptotic relative efficiency is 1. On the other hand, for $d \to \frac{1}{2}$, the limiting relative efficiency is 0.

10.4 ESTIMATION OF THE MEAN

A simple but illustrative example of a linear regression model is

$$y_t = \mu + \varepsilon_t$$

where μ is the unknown mean of the process y_t. In this case, the LSE is

$$\widehat{\mu} = \frac{1}{n}\sum_{t=1}^{n} y_t = \frac{1}{n}\mathbf{1}'\boldsymbol{y}_n,$$

where $\mathbf{1} = (1, 1, \ldots, 1)'$ and the BLUE is

$$\widetilde{\mu} = (\mathbf{1}'\Gamma^{-1}\mathbf{1})^{-1}\mathbf{1}'\Gamma^{-1}\boldsymbol{y}_n.$$

10.4.1 Consistency

In this case, $\sigma_n = n$ and condition (a) of Theorem 10.2 is equivalent to $n^{1-2d} \to \infty$ which is true for $d < \frac{1}{2}$. In addition, condition (b) requires that

$$\sum_{n=3}^{\infty}(n-1)^{-1}n^{2d-1}\log^2 n < \infty.$$

But,

$$\sum_{n=3}^{\infty}(n-1)^{-1}n^{2d-1}\log^2 n \leq 2\sum_{n=3}^{\infty} n^{2d-2}\log^2 n$$

$$\leq c\sum_{n=3}^{\infty} n^{2d-2+2\varepsilon} \leq c\sum_{n=3}^{\infty} n^{-\alpha-1},$$

where $\alpha = 1 - 2d - \varepsilon$ and $\varepsilon > 0$ and $c > 0$ is an appropriate constant. Consequently, for $d < \frac{1}{2}$ and ε sufficiently small we have

$$\sum_{n=3}^{\infty} n^{-\alpha-1} < \infty,$$

and (b) is satisfied. Thus, the LSE is strongly consistent.

Observe that we could also use Theorem 10.3 to establish the strong consistency of the mean. In this case $\|x_n\| = \sqrt{n}$, $R(0) = 1$ and

$$\liminf_{n\to\infty} \frac{\|x_n\|^2}{n^\delta} = \liminf_{n\to\infty} n^{1-\delta} > 0,$$

for any $\delta \leq 1$. By taking $\delta = 1$ we have $\delta > 2d$. Now, an application of Theorem 10.3 establishes that the LSE is strongly consistent.

10.4.2 Asymptotic Variance

On the other hand, the Grenander conditions are also satisfied: (1) $\|\boldsymbol{x}_n\| = \sqrt{n} \to \infty$ as $n \to \infty$; (2) $\|\boldsymbol{x}_{n+1}\|/\|\boldsymbol{x}_n\| = 1 + 1/\sqrt{n} \to 1$ as $n \to \infty$; (3) $R_{11}(h) = \lim_{n\to\infty}[1-|h|/n] = 1$ for all fixed h; and (4) $R(0) = 1$ is nonsingular. Furthermore, in this case

$$m^n(\lambda) = \frac{1}{2\pi n}\left|\sum_{t=1}^{n} e^{i\lambda t}\right|^2 = \frac{1}{2\pi n}\frac{|e^{in\lambda} - 1|^2}{|e^{i\lambda} - 1|^2} = \frac{1}{2\pi n}\frac{1 - \cos n\lambda}{1 - \cos\lambda},$$

which by Corollary 10.1 tends to the Dirac functional operator $\delta_D(\lambda)$ as $n \to \infty$. Therefore,

$$M(\lambda) = \int_{-\pi}^{\lambda} \delta_D(\omega)\, d\omega = \mathbf{1}_{[0,\pi)}(\lambda).$$

Hence, $M(\lambda)$ has a jump at the origin as shown in Figure 10.3, and therefore $s = 1$. In this case $D_n = n^{d+1/2}$ and therefore an application of Theorem 10.4(b) yields

$$n^{1-2d}\,\mathrm{Var}(\widehat{\mu}) \to 2\pi b,$$

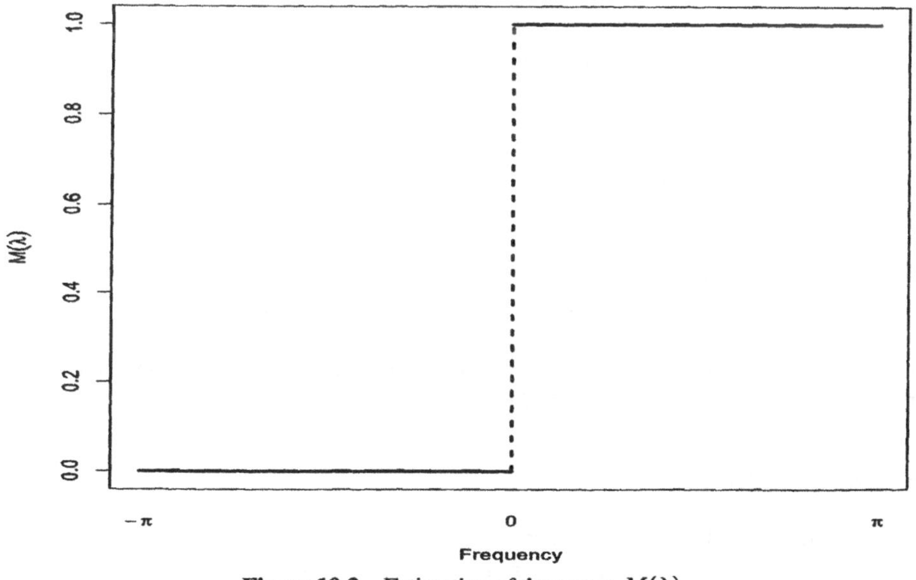

Figure 10.3 Estimation of the mean: $M(\lambda)$.

as $n \to \infty$ where

$$\begin{aligned} b &= \frac{1}{2\pi} f_0(0) \lim_{n\to\infty} n^{-1-2d} \int_{-\pi}^{\pi} |1-e^{i\lambda}|^{-2-2d} |1-e^{in\lambda}|^2 d\lambda \\ &= \frac{\Gamma(1-2d)}{\Gamma(d)\Gamma(1-d)} f_0(0) \int_0^1 \int_0^1 |x-y|^{2d-1} dx\, dy \\ &= \frac{\Gamma(1-2d)}{d(1+2d)\Gamma(d)\Gamma(1-d)} f_0(0). \end{aligned}$$

From expression (3.20), for an ARFIMA(p, d, q) model $f_0(0) = \dfrac{\sigma^2}{2\pi} \dfrac{|\theta(1)|^2}{|\phi(1)|^2}$. Consequently,

$$\operatorname{Var}(\widehat{\mu}) \sim \frac{c_\gamma n^{2d-1}}{d(1+2d)} = \sigma^2 \frac{|\theta(1)|^2}{|\phi(1)|^2} \left[\frac{\Gamma(1-2d)}{d(1+2d)\Gamma(d)\Gamma(1-d)} \right] n^{2d-1}. \quad (10.21)$$

10.4.3 Normality

Notice that

$$\begin{aligned} \int_{-\pi}^{\pi} \sqrt{m^n(\lambda)}\, d\lambda &= \frac{1}{\sqrt{2\pi n}} \int_{-\pi}^{\pi} \frac{\sqrt{1-\cos n\lambda}}{\sqrt{1-\cos\lambda}} d\lambda = \frac{2\sqrt{2}}{\sqrt{\pi n}} \int_0^{\pi/2} \left| \frac{\sin \lambda n}{\sin\lambda} \right| d\lambda \\ &= \frac{2\sqrt{2}}{\sqrt{\pi n}} \left[\int_0^{1/n} \left| \frac{\sin\lambda n}{\sin\lambda} \right| d\lambda + \int_{1/n}^{\pi/2} \left| \frac{\sin\lambda n}{\sin\lambda} \right| d\lambda \right] \\ &\leq \frac{c}{\sqrt{n}} \left[1 + \int_{1/n}^{\pi/2} \frac{1}{\lambda} d\lambda \right] \leq \frac{c}{\sqrt{n}} \left[1 + \log\left(\tfrac{\pi}{2}\right) + \log n \right]. \end{aligned}$$

Thus,

$$\int_{-\pi}^{\pi} \sqrt{m^n(\lambda)}\, d\lambda = \mathcal{O}\left(\frac{\log n}{\sqrt{n}} \right), \quad (10.22)$$

and therefore condition (10.18) is satisfied. Hence, for instance, if the noise sequence $\{v_t\}$ is independent and identically distributed, then $c_j(t_1, \ldots, t_j) = 0$ for all j; see Taniguchi and Kakizawa (2000, p. 303). Thus, condition (10.16) is satisfied and we conclude that $n^{1/2-d}(\widehat{\mu} - \mu)$ is asymptotically normal, as established by Theorem 10.5; see also Theorem 2 of Hosking (1996).

10.4.4 Relative Efficiency

Since in this case $M(\lambda)$ has a jump at zero frequency, the LSE (sample mean) is not asymptotically efficient as compared to the BLUE. However, we can analyze its relative efficiency with respect to the BLUE.

Following Adenstedt (1974), the BLUE of the mean of a fractional noise process with long-memory parameter d is given by

$$\widetilde{\mu} = \sum_{j=1}^{n} \alpha_j y_j,$$

where the coefficients α_j are

$$\alpha_j = \binom{n-1}{j-1} \frac{\Gamma(j-d)\Gamma(n+1-j-d)\Gamma(2-2d)}{\Gamma(n+1-2d)\Gamma(1-d)^2}.$$

The variance of this BLUE may be written as

$$\mathrm{Var}(\widetilde{\mu}) = \sigma^2 \frac{\Gamma(n)\Gamma(1-2d)\Gamma(2-2d)}{\Gamma(n+1-2d)\Gamma(1-d)^2} \sim \sigma^2 n^{2d-1} \frac{\Gamma(1-2d)\Gamma(2-2d)}{\Gamma(1-d)^2},$$

for large n. For an ARFIMA(p,d,q) process, the variance of the BLUE satisfies

$$\mathrm{Var}(\widetilde{\mu}) \sim \sigma^2 n^{2d-1} \frac{|\theta(1)|^2}{|\phi(1)|^2} \frac{\Gamma(1-2d)\Gamma(2-2d)}{\Gamma(1-d)^2}, \tag{10.23}$$

for large n. Now, from (10.21) and (10.23) we conclude that

$$r(d) = \lim_{n\to\infty} \frac{\mathrm{Var}(\widetilde{\mu})}{\mathrm{Var}(\widehat{\mu})} = \frac{(1+2d)\Gamma(1+d)\Gamma(2-2d)}{\Gamma(1-d)}.$$

Figure 10.4 shows the relative efficiency $r(d)$ for $d \in (0, \frac{1}{2})$. The minimal relative efficiency is approximately 0.9813 reached at $d = 0.318$.

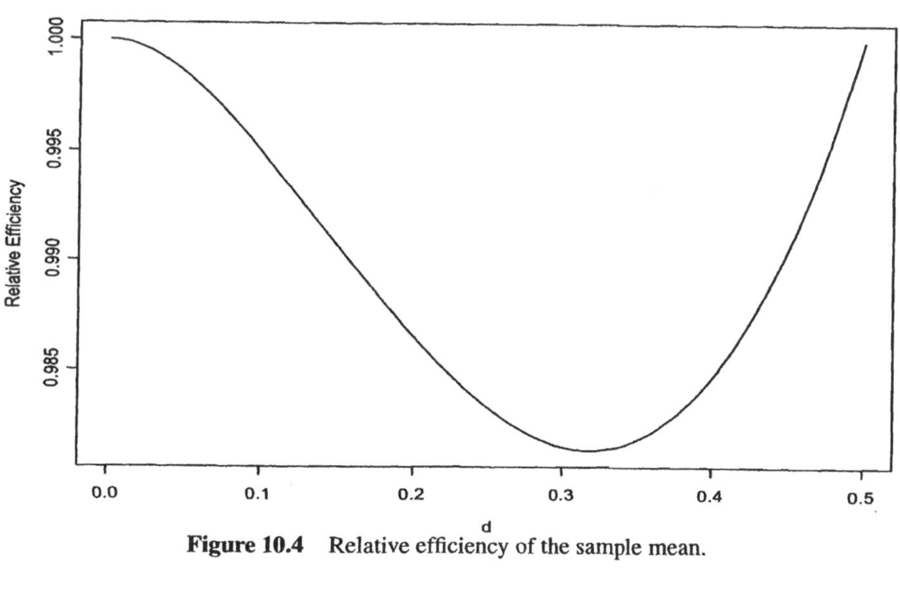

Figure 10.4 Relative efficiency of the sample mean.

10.5 POLYNOMIAL TREND

Extending the previous example, consider the polynomial regression

$$y_t = \beta_0 + \beta_1 t + \cdots + \beta_q t^q + \varepsilon_t,$$

where $\{\varepsilon_t\}$ satisfies (10.2). In this case, $p = q + 1$, $\boldsymbol{x}_n(i) = (1, 2^{i-1}, \ldots, n^{i-1})$ for $i = 1, \ldots, p$ and hence

$$\boldsymbol{x}_n(i)'\boldsymbol{x}_n(j) = \sum_{t=1}^{n} t^{i+j-2} = \frac{n^{i+j-1}}{i+j-1}[1 + o(n)],$$

for $i, j = 1, \ldots, p$.

Thus, $D_{ii} = \dfrac{n^{2i-1+d}}{2i-1}[1 + o(n)]$ and

$$m_{ij}^n(\lambda) = \frac{\sum_{t=1}^n \sum_{s=1}^n t^{i-1} s^{j-1} e^{i(s-t)\lambda}}{2\pi \|\boldsymbol{x}_n(i)\| \|\boldsymbol{x}_n(j)\|}.$$

By Lemma 10.1, this term converges weakly to

$$m_{ij}(\lambda) = \frac{\sqrt{2i-1}\sqrt{2j-1}}{i+j-1}\delta_D(\lambda).$$

Therefore,

$$\begin{aligned} M_{ij}(\lambda) &= \int_{-\pi}^{\lambda} \frac{\sqrt{2i-1}\sqrt{2j-1}}{i+j-1}\delta_D(\lambda)\, d\lambda \\ &= \frac{\sqrt{2i-1}\sqrt{2j-1}}{i+j-1}\mathbf{1}_{(0,\pi)}(\lambda), \end{aligned}$$

and

$$R_i j(h) = \int_{-\pi}^{\pi} e^{i\lambda h}\, dM_{ij}(\lambda) = \frac{\sqrt{2i-1}\sqrt{2j-1}}{i+j-1},$$

which actually does not depend on the lag h. For instance, for $p = 5$ we have that the matrix $R(h)$ is given by

$$R(h) = \begin{bmatrix} 1 & \frac{\sqrt{3}}{2} & \frac{\sqrt{5}}{3} & \frac{\sqrt{7}}{4} & \frac{\sqrt{9}}{5} \\ \frac{\sqrt{3}}{2} & 1 & \frac{\sqrt{15}}{4} & \frac{\sqrt{21}}{5} & \frac{\sqrt{27}}{6} \\ \frac{\sqrt{5}}{3} & \frac{\sqrt{15}}{4} & 1 & \frac{\sqrt{35}}{6} & \frac{\sqrt{45}}{7} \\ \frac{\sqrt{7}}{4} & \frac{\sqrt{21}}{5} & \frac{\sqrt{35}}{6} & 1 & \frac{\sqrt{63}}{8} \\ \frac{\sqrt{9}}{5} & \frac{\sqrt{27}}{6} & \frac{\sqrt{45}}{7} & \frac{\sqrt{63}}{8} & 1 \end{bmatrix}.$$

10.5.1 Consistency

Notice that

$$\frac{\|x_n(j)\|^2}{n^\delta} = \frac{n^{2j-1-\delta}}{2j-1}[1+o(n)].$$

Therefore, by taking $\delta = 1$ we have that

$$\liminf_{n\to\infty} \frac{\|x_n(j)\|^2}{n^\delta} > 0.$$

Furthermore, since $R(0)$ is nonsingular, we conclude by Theorem 10.3 that the LSE is strongly consistent.

10.5.2 Asymptotic Variance

Since $M(\lambda)$ has a jump at the origin, by virtue of Theorem 10.4(b) the asymptotic variance of $\widehat{\beta}_n$ satisfies

$$D_n^{-1}(x_n' x_n)\,\mathrm{Var}(\widehat{\beta}_n)(x_n' x_n)D_n^{-1} \to 2\pi B,$$

where

$$\begin{aligned} b_{ij} &= f_0(0) \lim_{n\to\infty} n^{-2d} \int_{-\pi}^{\pi} |1-e^{i\lambda}|^{-2d} m_{ij}^n(\lambda)\, d\lambda \\ &= f_0(0) \lim_{n\to\infty} \frac{n^{-2d}}{\|x_n(i)\|\|x_n(j)\|} \sum_{t=1}^{n}\sum_{s=1}^{n} t^{i-1}s^{j-1} \int_{-\pi}^{\pi} \frac{|1-e^{i\lambda}|^{-2d}}{2\pi} e^{i(t-s)\lambda}\, d\lambda \\ &= f_0(0) \lim_{n\to\infty} \frac{n^{-2d}}{\|x_n(i)\|\|x_n(j)\|} \sum_{t=1}^{n}\sum_{s=1}^{n} t^{i-1}s^{j-1}\gamma(t-s), \end{aligned}$$

where $\gamma(\cdot)$ is the ACF of a fractional noise process FN(d) with $\sigma^2 = 1$. Therefore,

$$\begin{aligned} b_{ij} &= f_0(0)c_\gamma \lim_{n\to\infty} \frac{n^{-2d}}{\|x_n(i)\|\|x_n(j)\|} \sum_{t=1}^{n}\sum_{s=1}^{n} t^{i-1}s^{j-1}|t-s|^{2d-1} \\ &= f_0(0)c_\gamma \sqrt{(2i-1)(2j-1)} \\ &\times \lim_{n\to\infty} n^{-2d} \sum_{t=1}^{n}\sum_{s=1}^{n} \left(\frac{t}{n}\right)^{i-1} \left(\frac{s}{n}\right)^{j-1} \left|\frac{t}{n}-\frac{s}{n}\right|^{2d-1} \frac{1}{n^2} \\ &= f_0(0)c_\gamma \sqrt{(2i-1)(2j-1)} \int_0^1\int_0^1 x^{i-1}y^{j-1}|x-y|^{2d-1}\, dx\, dy \\ &= f_0(0)c_\gamma \sqrt{(2i-1)(2j-1)} \frac{B(i,2d)+B(j,2d)}{i+j+2d-1}, \end{aligned}$$

see Problem 10.6 for finding the value of the double integral.

10.5.3 Normality

Analogously to (10.22), we can deduce that

$$\int_{-\pi}^{\pi} \sqrt{m_{jj}^{n}(\lambda)}\, d\lambda = \mathcal{O}(n^{-1/2} \log n),$$

so that condition (10.18) is satisfied. If the noise sequence $\{v_t\}$ is independent and identically distributed, then the cumulants satisfy $c_j(t_1, \ldots, t_j) = 0$ for all j; see Taniguchi and Kakizawa (2000, p. 303). Thus, condition (10.16) is also satisfied and we conclude that by Theorem 10.5, $\widehat{\beta}_n$ is asymptotically normally distributed.

10.5.4 Relative Efficiency

Analogous to the estimation of the mean case, since $M(\lambda)$ has a jump at the origin, the LSE of the polynomial regression is not asymptotically efficient. The relative efficiency for $q = 1$ is given by

$$r(d) = (9 - 4d^2)\left[\frac{(1+2d)\Gamma(1+d)\Gamma(3-2d)}{6\Gamma(2-d)}\right]^2.$$

Figure 10.5 shows $r(d)$ for $d \in (0, \frac{1}{2})$. The minimal relative efficiency is $\frac{8}{9}$ reached at $d = 0.5$.

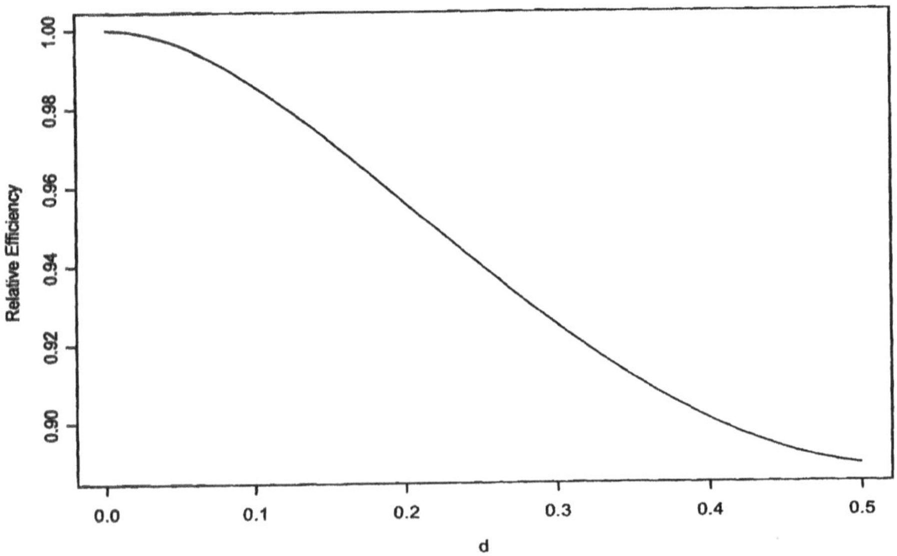

Figure 10.5 Relative efficiency of the polynomial regression with $q = 1$.

10.6 HARMONIC REGRESSION

Another important example is the harmonic regression

$$y_t = \boldsymbol{\beta}_1 e^{i\lambda_1 t} + \boldsymbol{\beta}_2 e^{i\lambda_2 t} + \cdots + \boldsymbol{\beta}_q e^{i\lambda_q t} + \varepsilon_t,$$

where $\{\varepsilon_t\}$ is a stationary process with spectral density (10.2) and $\lambda_j \neq 0$ for $j = 1, \ldots, q$ are known frequencies. In this case, $D_{ii} = \sqrt{n}$ for $i = 1, \ldots, q$, and by Corollary 10.1 we have that

$$\frac{\langle \boldsymbol{x}_{n,h}(i), \boldsymbol{x}_n(j)\rangle}{\|\boldsymbol{x}_{n,h}(i)\| \|\boldsymbol{x}_n(j)\|} = e^{i\lambda_j h}\frac{1}{n}\sum_{t=1}^{n} e^{i(\lambda_i - \lambda_j)t} \to e^{i\lambda_j h}\delta(\lambda_j - \lambda_i),$$

as $n \to \infty$. Hence, $R_{ij}(h) = e^{i\lambda_j h}\delta(\lambda_j - \lambda_i)$ and from equation (10.9) we conclude that $dM_{ij}(\lambda) = \delta(\lambda_j - \lambda_i)\delta_D(\lambda - \lambda_j)\, d\lambda$. Therefore, $M_{ij}(\lambda)$ does not have a jump at the origin for any $\lambda_j \neq 0$.

10.6.1 Consistency

Observe that $R(0) = I_n$ where I_n is the $n \times n$ identity matrix. Consequently, it satisfies the Grenander condition (4), that is, $R(0)$ is nonsingular. Besides,

$$\frac{\|\boldsymbol{x}_n(j)\|^2}{n^\delta} = n^{1-\delta}.$$

Therefore, by taking $\delta = 1$ in Theorem 10.3 we conclude that the LSE is strongly consistent.

10.6.2 Asymptotic Variance

Since $R_{ij}(0) = \delta(i - j)$ and $D_{ii} = \sqrt{n}$ we have that

$$\int_{-\pi}^{\pi} f(\lambda)\, dM_{ij}(\lambda) = \int_{-\pi}^{\pi} f(\lambda)\delta_D(\lambda - \lambda_j)\, d\lambda = f(\lambda_j)\delta(i - j).$$

Now, an application of Theorem 10.4(a) yields

$$\lim_{n\to\infty} n \operatorname{Var}(\widehat{\boldsymbol{\beta}}_n) = 2\pi \begin{bmatrix} f(\lambda_1) & 0 & 0 & \cdots & 0 \\ 0 & f(\lambda_2) & 0 & \cdots & 0 \\ \vdots & & \ddots & & \vdots \\ 0 & \cdots & 0 & f(\lambda_{q-1}) & 0 \\ 0 & \cdots & 0 & 0 & f(\lambda_q) \end{bmatrix}.$$

10.6.3 Normality

Observe that for the harmonic regression we have

$$m_{jj}^n(\lambda) = \frac{1}{2\pi n}\left|\sum_{t=1}^{n} e^{i(\lambda - \lambda_j)t}\right|^2 = \frac{1}{2\pi n}\left|\frac{\sin\frac{n(\lambda-\lambda_j)}{2}}{\sin\frac{(\lambda-\lambda_j)}{2}}\right|^2.$$

Thus, for small $\delta > 0$ and $\lambda_j \neq 0$, $m_{jj}^n(\lambda)$ is bounded by

$$\max_{\lambda \leq \delta} m_{jj}^n(\lambda) \leq \frac{c}{n},$$

for some positive constant c. Therefore, $m_{jj}^n(\lambda)$ satisfies condition (10.17) for any $d \in (0, \frac{1}{2})$.

On the other hand, by an argument analogous to Subsection 10.4.3 for obtaining (10.22), we have

$$\int_{-\pi}^{\pi} \sqrt{m_{jj}^n(\lambda)} d\lambda = \mathcal{O}\left(\frac{\log n}{\sqrt{n}}\right).$$

Consequently, condition (10.18) holds for this harmonic regression. Furthermore, if $\{\nu_t\}$ is a strict white noise sequence, condition (10.16) is also satisfied. Therefore, by virtue of Theorem 10.5, the LSE estimator $\widehat{\beta}_n$ is asymptotically normal.

10.6.4 Efficiency

As shown in Figure 10.6, $M_{jj}(\lambda)$ displays one jump of size one at frequency λ_j. Thus, $M(\lambda)$ increases at q frequencies each of rank one; so according to Theorem 10.7, the LSE is asymptotically efficient.

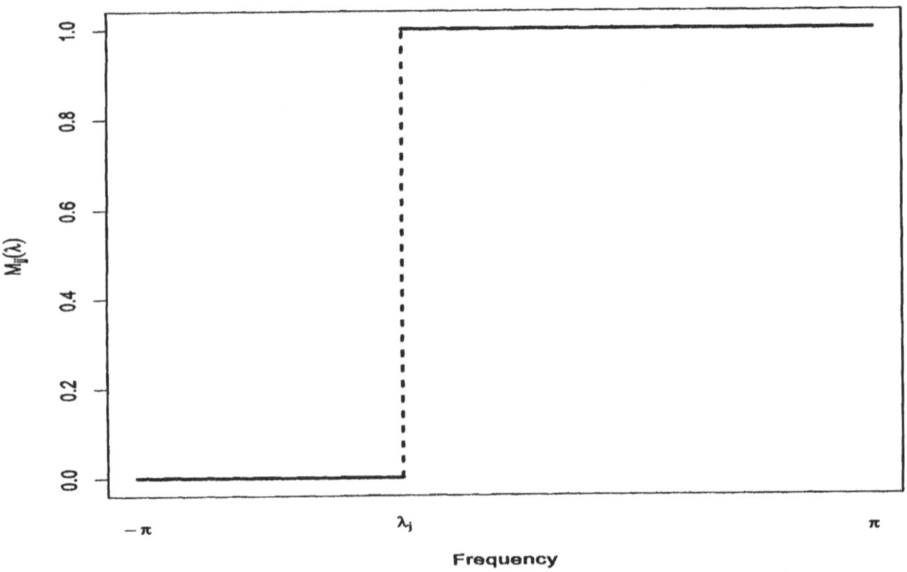

Figure 10.6 Harmonic regression: $M_{jj}(\lambda)$.

10.7 ILLUSTRATION: AIR POLLUTION DATA

As an illustration of the long-memory regression techniques discussed in this chapter consider the following air pollution data exhibited in Figure 10.7. This time series consists of 4014 daily observations of fine particulate matter with diameter less than 2.5 μm (PM2.5) measured in Santiago, Chile, during the period 1989–1999; see Section 10.8 for further details about these data.

In order to stabilize the variance of these data, a logarithmic transformation has been made. The resulting series is shown in Figure 10.8. This series displays a clear seasonal component and a possible downward linear trend. Consequently, the following model for the log-PM25 is proposed,

$$y_t = \beta_0 + \beta_1 t + \sum_{j=1}^{k} [a_j \sin(\omega_j t) + c_j \cos(\omega_j t)] + \varepsilon_t, \tag{10.24}$$

where $\varepsilon_t \sim (0, \sigma_\varepsilon^2)$.

An analysis of the periodogram of the detrended data and the ACF reveals the presence of three plausible seasonal frequencies, $\omega_1 = 2\pi/7$, $\omega_2 = 2\pi/183$, and $\omega_3 = 2\pi/365$.

The least squares fitting assuming uncorrelated errors is shown in Table 10.1. Observe that according to this table all the regression coefficients in model (10.24) are significant at the 5% level, excepting c_1. In particular, even though the LS estimate

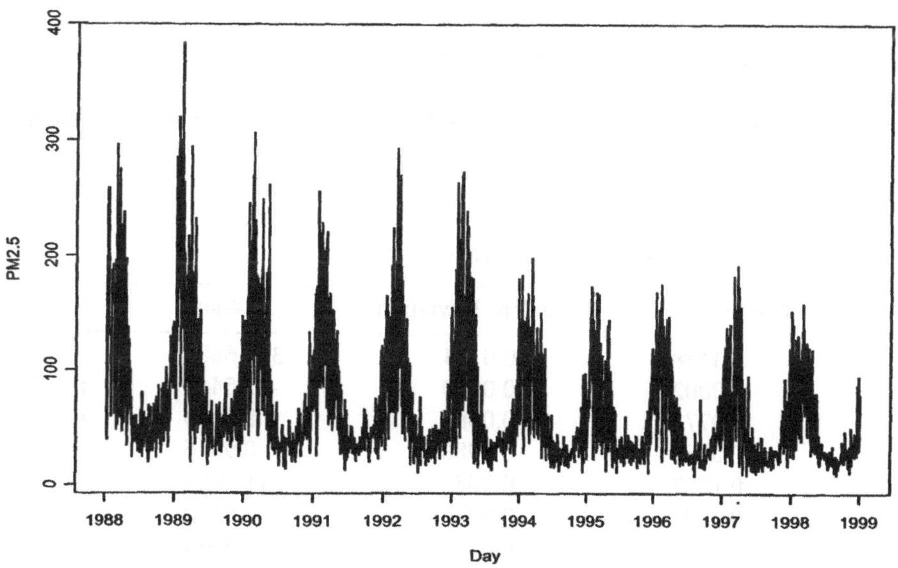

Figure 10.7 Air pollution data: Particulate matter with diameter less than 2.5 μm (PM2.5).

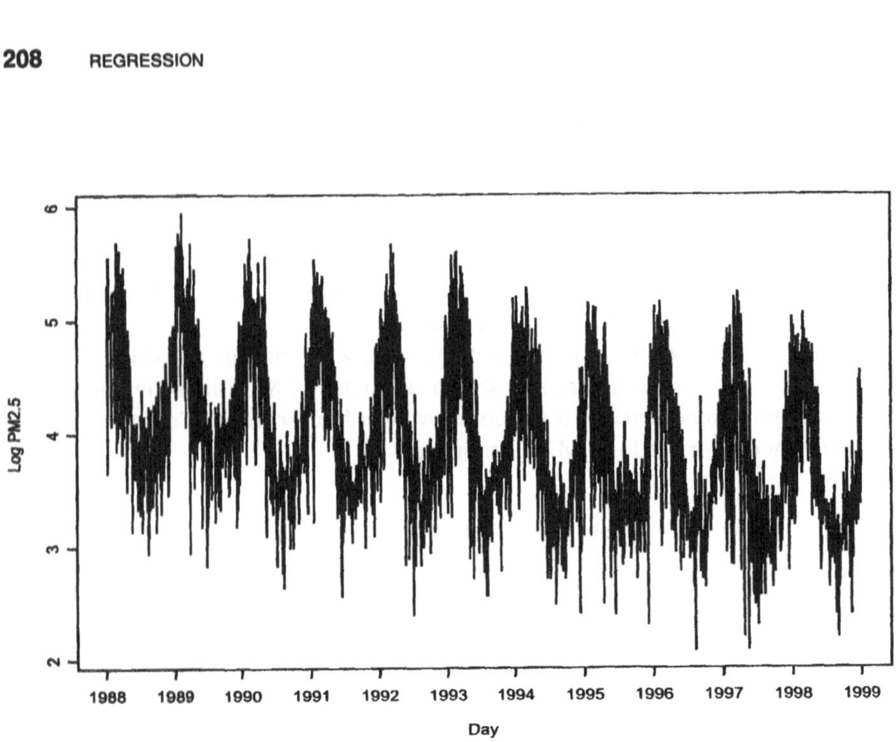

Figure 10.8 Air pollution data: Log PM2.5.

of the linear trend coefficient β_1 is very small, $\widehat{\beta}_1 = -0.0002$, it is significant at the 5% level. The sample autocorrelation function of the residuals from the LSE fit is shown in Figure 10.9. As observed in this plot, the components of the autocorrelation function are significant even after 30 lags. Additionally, the variance plot (see Subsection 4.5.3) displayed in Figure 10.10 indicates the possible presence of long-range dependence in the data. As a result from these two plots, it seems that the disturbances ε_t in the linear regression model (10.24) may have long-memory correlation structure and the LSE fitting may not be adequate.

Table 10.1 Air Pollution Data: Least Squares Fit

	Coefficient	Standard Deviation	t-stat	P-value
β_0	4.3148	0.0124	347.6882	0.0000
β_1	-0.0002	0.0000	-35.5443	0.0000
a_1	0.0775	0.0088	8.8556	0.0000
c_1	-0.0083	0.0088	-0.9446	0.3449
a_2	0.1007	0.0087	11.5086	0.0000
c_2	-0.0338	0.0088	-3.8590	0.0001
a_3	0.4974	0.0088	56.6914	0.0000
c_3	0.4479	0.0088	51.1620	0.0000

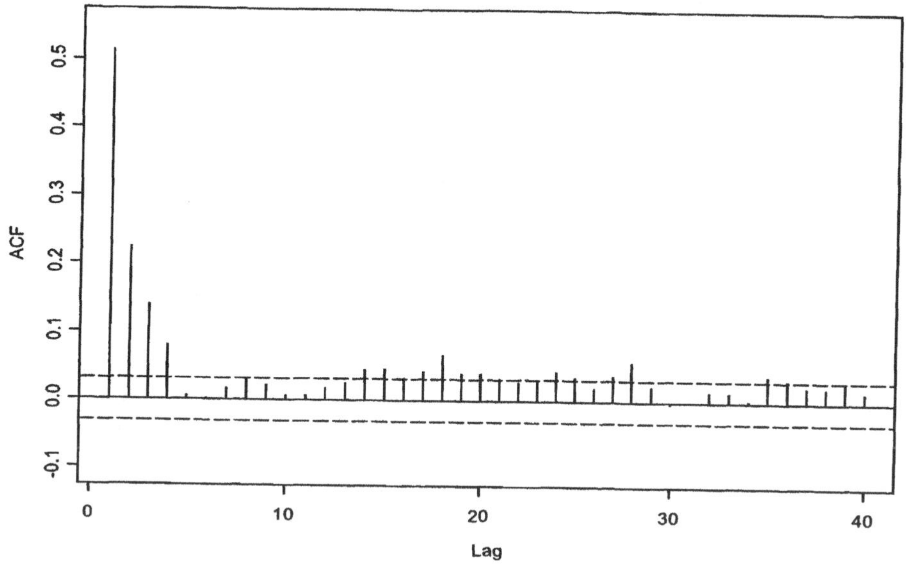

Figure 10.9 Air pollution data: Sample autocorrelation function of the residuals from the least squares fit.

To account for the possible long-memory behavior of the errors, the following ARFIMA(p, d, q) model is proposed for the regression disturbances $\{\varepsilon_t\}$:

$$\phi(B)\varepsilon_t = \theta(B)(1 - B)^{-d}\eta_t,$$

where $\{\eta_t\}$ is a white noise sequence with variance σ^2. The model selected according to the Akaike's information criterion (AIC) is the ARFIMA$(0, d, 0)$, with $\widehat{d} = 0.4252$, $t_d = 34.37$, and $\widehat{\sigma}_\eta = 0.3557$. Table 10.2 shows the results from the least squares fit

Table 10.2 Air Pollution Data: Least Squares Fit with Long-Memory Errors

	Coefficient	Standard Deviation	t-stat	P-value
β_0	4.3148	0.5969	7.2282	0.0000
β_1	-0.0002	0.0003	-0.7405	0.4590
a_1	0.0775	0.0084	9.1837	0.0000
c_1	-0.0083	0.0084	-0.9803	0.3270
a_2	0.1007	0.0363	2.7713	0.0056
c_2	-0.0338	0.0335	-1.0094	0.3129
a_3	0.4974	0.0539	9.2203	0.0000
c_3	0.4479	0.0447	10.0267	0.0000

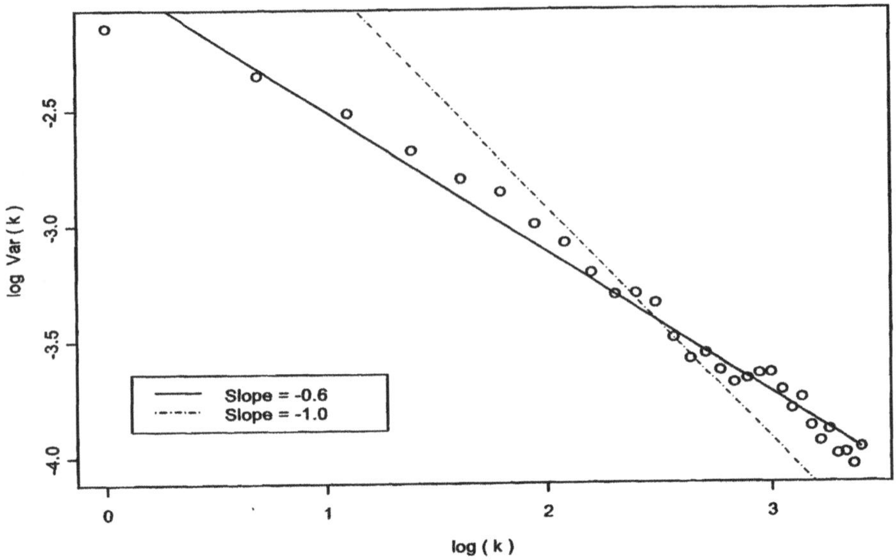

Figure 10.10 Air pollution data: Variance plot of the residuals from the LSE fit.

with ARFIMA errors. The standard deviations of the regression coefficients and the t tests have been calculated following Theorem 10.4 and Theorem 10.5. From this table, observe that the linear trend coefficient β_1 is no longer significant at the 5% level. Similarly, the coefficients c_1 and c_2 are not significant at that level.

10.8 BIBLIOGRAPHIC NOTES

Chapter 7 of Grenander and Rosenblatt (1957), Section VII.4 of Ibragimov and Rozanov (1978), and Chapter VIII of Hannan (1970) are excellent references for the analysis of regression with correlated errors. However, most of their results do not apply directly to strongly dependent processes. On the other hand, there is an extensive literature about regression data with long-memory disturbances; see, for example, Künsch (1986), Yajima (1988, 1991), Dahlhaus (1995), Sibbertsen (2001), and Choy and Taniguchi (2001), among others. Yajima (1991) established many asymptotic results for least squares error estimates and best linear unbiased estimates. As described in this chapter and according to Yajima, the convergence rates for the variance of these estimates as the sample size increases, depend on the structure of the characteristic function of the design matrix. The Grenander conditions were introduced by Grenander (1954). All the theorems stated in this chapter are proved in Yajima (1988, 1991). The problem of random design matrix in the context of analysis of variance is studied by Künsch, Beran, and Hampel (1993). Stochastic regressors are further investigated by Robinson and Hidalgo (1997) where the components of the design matrix follow a stationary process. Technical aspects about the weak convergence of characteristic functions can be found, for example, in Chapter IX of Kawata (1972).

The air pollution data discussed in Section 10.7 are provided by the Ambient Air Quality Monitoring Network (MACAM in Spanish, www.sesma.cl) in Santiago, Chile. Further details about the fine particulate matter data and other ambient variables can be found in Iglesias, Jorquera, and Palma (2006).

Problems

10.1 Assume that x_j denotes the jth column of the $n \times p$ design matrix $\boldsymbol{x}_n$, that is, $x_{tj} = t^{j-1}$ and let $D_n = \text{diag}(\|x_1\|_n, \|x_2\|_n, \dots, \|x_p\|_n)$ where $\|x_j\|_n = [\sum_{t=1}^n x_{tj}^2]^{1/2}$.

a) Show that

$$\lim_{n\to\infty} D_n^{-1}\boldsymbol{x}_n'\boldsymbol{x}_n D_n^{-1} = M,$$

where $M = (m_{ij})_{i,j=1,\dots,p}$ with $m_{ij} = \dfrac{\sqrt{2i-1}\sqrt{2j-1}}{i+j-1}$.

b) Verify that

$$\lim_{n\to\infty} n^{-2d} D_n^{-1}\boldsymbol{x}_n'\Gamma\boldsymbol{x}_n D_n^{-1} = 2\pi f_0(0) H,$$

where

$$h_{ij} = \frac{\sqrt{2i-1}\sqrt{2j-1}\,\Gamma(1-2d)}{\Gamma(d)\Gamma(1-d)} \int_{-1}^{1}\int_{-1}^{1} x^{i-1}y^{j-1}|x-y|^{2d-1}\,dx\,dy.$$

10.2 Consider the harmonic regression

$$y_t = \alpha_1 \sin(\lambda_0 t) + \alpha_2 \cos(\lambda_0 t) + \varepsilon_t,$$

for $t = 1, \dots, n$ where ε_t is a stationary long-memory process with spectral density (10.2).

a) Show that y_t may be written as

$$y_t = \beta_1 e^{i\lambda_0 t} + \beta_2 e^{-i\lambda_0 t} + \varepsilon_t,$$

and find expressions for α_1 and α_2 in terms of β_1 and β_2.

b) Verify that the LSE of α_1 and α_2 are

$$\widehat{\alpha}_1 = \frac{1}{2i}(i\widehat{\beta}_2 + \widehat{\beta}_1),$$

and

$$\widehat{\alpha}_2 = \frac{1}{2i}(i\widehat{\beta}_2 - \widehat{\beta}_1),$$

where $\widehat{\beta}_1$ and $\widehat{\beta}_2$ are the LSE of β_1 and β_2, respectively.

c) Show that

$$\lim_{n\to\infty} n \operatorname{Var}(\widehat{\alpha}_1) = \lim_{n\to\infty} n \operatorname{Var}(\widehat{\alpha}_2) = \pi f(\lambda_0).$$

10.3 Consider the linear regression model

$$y_t = \beta t^p e^{i\lambda_0 t} + \varepsilon_t,$$

where $p \in \{0, 1, 2, \ldots\}$ and ε_t is a stationary long-memory process with spectral density (10.2).

a) Is the LSE of β strongly consistent?

b) Are the Grenander conditions satisfied in this case?

c) Is the LSE asymptotically normal?

d) Is the LSE asymptotically efficient in this case?

10.4 Let $h(\lambda) \geq 0$ and define the $n \times n$ matrix Γ with elements

$$\gamma_{ij} = \int_{-\pi}^{\pi} h(\lambda) e^{i(i-j)\lambda} d\lambda.$$

a) Verify that Γ is Hermitian, that is, $\gamma_{ij} = \overline{\gamma_{ji}}$.

b) Show that Γ is positive semidefinite, that is, for any $x \in \mathbb{C}^n$

$$x^* \Gamma x \geq 0.$$

c) Observe that since Γ is Hermitian, it may be written as

$$\Gamma = UDU^*,$$

where U is a nonsingular $n \times n$ matrix and $D = \operatorname{diag}(d_i)$ with $d_i \geq 0$ for $i = 1, \ldots, n$. Using this fact and the Cauchy-Schwartz inequality

$$|x^* y| \leq \|x\| \|y\|,$$

show that for all $u, v \in \mathbb{C}^n$

$$2|u^* \Gamma v| \leq u^* \Gamma u + v^* \Gamma v. \tag{10.25}$$

d) Verify that

$$\int_{-\pi}^{\pi} 2\pi h(\lambda) m_{ij}^n(\lambda)\, d\lambda = \sum_{t=1}^{n} \sum_{s=1}^{n} \frac{\langle x_{ti}, x_{sj} \rangle}{\|x_n(i)\| \|x_n(j)\|} \gamma_{ij} = u^* \Gamma v$$

where $u = x_n(i)/\|x_n(i)\|$ and $v = x_n(j)/\|x_n(j)\|$.

e) Using (10.25) show that

$$\left| \int_{-\pi}^{\pi} h(\lambda) m_{ij}^n(\lambda)\, d\lambda \right| \leq \int_{-\pi}^{\pi} h(\lambda) m_{ii}^n(\lambda)\, d\lambda + \int_{-\pi}^{\pi} h(\lambda) m_{jj}^n(\lambda)\, d\lambda.$$

f) Deduce that for any $f(\lambda) \geq 0$,

$$\left| \int_A f(\lambda)\, dM_{ij}(\lambda) \right| \leq \int_A f(\lambda)\, dM_{ii}(\lambda) + \int_A f(\lambda)\, dM_{jj}(\lambda),$$

cf., Yajima (1991, p. 162).

10.5 Consider the following *trend break* regression

$$y_t = \beta_0 + \beta_1 x_t + \varepsilon_t,$$

for $t = 1, \dots, n$, where the covariate x_t is defined by

$$x_t = \begin{cases} 0 & \text{if } \ t \leq \dfrac{n}{2}, \\ 1 & \text{if } \ \dfrac{n}{2} < t \leq n. \end{cases}$$

and for simplicity we assume that n is even. Let $\boldsymbol{x}_n - [\boldsymbol{x}_n(1), \boldsymbol{x}_n(2)]$ the design matrix.

a) Show that $\|\boldsymbol{x}_n(1)\| = \sqrt{n}$ and $\|\boldsymbol{x}_n(2)\| = \sqrt{n/2}$.

b) Following expression (10.8) verify that

$$\frac{\langle \boldsymbol{x}_{n,h}(1), \boldsymbol{x}_n(2) \rangle}{\|\boldsymbol{x}_{n,h}(1)\| \|\boldsymbol{x}_n(2)\|} \to \frac{1}{\sqrt{2}},$$

as $n \to \infty$ and that

$$R(h) = \begin{bmatrix} 1 & \dfrac{1}{\sqrt{2}} \\ \dfrac{1}{\sqrt{2}} & 1 \end{bmatrix}.$$

c) Are the Grenander conditions fulfilled in this case?

d) Prove that the matrix $M(\lambda)$ may be written as

$$M(\lambda) = \delta_D(\lambda) \begin{bmatrix} 1 & \dfrac{1}{\sqrt{2}} \\ \dfrac{1}{\sqrt{2}} & 1 \end{bmatrix}.$$

e) Let $\widehat{\boldsymbol{\beta}}_n$ be the LSE of $\boldsymbol{\beta} = (\boldsymbol{\beta}_1, \boldsymbol{\beta}_2)'$. Is this estimator consistent?

f) Find an expression for the asymptotic variance of $\widehat{\boldsymbol{\beta}}_n$.

10.6 Show that

$$\int_0^1 \int_0^1 x^\alpha y^\beta |x - y|^\gamma \, dx\, dy = \frac{B(\alpha + 1, \gamma + 1) + B(\beta + 1, \gamma + 1)}{\alpha + \beta + \gamma + 2},$$

for $\alpha > -1$, $\beta > -1$ and $\gamma > -1$, where $B(\cdot, \cdot)$ is the beta function. Hint: Try the change of variables $x = uv$ and $y = v$.

10.7 Consider the liner regression model

$$y_t = \beta t^\alpha + \varepsilon_t,$$

for $t = 1, 2, \ldots, n$, where α is known and $\{\varepsilon_t\}$ is a long-memory stationary process with spectral density satisfying

$$f(\lambda) \sim c_f |\lambda|^{-2d},$$

as $|\lambda| \to 0$ with $0 < d < \frac{1}{2}$. Let $\widehat{\boldsymbol{\beta}}_n$ be the LSE of $\boldsymbol{\beta}$ and let $\boldsymbol{x}_n = (1, 2^\alpha, \ldots, n^\alpha)'$.

a) Verify that $\|\boldsymbol{x}_n\|^2 \sim \dfrac{n^{2\alpha+1}}{2\alpha+1}$, for $\alpha > -\frac{1}{2}$.

b) Show that if $\alpha > d - \frac{1}{2}$, then $\widehat{\boldsymbol{\beta}}_n$ is strongly consistent.

c) Prove that

$$\frac{\langle \boldsymbol{x}_{n,h}, \boldsymbol{x}_n \rangle}{\|\boldsymbol{x}_{n,h}\| \|\boldsymbol{x}_n\|} \to R(h),$$

as $n \to \infty$, where $R(h) = 1$ for all $h \in \mathbb{Z}$.

d) Are the Grenander conditions satisfied in this case?

e) Show that the variance of $\widehat{\boldsymbol{\beta}}_n$ satisfies

$$\mathrm{Var}(\widehat{\boldsymbol{\beta}}_n) \sim \frac{4\pi c_f (2\alpha+1)^2}{\Gamma(d)\Gamma(1-d)} \frac{B(\alpha+1, 2d)}{2\alpha+2d+1} n^{2d-1-2\alpha},$$

as $n \to \infty$.

f) Assume that the disturbances $\{\varepsilon_t\}$ are independent and identically distributed. Is the LSE $\widehat{\boldsymbol{\beta}}_n$ asymptotically normal?

CHAPTER 11

MISSING DATA

The methodologies studied so far assume complete data. However, in many practical situations only part of the data may be available. As a result, in these cases we are forced to fit models and make statistical inferences based on partially observed time series. In this chapter we explore the effects of data gaps on the analysis of long-memory processes and describe some methodologies to deal with this situation. As in many other areas of statistics, there are several ways to deal with incomplete time series data. A fairly usual approach, especially when there are only a small number of missing values, is to replace them by zero or the sample mean of the series. Another approach is to use some imputation technique such as cubic splines and other interpolation methods, including, for example, the repetition of previous values. In this chapter we discuss some of these techniques. Special attention is focused on the method of integrating out the missing values from the likelihood function. As discussed in Section 11.1, this approach does not introduce artificial correlation in the data and avoids some dramatic effects on parameter estimates produced by some imputation techniques.

The definition of an appropriate likelihood function to account for unobserved values is studied in Section 11.2. In particular, it is shown that if the distribution of the location of the missing data is independent of the distribution of the time series, then the statistical inferences can be carried out by ignoring the missing observations. This section also discusses the computation of the likelihood function for a general class of time series models and modifications to the Kalman filter equations to account for missing values. Sections 11.3 and 11.4 are dedicated to investigate the effects of data gaps on estimates and predictions, respectively, establishing some general theoretical results. A number of illustrations of these results are discussed in Section 11.5. Estimation of missing values via interpolation and Bayesian imputation is discussed in Section 11.6 while suggestions for further reading on this topic are given in Section 11.7. This chapter concludes with a list of proposed problems.

11.1 MOTIVATION

Missing data is an important problem for the analysis of time series. In particular, the parameter estimates may suffer serious distortions if the missing values are not appropriately treated. This point is illustrated by Table 11.1 which displays the maximum-likelihood estimates of the long-memory parameter d of a fractional noise process, along with their mean-squared errors. The results are based on 1000 simulations of fractional processes with $d = 0.4$ with 1000 observations, for different percentages of missing values. Three maximum-likelihood estimates are considered, one based on the full sample (*full*), one based on the series with imputed missing values using a spline interpolation method (*imputed*), and one based only on the available data (*NAs*).

Observe from Table 11.1 that the imputation method changes dramatically the estimated parameter d, taking it close to the upper bound of stationarity ($d = \frac{1}{2}$). Besides, the mean-squared errors are greatly increased when compared to the MLE based on the full sample. On the other hand, when the missing values are adequately accounted for (MLE with NAs), the estimates are quite close to the true value of the long-memory parameter and the mean-squared errors are marginally increased when compared to the full sample MLE.

As discussed at the beginning of this chapter, there are many ways to deal with missing data in the context of time series. For example, the missing values can be *imputed*, the data can be *repeated* as in the Nile River series (see Section 11.3),

Table 11.1 Maximum-Likelihood Estimates for Simulated FN(d) Processes Based on 1000 Observations.[a]

%		$\widehat{d}$			MSE	
Missing	Full	Imputed	NAs	Full	Imputed	NAs
10	0.3945	0.4380	0.3994	0.0251	0.0445	0.0269
20	0.3927	0.4795	0.3975	0.0250	0.0806	0.0271
30	0.3936	0.4951	0.3991	0.0256	0.0951	0.0275

[a] The three estimates correspond to the full, imputed, and incomplete data, respectively.

replaced by zeroes or other values, *ignored* or *integrated out*, among many other techniques. In the next section, we study some techniques for obtaining a likelihood function based only on the available data.

11.2 LIKELIHOOD FUNCTION WITH INCOMPLETE DATA

An important issue when defining the likelihood function with missing data is the distribution of the *location* of the missing values. For example, if there is *censoring* in the collection of the data and any value above or below a given threshold is not reported, then the location of those missing observations will depend on the distribution of the time series. In other situations, the location may not depend on the values of the stochastic process but it may depend on time. For example, daily economical data may miss Saturday and Sunday observations, producing a systematic pattern of missing data. If the distribution of the location of missing data does not depend on the distribution of the time series—*observed* and *unobserved*—values, then maximum-likelihood estimations and statistical inferences can be carried out by ignoring the exact location of the missing values as described below.

11.2.1 Integration

Let the *full data* be the triple $(y_{\text{obs}}, y_{\text{mis}}, \eta)$, where y_{obs} denotes the observed values, y_{mis} represents the unobserved values, and η is the location of the missing values. In this analysis, η may be regarded as a random variable. The observed data is then given by the pair (y_{obs}, η), that is, the available information consists of the observed values and the pattern of the missing observations. The distribution of the observed data may be obtained by integrating the missing data out of the joint distribution of $(y_{\text{obs}}, y_{\text{mis}}, \eta)$:

$$\begin{aligned} f_\theta(y_{\text{obs}}, \eta) &= \int f_\theta(y_{\text{obs}}, y_{\text{mis}}, \eta)\, dy_{\text{mis}} \\ &= \int f_\theta(\eta | y_{\text{obs}}, y_{\text{mis}}) f_\theta(y_{\text{obs}}, y_{\text{mis}})\, dy_{\text{mis}}. \end{aligned}$$

If the distribution of η does not depend on $(y_{\text{obs}}, y_{\text{mis}})$, then

$$\begin{aligned} f_\theta(y_{\text{obs}}, \eta) &= \int f_\theta(\eta) f_\theta(y_{\text{obs}}, y_{\text{mis}})\, dy_{\text{mis}} \\ &= f_\theta(\eta) \int f_\theta(y_{\text{obs}}, y_{\text{mis}})\, dy_{\text{mis}}. \end{aligned}$$

Furthermore, if the location η does not depend upon the parameter θ, then

$$f_\theta(y_{\text{obs}}, \eta) = f(\eta) \int f_\theta(y_{\text{obs}}, y_{\text{mis}})\, dy_{\text{mis}}.$$

In this case, the likelihood function for θ satisfies

$$\mathcal{L}(\theta|y_{\text{obs}},\eta) \quad \propto \quad \int f_\theta(y_{\text{obs}},y_{\text{mis}})\,dy_{\text{mis}} = f_\theta(y_{\text{obs}}) = \mathcal{L}(\theta|y_{\text{obs}}).$$

Thus, maximum-likelihood estimates do not depend on the location of the missing data. Observe that this result is still valid if η depends upon time.

11.2.2 Maximization

An alternative way of dealing with unobserved data is through the maximization of the likelihood function $\mathcal{L}(\theta|y_{\text{obs}},y_{\text{mis}})$ with respect to y_{mis}. Under this approach, let the function $\widetilde{\mathcal{L}}$ be defined by

$$\widetilde{\mathcal{L}}(\theta|y_{\text{obs}}) = \max y_{\text{mis}}\mathcal{L}(\theta|y_{\text{obs}},y_{\text{mis}}).$$

Consider, for instance, the class of random variables $y = (y_1,\ldots,y_n)' \in \mathbb{R}^n$, with zero mean and joint density given by

$$f_\theta(y) = |\Sigma|^{-1/2}h(y'\Sigma^{-1}y), \tag{11.1}$$

where h is a positive real function and Σ is a symmetric definite positive matrix. Several well-known distributions are described by (11.1), including the multivariate normal, mixture of normal distributions and multivariate t, among others.

For this family of distributions we find that

$$\begin{aligned}\mathcal{L}(\theta|y_{\text{obs}},y_{\text{mis}}) &= |\Sigma|^{-1/2}h[(y_{\text{obs}},y_{\text{mis}})'\Sigma^{-1}(y_{\text{obs}},y_{\text{mis}})]\\ &= |A|^{1/2}h[(y_{\text{obs}},y_{\text{mis}})'A(y_{\text{obs}},y_{\text{mis}})],\end{aligned}$$

where

$$A = \begin{bmatrix} A_{11} & A_{12} \\ A_{21} & A_{22}\end{bmatrix} = \Sigma^{-1} = \begin{bmatrix} \Sigma_{11} & \Sigma_{12} \\ \Sigma_{21} & \Sigma_{22}\end{bmatrix}^{-1},$$

with $\Sigma_{11} = \text{Var}[y_{\text{obs}}]$, $\Sigma_{22} = \text{Var}[y_{\text{mis}}]$, $\Sigma_{12} = \text{Cov}[y_{\text{obs}},y_{\text{mis}}] = \Sigma_{21}'$.

Thus,

$$\frac{\partial\mathcal{L}(\theta|y_{\text{obs}},y_{\text{mis}})}{\partial y_{\text{mis}}} = 2|A|^{1/2}h'(s)[y_{\text{obs}}'A_{12} + y_{\text{mis}}'A_{22}] = 0,$$

where $s = (y_{\text{obs}},y_{\text{mis}})A(y_{\text{obs}},y_{\text{mis}})' \geq 0$. Consequently,

$$\widehat{y}_{\text{mis}} = -A_{22}^{-1}A_{21}y_{\text{obs}} = \Sigma_{21}\Sigma_{11}^{-1}y_{\text{obs}},$$

is a critical point of $\mathcal{L}(\theta|y_{\text{obs}},y_{\text{mis}})$ as a function of y_{obs}. Furthermore, the Hessian is

$$H(\widehat{y}_{\text{mis}}) = 2|A|^{1/2}h'(s)A_{22},$$

which is negative definite if $h'(s) < 0$. This occurs for several distributions. For example, in the Gaussian case, $h(s) = \exp(-s^2/2)$, and hence its derivative satisfies

$$h'(s) = -se^{-s^2/2} < 0,$$

for any $s > 0$. Therefore, in this situation $\widehat{y}_{\text{mis}}$ is indeed a maximum and

$$\begin{aligned}\widetilde{\mathcal{L}}(\theta|y_{\text{obs}}) &= |\Sigma|^{-1/2}h(y'_{\text{obs}}\Sigma_{11}^{-1}y_{\text{obs}}) \\ &= |\Sigma_{22} - \Sigma_{21}\Sigma_{11}^{-1}\Sigma_{12}|^{-1/2}|\Sigma|_{11}^{-1/2}h(y'_{\text{obs}}\Sigma_{11}^{-1}y_{\text{obs}}) \\ &= |\Sigma_{22} - \Sigma_{21}\Sigma_{11}^{-1}\Sigma_{12}|^{-1/2}\mathcal{L}(\theta|y_{\text{obs}}).\end{aligned}$$

Hence, by defining $g(\theta) = |\Sigma_{22} - \Sigma_{21}\Sigma_{11}^{-1}\Sigma_{12}|^{-1/2}$, we conclude that

$$\widetilde{\mathcal{L}}(\theta|y_{\text{obs}}) = g(\theta)\mathcal{L}(\theta|y_{\text{obs}}).$$

Thus, the maximization of $\widetilde{\mathcal{L}}(\theta|y_{\text{obs}})$ with respect to θ may give a different result than the maximization of $\mathcal{L}(\theta|y_{\text{obs}})$.

11.2.3 Calculation of the Likelihood Function

Naturally, the actual calculation of the likelihood function depends on the joint distribution of the process. For example, for the class of distributions described by (11.1), the log-likelihood function is given by

$$\mathcal{L}(\theta) = -\tfrac{1}{2}\log\det\Sigma + \log h(y'\Sigma^{-1}y). \tag{11.2}$$

The matrix Σ can be diagonalized as in Chapter 4. That is, since Σ is assumed to be definite positive and symmetric, there is a lower triangular matrix L with ones in the diagonal and a diagonal matrix $D = \text{diag}\{d_1, \ldots, d_n\}$ such that

$$\Sigma = L'DL.$$

Define $e = Ly$. Hence, $y'\Sigma^{-1}y = e'De$ and $\det\Sigma = \det D$. Consequently,

$$\mathcal{L}(\theta) = -\frac{1}{2}\sum_{t=1}^{n}\log d_t + \log h\left(\sum_{t=1}^{n}\frac{e_t^2}{d_t}\right).$$

As discussed in the next section, in the presence of data gaps the likelihood function of this class of distributions may be calculated from the output of the Kalman filter equations.

11.2.4 Kalman Filter with Missing Observations

Here we analyze with greater detail the state space approach for dealing with missing observations that was outlined in Subsection 2.3.4. Consider the state space system

$$\begin{cases} x_{t+1} = Fx_t + H\varepsilon_t, \\ y_t = Gx_t + \varepsilon_t, \end{cases}$$

where x_t is the state, y_t is the observation, F and G are system matrices, and the observation noise variance is $\mathrm{Var}(\varepsilon_t) = \sigma^2$.

The following theorem summarizes the modifications to the Kalman filter equations in order to account for the missing observations.

Theorem 11.1. *Let $\widehat{x}_t$ be the projection of the state x_t onto $\mathcal{P}_{t-1} = \overline{\mathrm{sp}}\{observed\ y_s : 1 \le s \le t-1\}$ and let $\Omega_t = E[(x_t - \widehat{x}_t)(x_t - \widehat{x}_t)']$ be the state error variance, with $\widehat{x}_1 = 0$ and $\Omega_1 = E[x_1 x_1']$. Then, the state predictor $\widehat{x}_t$ is given by the following recursive equations for t $\ge$ 1:*

$$
\begin{aligned}
\Delta_t &= G\Omega_t G' + \sigma^2, \\
K_t &= (F\Omega_t G' + \sigma^2 H)\Delta_t^{-1}, \\
\Omega_{t+1} &= \begin{cases} F\Omega_t F' + \sigma^2 HH' - \Delta_t K_t K_t', & y_t \text{ observed}, \\ F\Omega_t F' + \sigma^2 HH', & y_t \text{ missing}, \end{cases} \\
\nu_t &= \begin{cases} y_t - G\widehat{x}_t, & y_t \text{ observed}, \\ 0, & y_t \text{ missing}, \end{cases} \\
\widehat{x}_{t+1} &= F\widehat{x}_t + K_t \nu_t, \\
\widehat{y}_{t+1} &= G\widehat{x}_{t+1}.
\end{aligned}
$$

Proof. The modifications of the state covariance matrix equation and the state prediction are obtained by noting that if y_t is missing, then $\mathcal{P}_t = \mathcal{P}_{t-1}$ and $\widehat{x}_{t+1} = E[x_{t+1}|\mathcal{P}_t] = E[Fx_t + H\varepsilon_t|\mathcal{P}_t] = F\,E[x_t|\mathcal{P}_t] + E[H\varepsilon_t|\mathcal{P}_t] = F\,E[x_t|\mathcal{P}_{t-1}] = F\widehat{x}_t$. Thus, $x_{t+1} - \widehat{x}_{t+1} = F(x_t - \widehat{x}_t) + H\varepsilon_t$ and $\Omega_{t+1} = F\Omega_t F' + HH'\sigma^2$.

On the other hand, since y_t is missing, the innovation is null, so that $\nu_t = 0$. Besides, let the observation y_t be missing. Hence, by definition

$$\widehat{y}_t = E[y_t|\mathcal{P}_{t-1}].$$

From the state space system equation we have

$$y_t = Gx_t + \varepsilon_t,$$

and therefore,

$$\widehat{y}_t = E[Gx_t + \varepsilon_t|\mathcal{P}_{t-1}] = G\widehat{x}_t + E[\varepsilon_t|\mathcal{P}_{t-1}] = G\widehat{x}_t. \tag{11.3}$$

□

Remark 11.1. Notice that the equation linking the state and the observation estimation (11.3) is the same, with or without missing data.

Based on the modified Kalman equations specified in Theorem 11.1, the log-likelihood function of a stochastic process with missing data and probability distribution (11.1) is given by

$$\mathcal{L}(\theta) = -\frac{1}{2}\sum \log \Delta_t + \log h\left(\sum \frac{\nu_t^2}{\Delta_t}\right),$$

where the sum is over all the observed values of y_t.

■ **EXAMPLE 11.1**

Let y_t be a stationary process with autocorrelations ρ_k and variance σ_y^2 and suppose that we have observed the values $y_1, y_2, y_4, \ldots$, but the value y_3 is missing. By applying the modified Kalman equations described above we conclude that $\Delta_1 = \sigma_y^2$, $\Delta_2 = \sigma_y^2(1-\rho_1^2)$, $\Delta_3 = \sigma_y^2(1-\rho_2^2)$, $\Delta_4 = \sigma_y^2(1-\rho_1^2-\rho_2^2-\rho_3^2-2\rho_1\rho_2\rho_3)/(1-\rho_2^2), \ldots$ Thus, the magnitude of the jump in the prediction error variance from Δ_2 to Δ_3 is $\rho_2^2-\rho_1^2$.

11.3 EFFECTS OF MISSING VALUES ON ML ESTIMATES

In Section 11.1 we discussed some of the effects of imputation on ML estimates. To further illustrate the effects of repeating values to fill data gaps consider the following example involving the Nile River time series, which is exhibited in Figure 11.1. Notice that this series in longer than the Nile River data analyzed in Subsection 8.5.

Panel (a) of Figure 11.1 displays the original data. From this plot, it seems that during several periods the data was repeated year after year in order to complete the series. Panel (b) shows the same series, but without those repeated values (filtered series).

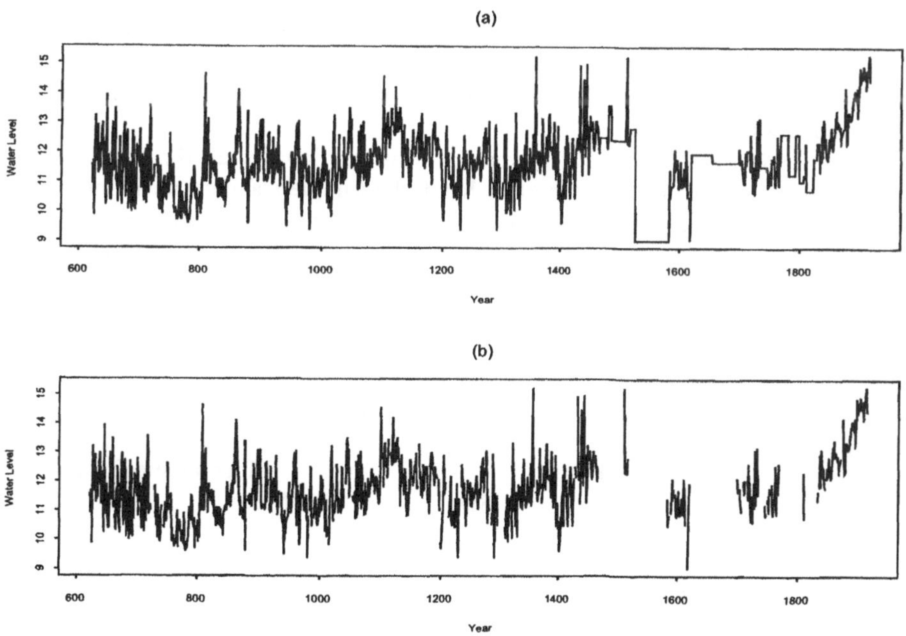

Figure 11.1 Nile River data (A.D. 622 to A.D. 1921). (a) Original data and (b) filtered data.

Table 11.2 Nile River Data Estimates

Series	$\widehat{d}$	$t_{\widehat{d}}$	$\widehat{\sigma}$
Original data	0.499	25.752	0.651
Filtered data	0.434	19.415	0.718

Table 11.2 shows the fitted parameters of an ARFIMA$(0, d, 0)$ using an AR(40) approximation along the modified Kalman filter equations of Subsection 11.2.4 for both the original and the filtered data.

Observe that for the original data, the estimate of the long-memory parameter is 0.5, indicating that the model has reached the nonstationary boundary. On the other hand, for the filtered data, the estimate of d is inside the stationary region.

Thus, in this particular case, the presence of data irregularities such as the replacement of missing data with repeated values induces nonstationarity. On the other hand, when the missing data is appropriately taken care of, the resulting model is stationary.

11.3.1 Monte Carlo Experiments

Table 11.3 displays the results from Monte Carlo simulations of Kalman maximum-likelihood estimates for a fractional noise ARFIMA$(0, d, 0)$ with missing values at random.

The sample size is $n = 400$ and the Kalman moving-average (MA) truncation uses $m = 40$. The long-memory parameters are $d = 0.1, 0.2, 0.3, 0.4$, and $\sigma = 1$. The number of missing values is 80 (20% of the sample) which have been selected randomly for each sample. The sample mean and standard deviations are based on 1000 repetitions.

From Table 11.3, notice that the bias is relatively low for all the values of the long-memory parameter. On the other hand, the sample standard deviation of the estimates seems to be close to the expected values. The expected standard deviation is $\sqrt{6/400\pi^2} = 0.0390$ for the full sample and $\sqrt{6/320\pi^2} = 0.04359$ for the incomplete data case.

Table 11.3 Maximum-Likelihood Estimates for Simulated FN(d) Processes Based on 1000 Observations for Different Values of the Long-Memory Parameter d

	Full Data		Incomplete Data	
d	Mean	SD	Mean	SD
0.1	0.0913	0.0387	0.0898	0.0446
0.2	0.1880	0.0396	0.1866	0.0456
0.3	0.2840	0.0389	0.2862	0.0484
0.4	0.3862	0.0388	0.3913	0.0508

Table 11.4 Finite Sample Behavior of Maximum-Likelihood Estimates with Missing Values.[a]

	Full Data			Incomplete Data		
	d	ϕ	θ	d	ϕ	θ
Mean	0.2559	-0.5102	0.1539	0.2561	-0.5086	0.1639
SD	0.0873	0.0729	0.1394	0.1070	0.0862	0.1795
Theoretical SD	0.0790	0.0721	0.1305	0.0883	0.0806	0.1459

[a] ARFIMA$(1, d, 1)$ model with sample size $n = 400$ and parameters $d = 0.3$, $\phi = -0.5$, and $\theta = 0.2$. The incomplete data have 80 missing observations selected randomly.

Table 11.4 shows the results from Monte Carlo simulations of Kalman maximum-likelihood estimates for ARFIMA$(1, d, 1)$ processes with missing values at random. The sample size of the full data is $n = 400$ and the parameters are $d = 0.3$, $\phi = -0.5$, $\theta = 0.2$, and $\sigma^2 = 1$. The incomplete data have 80 missing observations.

From Table 11.4, the estimates of the long-memory parameter d seem to be slightly downward biased for both full and incomplete data.

On the other hand, the sample standard deviations are slightly higher than expected, for the estimates of d. The theoretical standard deviations reported in this table are based on formula (5.7), with $n = 400$ for the full data sample and $n = 320$ for the incomplete data case.

11.4 EFFECTS OF MISSING VALUES ON PREDICTION

In this section we turn our attention to the evolution of the one-step mean-squared prediction error, $E[y_t - \widehat{y}_t]^2$, for ARFIMA models, during and after a block of missing data. For simplicity, the analysis will be conducted by taking into account the full past instead of the finite past of the time series.

Throughout this section we use the concepts of exponential and hyperbolic rates of convergence of a sequence $\{y_k\}$ to its limit y as $k \to \infty$. *exponential convergence* to y means that $|y_k - y| \leq ca^{-k}$ for large k, $|a| < 1$, and a positive constant c. On the other hand, *hyperbolic convergence* to y means that $|y_k - y| \leq ck^{-\alpha}$ for some positive constant c and $\alpha > 0$.

Let y_t be a stationary invertible process with AR(∞) representation $y_t = \varepsilon_t - \sum_{j=1}^{\infty} \pi_j y_{t-j}$ and MA(∞) decomposition $y_t = \sum_{j=1}^{\infty} \psi_j \varepsilon_{t-j}$, where the white noise sequence $\{\varepsilon_t\}$ has variance σ^2. Assume that the observations $y_0, \ldots, y_{m-1}$ are missing and define

$$\mathcal{H}_{k,m} = \overline{\text{sp}} \{y_s, s < k, s \neq 0, 1, \ldots, m-1\}.$$

Let $e(t, m, k)$ be the error of the best linear predictor of y_t given $\mathcal{H}_{k,m}$, that is, given all the available information before time k, and let $\sigma^2(t, m, k)$ be its variance. The following theorems characterize the behavior of $\sigma^2(t, m, k)$ during and after the data gap and specify the convergence rates.

Theorem 11.2. *The variance $\sigma^2(k,m,k)$ satisfies*

(a) *for* $k = 0,\ldots,m$, $\sigma^2(k,m,k) = \sigma^2 \sum_{j=0}^{k} \psi_j^2$,

(b) *for* $k > m$, $\sigma^2(k,m,k) - \sigma^2 \le \sigma_y^2 m^2 \max_{\{j \ge k-m\}} \pi_j^2$,

(c) $\lim_{k\to\infty} \sigma^2(k,m,k) = \sigma^2$.

Proof. Part (a). This is a standard result for the k steps ahead error variance; see Subsection 9.2. Part (b). Let $a(k,m) = \sum_{j=k-m+1}^{k} \pi_j e(k-j,m,k)$ and take $t = k$ in the AR(∞) decomposition and subtract from both sides the best linear predictor. Then all terms vanish, except for y_k and those associated with the missing observations, which yields the identity

$$e(k,m,k) = \varepsilon_k - a(k,m). \tag{11.4}$$

By the orthogonality of ε_k to all previous observations,

$$\sigma^2(k,m,k) = \sigma^2 + \mathrm{Var}[a(k,m)],$$

for $m \ge 1$. Now, by noting that

$$\begin{aligned} \|a(k,m)\| &\le \sum_{j=k-m+1}^{k} |\pi_j| \|e(k-j,m,k)\| \\ &\le \max_{\{j\ge k-m\}} |\pi_j| \sum_{j=k-m+1}^{k} \|e(k-j,m,k)\| \\ &\le \max_{\{j\ge k-m\}} |\pi_j| m\sigma_y, \end{aligned}$$

part (b) is obtained. Part (c) is a direct consequence of part (b) since $\pi_j \to 0$ as $j \to \infty$ and $\sigma^2(k,m,k) \ge \sigma^2$ for all k. □

Theorem 11.2 shows that, as expected, during the data gap the mean-squared prediction error increases monotonically up to time m, since no new information is being added. In contrast, after time m the variance of the prediction error decreases as new observations are incorporated. The next result specifies the rates at which these error variances increase or decrease during and after the gap.

Theorem 11.3. *For an ARFIMA model and fixed data gap length m we have*

(a) $\sigma_y^2 - \sigma^2(k,m,k) \sim ck^{2d-1}$ *for some constant* $c > 0$ *and large* k, $k \le m$,

(b) $\sigma^2(k,m,k) - \sigma^2 \sim ck^{-2d-2}$ *for some constant* $c > 0$ *and large* k, $k \gg m$.

Proof. Part (a). Notice that for $k \le m$ we have that $\sigma^2(k,m,k) - \sigma_y^2 = \sigma^2 \sum_{j=k}^{\infty} \psi_j^2$. Since $\psi_j \sim c_0 j^{d-1}$ for large j, $\sigma^2 \sum_{j=k}^{\infty} \psi_j^2 \sim ck^{2d-1}$ for large k. Part (b). Observe that from equation (11.4) we have for $k \gg m$

$$\begin{aligned}\pi_{k-m+1}^{-2}\{\sigma^2(k,m,k)-\sigma^2\} &= \operatorname{Var}\left[\sum_{j=k-m+1}^{k}\frac{\pi_j}{\pi_{k-m+1}}e(k-j,m,k)\right],\\ &= \operatorname{Var}\left[\sum_{j=k-m+1}^{k} e(k-j,m,k)\right.\\ &\quad\left.+\sum_{j=k-m+1}^{k} b_{kj}e(k-j,m,k)\right],\end{aligned}$$

where $b_{kj} = \pi_j/\pi_{k-m+1} - 1$ for $j = k-m+1,\ldots,k$. For m fixed and large k this term is bounded by

$$|b_{kj}| \le \frac{c_1}{k},$$

for $j = k-m+1,\ldots,k$ and $c_1 > 0$. Since $\|e(k-j,m,k)\| \le \sigma_y$, we conclude that

$$\pi_{k-m+1}^{-2}[\sigma^2(k,m,k)-\sigma^2] = \operatorname{Var}\left[\sum_{j=k-m+1}^{k} e(k-j,m,k)\right] + O\left(\frac{1}{k}\right),$$

for large k. Now, by defining the following random variable:

$$z_k = \sum_{j=0}^{m-1} y_j - E\left[\sum_{j=0}^{m-1} y_j | y_k,\ldots,y_m,y_{-1},y_{-2},\ldots\right],$$

we have

$$\pi_{k-m+1}^{-2}\left\{\sigma^2(k,m,k)-\sigma^2\right\} = \operatorname{Var}[z_k] + O\left(\frac{1}{k}\right).$$

But, $0 < \operatorname{Var}[z_\infty] \le \operatorname{Var}[z_k] \le \operatorname{Var}[z_m] < \infty$. Therefore $\sigma^2(k,m,k)-\sigma^2 \sim c_2\pi_{k-m+1}^2 \sim ck^{-2d-2}$ for $k \gg m$. □

■ EXAMPLE 11.2

According to Theorem 11.3, there are two different hyperbolic rates for the mean-squared prediction error during and after the data gap. For instance, if $d = 0.3$, then the variance of the prediction error during the gap increases at rate $O(k^{-0.4})$, whereas it decreases to σ^2 at rate $O(k^{-2.6})$. Thus, the information is lost during the block of missing observations at a much slower rate than it is gained after the data gap.

Theorem 11.2 and Theorem 11.3 assume the data gap length to be fixed. However, if the length of the gap increases to infinity, then the prediction process is statistically equivalent to the one during the transition period at the beginning of the time series, with no previous observation. The following result characterizes the convergence of the prediction error in this case.

Theorem 11.4. *For an ARFIMA(p, d, q) process we have*

$$\sigma^2(k, \infty, k) - \sigma^2 \sim \frac{d^2}{k},$$

as $k \to \infty$,

Proof. By observing that $\sigma^2(k, \infty, k) - \sigma^2 = \delta_k$, this is a direct consequence of Theorem 9.3. □

Theorem 11.4 sheds some light on the relationship between predicting with finite and infinite past. Let $\widetilde{e}(t)$ be the error of the best linear predictor of y_t based on the finite past $(y_{t-1}, y_{t-2}, ..., y_1)$ and let $\widetilde{\sigma}^2(t)$ be its variance. Then $\widetilde{\sigma}^2(t) = \sigma_y^2 \prod_{i=1}^{t-1}(1 - \phi_{ii}^2) = \sigma^2(t, \infty, t)$. According to Theorem 11.4, this term converges hyperbolically to σ^2 which is the error variance of the best linear predictor of y_t based on the infinite past $(y_{t-1}, y_{t-2}, ...)$.

The following result describes the behavior of the prediction error variance for the case where the data gap corresponds to only one missing value.

Theorem 11.5. *Under the assumptions of Theorem 11.2, the mean-squared prediction error of an isolated missing observation behaves as follows:*

(a) *for* $k > 0$, $\sigma^2(k, 1, k) - \sigma^2 = \pi_k^2 \sigma^2(0, 1, k)$,

(b) *for* $k > 0$, *if* π_k *is a monotonically decreasing sequence, then* $\sigma^2(k, 1, k) - \sigma^2$ *is a monotonically decreasing sequence converging to zero.*

Proof. Taking $m = 1$ in (11.4) yields part (a). Part (b) follows from the monotonic behavior of $\sigma^2(0, 1, r)$, which decreases from σ^2 to $\sigma^2(0, 1, \infty)$, the variance of the interpolation error given all observations but the missing one; see equation (11.7). □

■ EXAMPLE 11.3

According to Theorem 11.5(a), the magnitude of the jump in the mean-squared prediction error after a missing value is $\pi_1^2 \sigma_\varepsilon^2$. In some cases, $\pi_1 = 0$ and, as a consequence, there is no jump at all. For instance, in an ARFIMA$(1, d, 1)$ process $\pi_1 = \theta - \phi - d$, hence there is no jump in the one-step mean-squared prediction error whenever $d = \theta - \phi$.

The monotonicity condition in Theorem 11.5(b) is shared by all fractional noise processes with long-memory parameter d. In fact, $\pi_j = \prod_{k=1}^{j}(k - 1 - d)/k > 0$ and so $\pi_k/\pi_{k+1} = (k + 1)/(k - d)$. Thus, for $d \in (-1, 1)$ and $k \geq 1$, $(k + 1)/(k - d) > 1$, proving that $\pi_k/\pi_{k+1} > 1$.

11.5 ILLUSTRATIONS

Figure 11.2 to Figure 11.4 show the evolution of the one-step prediction error variance for a fractional noise process with $d = 0.4$ and $\sigma = 1$, from $t = 1$ to $t = 200$.

In Figure 11.2, the heavy line represents the evolution for the full sample (no missing values) and the broken line indicates the evolution in the presence of a data gap from $t = 20$ to $t = 50$. It is clear that the prediction error variance increases from the beginning of the data gap up to $t = 50$ and then decays rapidly to 1. According to Theorem 11.2 and Theorem 11.3, the one-step prediction error variance increases at the rate $\mathcal{O}(k^{-0.2})$ during the data gap. In turn, according to Theorem 11.4 after the gap the one-step prediction error variance decays to 1 at rate $\mathcal{O}(k^{-1})$.

The effect of data gaps on out-of-sample forecasts can be analyzed with the help of Figure 11.3 and Figure 11.4. Unlike Figure 11.2 where the data from $t = 101$ to $t = 200$ were available, Figure 11.3 and Figure 11.4 depict the evolution of the prediction error variance in the case where the last observation is made a time $t = 100$ and no new data are available from $t = 101$ to $t = 200$. Naturally, this situation is similar to observing values from $t = 1$ to $t = 100$ and then predicting the future values from $t = 101$ to $t = 200$.

Observe in Figure 11.3 that the prediction error variance after $t = 100$ is not greatly affected by the presence of the data gap. On the other hand, a data gap of the

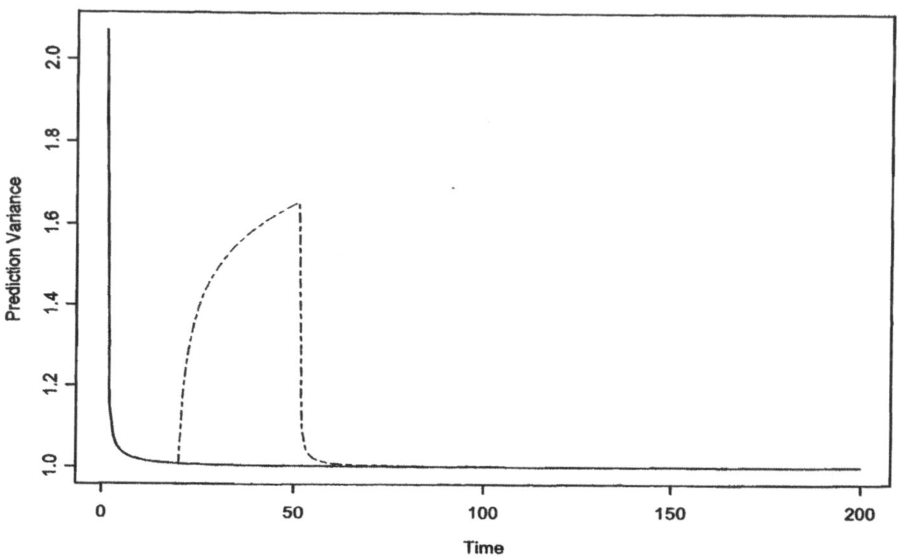

Figure 11.2 Prediction error variance for a fractional noise process with $d = 0.4$, $\sigma^2 = 1$, and 200 observations. Heavy line: full sample. Broken line: Data gap from $t = 20$ to $t = 50$.

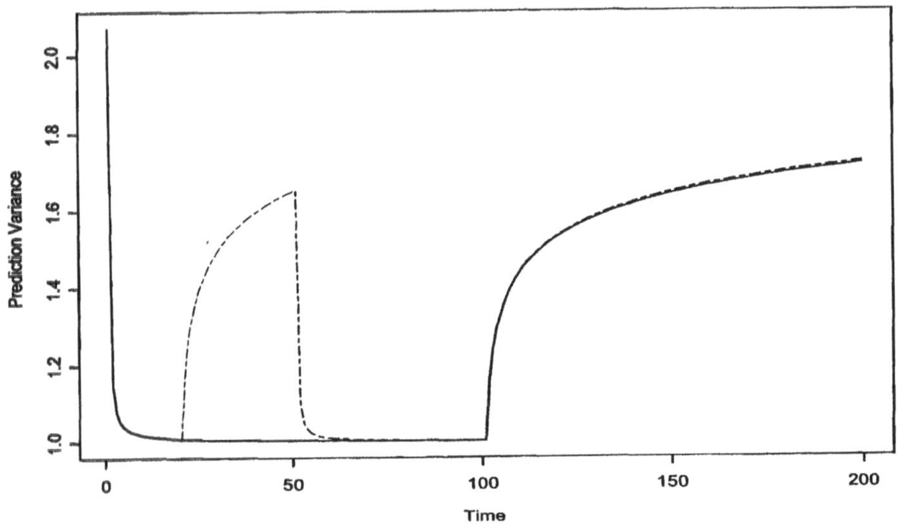

Figure 11.3 Prediction error variance for a fractional noise process with $d = 0.4$, $\sigma^2 = 1$, and 200 observations. Heavy line: Data gaps from $t = 20$ to $t = 50$ and from $t = 100$ to $t = 200$. Broken line: Data gaps from $t = 20$ to $t = 50$ and from $t = 100$ to $t = 200$.

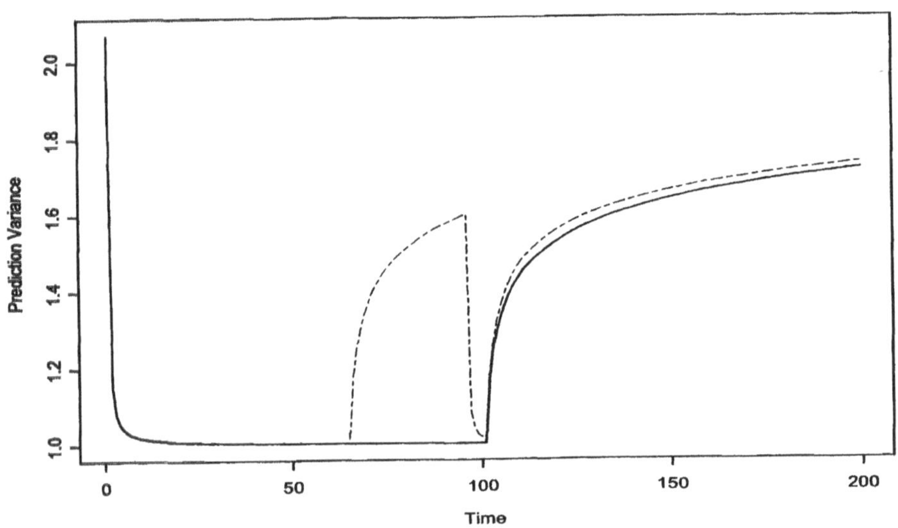

Figure 11.4 Prediction error variance for a fractional noise process with $d = 0.4$, $\sigma^2 = 1$, and 200 observations. Heavy line: Data gaps from $t = 20$ to $t = 50$ and from $t = 100$ to $t = 200$. Broken line: Data gaps from $t = 65$ to $t = 95$ and from $t = 100$ to $t = 200$.

same length but located closer to the end of the series may have a slightly greater effect on the variance of the forecasts, as suggested by Figure 11.4.

11.6 INTERPOLATION OF MISSING DATA

We now focus our attention on the problem of finding estimates for the missing values of a long-memory linear process. To this end, let $\{y_t : t \in \mathbb{Z}\}$ be a stationary process with spectral density $f(\lambda)$ and autoregressive representation

$$\sum_{j=0}^{\infty} \pi_j y_{t-j} = \varepsilon_t, \tag{11.5}$$

where $\pi_0 = 1$ and $\{\varepsilon_t\}$ is a white noise sequence with variance σ^2.

Let $\mathcal{M} = \overline{\text{sp}}\{y_{\text{obs}}\}$ be the Hilbert space generated by the observed series and let y_{mis} be a missing value. According to the projection theorem of Chapter 1, the best linear interpolator of y_{mis} based on the observed data y_{obs} is given by

$$\widetilde{y}_{\text{mis}} = E(y_{\text{mis}} | \mathcal{M}).$$

In particular, if $y_{\text{mis}} = y_0$ and the full past and full future is available, that is, $y_{\text{obs}} = \{y_t, t \neq 0\}$, then the interpolator may be expressed as

$$\widetilde{y}_{\text{mis}} = -\sum_{j=1}^{\infty} \alpha_j (y_j + y_{-j}), \tag{11.6}$$

where the coefficients α_j are given by

$$\alpha_j = \frac{\sum_{i=0}^{\infty} \pi_i \pi_{i+j}}{\sum_{i=0}^{\infty} \pi_i^2},$$

for $j \geq 1$. In this case, the interpolation error variance is

$$\sigma_{\text{int}}^2 = 4\pi^2 \left[\int_{-\pi}^{\pi} \frac{d\lambda}{f(\lambda)} \right]^{-1}. \tag{11.7}$$

On the other hand, if $y_{\text{mis}} = y_0$ and the full past and part of the future is available, that is, $y_{\text{obs}} = \{y_t, t \neq 0, t \leq n\}$, then the best linear interpolator of y_0 may be written as

$$\widetilde{y}_{\text{mis}} = \widehat{y}_0 - \sum_{j=1}^{n} \beta_{j,n} (y_j - \widehat{y}_j),$$

where the coefficients $\beta_{j,n}$ are given by

$$\beta_{j,n} = \frac{\sum_{i=0}^{n-j} \pi_i \pi_{i+j}}{\sum_{i=0}^{n} \pi_i^2},$$

for $j = 1, 2, \ldots, n$ and $\widehat{y}_j$ is the best linear predictor of y_j based on $\overline{\text{sp}}\{y_t, t < 0\}$. In this case, the interpolation error variance is

$$\sigma^2_{\text{int}}(n) = \frac{\sigma^2}{\sum_{i=0}^{n} \pi_i^2}. \tag{11.8}$$

Notice that

$$\sigma^2_{\text{int}}(n) \to \frac{\sigma^2}{\sum_{i=0}^{\infty} \pi_i^2},$$

as $n \to \infty$. But, recalling that the spectral density of process (11.5) is given by

$$f(\lambda) = \frac{\sigma^2}{2\pi} \left| \sum_{j=0}^{\infty} \pi_j e^{i\lambda j} \right|^{-2},$$

we conclude that

$$\begin{aligned}
\int_{-\pi}^{\pi} \frac{d\lambda}{f(\lambda)} &= \frac{2\pi}{\sigma^2} \int_{-\pi}^{\pi} \left| \sum_{j=0}^{\infty} \pi_j e^{i\lambda j} \right|^2 d\lambda, \\
&= \frac{2\pi}{\sigma^2} \sum_{i=0}^{\infty} \sum_{j=0}^{\infty} \pi_i \pi_j \int_{-\pi}^{\pi} e^{i\lambda(i-j)} d\lambda \\
&= \frac{(2\pi)^2}{\sigma^2} \sum_{i=0}^{\infty} \pi_i^2.
\end{aligned}$$

Therefore,

$$4\pi^2 \left[\int_{-\pi}^{\pi} \frac{d\lambda}{f(\lambda)} \right]^{-1} = \frac{\sigma^2}{\sum_{i=0}^{\infty} \pi_i^2}.$$

Hence, as expected,

$$\sigma^2_{\text{int}}(n) \to \sigma^2_{\text{int}},$$

as $n \to \infty$.

Finally, if y_{obs} is a finite observed trajectory of a stationary time series, then we may use the projection theorem discussed in Chapter 1 directly to calculate the best linear interpolator of y_{mis} by means of the formula:

$$\widetilde{y}_{\text{mis}} = E(y_{\text{mis}} y'_{\text{obs}})[E(y_{\text{obs}} y'_{\text{obs}})]^{-1} y_{\text{obs}}. \tag{11.9}$$

Naturally, the calculation of this general expression implies obtaining the inverse of the variance-covariance matrix of the observed data and this procedure could be computationally demanding in the long-memory case.

EXAMPLE 11.4

Consider the fractional noise process FN(d). For this model we have that

$$\alpha_j = \frac{\Gamma(j-d)\Gamma(1+d)}{\Gamma(j+d+1)\Gamma(-d)},$$

for $j = 1, 2, \ldots$ Therefore, the best linear interpolator of y_0 based on the full past and full future is given by

$$\widetilde{y}_0 = \frac{d\Gamma(1+d)}{\Gamma(1-d)} \sum_{j=1}^{\infty} \frac{\Gamma(j-d)}{\Gamma(j+d+1)}(y_j + y_{-j}),$$

and its interpolation error variance equals

$$\sigma^2_{\text{int}}(d) = \sigma^2 \frac{\Gamma(1+d)^2}{\Gamma(1+2d)}. \tag{11.10}$$

Figure 11.5 displays the coefficients α_j for $j = 1, 2, \ldots, 40$ and $d = 0.4$. Observe that they approach zero rapidly. However, this decay rate may be much faster for short-memory processes. For instance, for an AR(1) model, $\alpha_j = 0$ for $j > 1$; see Problem 11.11.

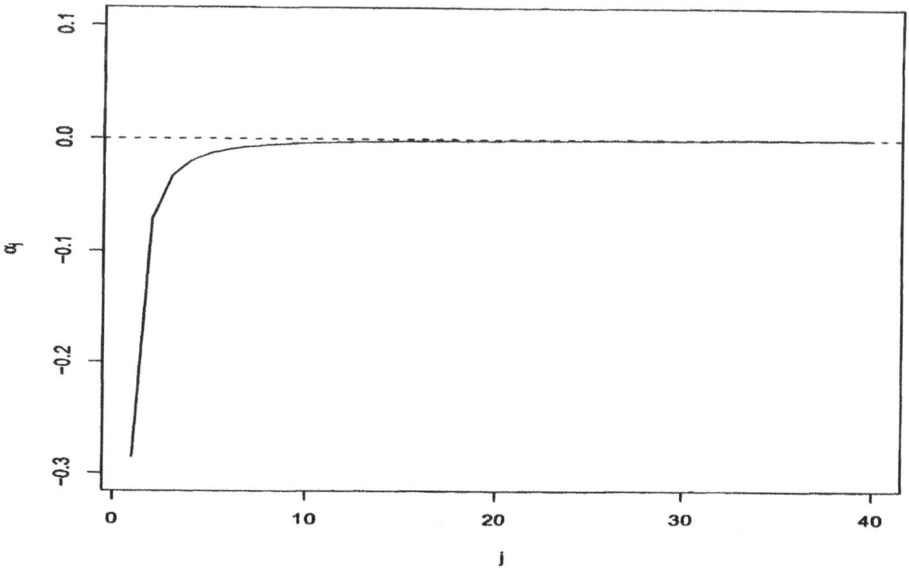

Figure 11.5 Example 11.4: Behavior of the interpolation coefficients α_j for a fractional noise process with $d = 0.4$ and $j = 1, 2, \ldots, 40$.

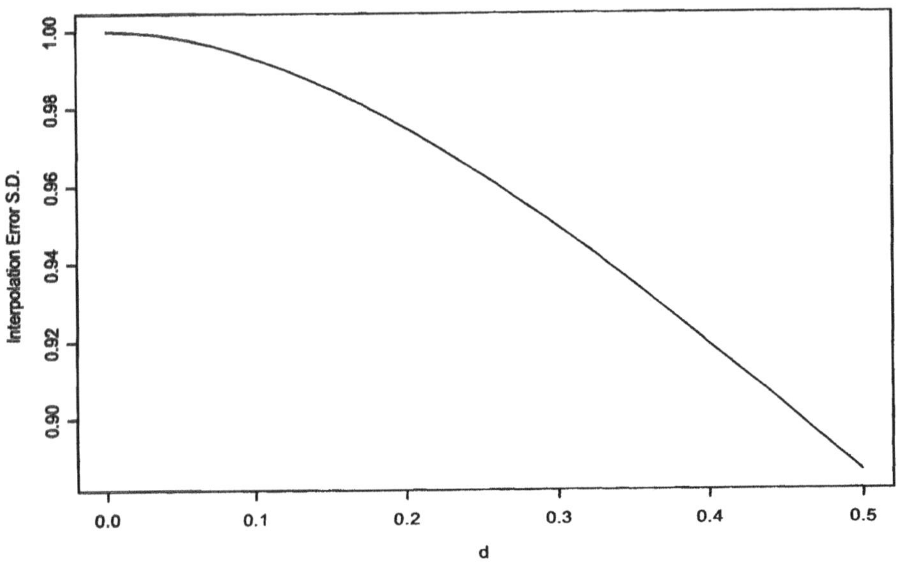

Figure 11.6 Example 11.4: Interpolation error standard deviation of a fractional noise process based on the full past and full future, $\sigma_{\text{int}}(d)$ for different values of the long-memory parameter d.

The evolution of $\sigma_{\text{int}}(d)$ as d moves from 0 to 0.5 is depicted in Figure 11.6. In this case, $\sigma_{\text{int}}(0) = 1$ and $\sigma_{\text{int}}(0.5) = 0.8862$.

EXAMPLE 11.5

Extending the previous example, let $\{y_t : t \in \mathbb{Z}\}$ be an ARFIMA$(1, d, 0)$ process

$$(1 + \phi B)(1 - B)^d y_t = \varepsilon_t, \tag{11.11}$$

where $|\phi| < 1$, $d \in (-1, \frac{1}{2})$ and $\{\varepsilon_t\}$ is a white noise process with variance σ^2. Let η_j be the coefficients of the differencing operator $(1 - B)^d$,

$$(1 - B)^d = \sum_{j=0}^{\infty} \eta_j B^j.$$

Consequently, the coefficients of the autoregressive expansion (11.11)

$$(1 + \phi B)(1 - B)^d = \sum_{j=0}^{\infty} \pi_j B^j$$

are given by

$$\pi_j = \begin{cases} 1, & j = 0, \\ \eta_j + \phi\eta_{j-1}, & j \geq 1. \end{cases}$$

Hence,

$$\begin{aligned} \sum_{j=0}^{\infty} \pi_j^2 &= (1+\phi^2)\sum_{j=0}^{\infty} \eta_j^2 + 2\phi\sum_{j=0}^{\infty} \eta_j\eta_{j+1} \\ &= \left[1+\phi^2 - \frac{2\phi d}{1+d}\right]\frac{\Gamma(1+2d)}{\Gamma(1+d)^2}. \end{aligned}$$

Finally, the interpolation error of y_0 based on $\overline{\mathrm{sp}}\{y_t, t \neq 0\}$ for this ARFIMA$(1, d, 0)$ model is given by

$$\sigma^2_{\mathrm{int}} = \sigma^2 \left[\frac{1+d}{1+\phi^2+d(1-\phi)^2}\right]\frac{\Gamma(1+d)^2}{\Gamma(1+2d)}.$$

■ EXAMPLE 11.6

Consider an ARFIMA(p, d, q) process with autocovariance function $\gamma_h(d, \phi, \theta)$ for lag h. By analyzing the interpolation error variances (11.7) and (11.8) we conclude that

$$\sigma^2_{\mathrm{int}}(n) - \sigma^2_{\mathrm{int}} = \sigma^2 \frac{\sum_{j=n+1}^{\infty}\pi_j^2}{\sum_{j=0}^{n}\pi_j^2\sum_{j=0}^{\infty}\pi_j^2}.$$

But, from (3.18) the AR(∞) coefficients of the ARFIMA process satisfy

$$\pi_j \sim \frac{\phi(1)}{\theta(1)\Gamma(-d)} j^{-d-1},$$

for large j. As a result,

$$\sigma^2_{\mathrm{int}}(n) - \sigma^2_{\mathrm{int}} \sim \left[\frac{\sigma^2\phi(1)^2}{\theta(1)^2\Gamma(-d)^2\sum_{j=0}^{n}\pi_j^2\sum_{j=0}^{\infty}\pi_j^2}\right]\sum_{j=n+1}^{\infty} j^{-2d-2}.$$

Now, an application of Lemma 3.2 yields

$$\sigma^2_{\mathrm{int}}(n) - \sigma^2_{\mathrm{int}} \sim \left[\frac{\sigma^2\phi(1)^2}{(2d+1)\theta(1)^2\Gamma(-d)^2(\sum_{j=0}^{\infty}\pi_j^2)^2}\right] n^{-2d-1}.$$

Finally, by taking into account that

$$\sum_{j=0}^{\infty}\pi_j^2 = \frac{\gamma_0(-d, \theta, \phi)}{\sigma^2},$$

we conclude

$$\sigma_{\text{int}}^2(n) - \sigma_{\text{int}}^2 \sim \left[\frac{\sigma^6 \phi(1)^2}{(2d+1)\theta(1)^2\Gamma(-d)^2\gamma_0(-d,\theta,\phi)^2} \right] n^{-2d-1}.$$

11.6.1 Bayesian Imputation

Finding estimates for missing values can be also achieved by means of Bayesian methods. In this case, we may obtain estimates for y_{mis} based on the *predictive* distribution

$$f(y_{\text{mis}}|y_{\text{obs}}) = \int_{\Theta} f(y_{\text{mis}}|y_{\text{obs}}, \boldsymbol{\theta})\pi(\boldsymbol{\theta}|y_{\text{obs}})\, d\boldsymbol{\theta}.$$

A simple method for drawing samples from this distribution is by adding an extra step to iterative techniques such as the Gibbs sampler or the Metropolis-Hastings algorithm, both described in Chapter 8.

Consider, for example, that at step m of the algorithms described in Section 8.2 we have the sample $\theta^{(m)}$ of the parameter vector. Then, we may obtain a sample $y_{\text{obs}}^{(m)}$ by drawing from the distribution $f(y_{\text{mis}}|y_{\text{obs}}, \theta^{(m)})$.

■ EXAMPLE 11.7

Suppose that both y_{mis} and y_{obs} are normally distributed:

$$\begin{aligned} y_{\text{mis}} &\sim N[\mu_{\text{mis}}(\boldsymbol{\theta}), \sigma_{\text{mis}}^2(\boldsymbol{\theta})], \\ y_{\text{obs}} &\sim N[\mu_{\text{obs}}(\boldsymbol{\theta}), \Sigma_{\text{obs}}(\boldsymbol{\theta})], \end{aligned}$$

and $\langle y_{\text{mis}}, y_{\text{obs}} \rangle = \Sigma_{\text{mis,obs}}(\boldsymbol{\theta})$, where $\Sigma_{\text{obs}}(\boldsymbol{\theta})$ is a $n \times n$ matrix and $\Sigma_{\text{mis,obs}}(\boldsymbol{\theta})$ is a $n \times 1$ vector. Thus, the conditional distribution of y_{mis} given y_{obs} and $\boldsymbol{\theta}$ is

$$y_{\text{mis}}|y_{\text{obs}}, \boldsymbol{\theta} \sim N[\mu_{\text{mis|obs}}(\boldsymbol{\theta}), \sigma_{\text{mis|obs}}^2(\boldsymbol{\theta})],$$

where

$$\begin{aligned} \mu_{\text{mis|obs}}(\boldsymbol{\theta}) &= \mu_{\text{mis}}(\boldsymbol{\theta}) + \Sigma_{\text{mis,obs}}'(\boldsymbol{\theta})\Sigma_{\text{obs}}^{-1}(\boldsymbol{\theta})(y_{\text{obs}} - \mu_{\text{obs}}(\boldsymbol{\theta})), \\ \sigma_{\text{mis|obs}}^2(\boldsymbol{\theta}) &= \sigma_{\text{mis}}^2(\boldsymbol{\theta}) + \Sigma_{\text{mis,obs}}'(\boldsymbol{\theta})\Sigma_{\text{obs}}^{-1}(\boldsymbol{\theta})\Sigma_{\text{mis,obs}}(\boldsymbol{\theta}). \end{aligned}$$

Notice that the matrix $\Sigma_{\text{obs}}(\boldsymbol{\theta})$ is not necessarily Toeplitz due to the data gaps. Consequently, the Durbin-Levinson algorithm cannot be used to facilitate these calculations. An alternative method for obtaining the quantities $\mu_{\text{mis}}(\boldsymbol{\theta})$ and $\sigma_{\text{mis|obs}}^2(\boldsymbol{\theta})$ is provided by the Kalman filter equations of Chapter 2 with the missing data modifications discussed in Subsection 2.3.4.

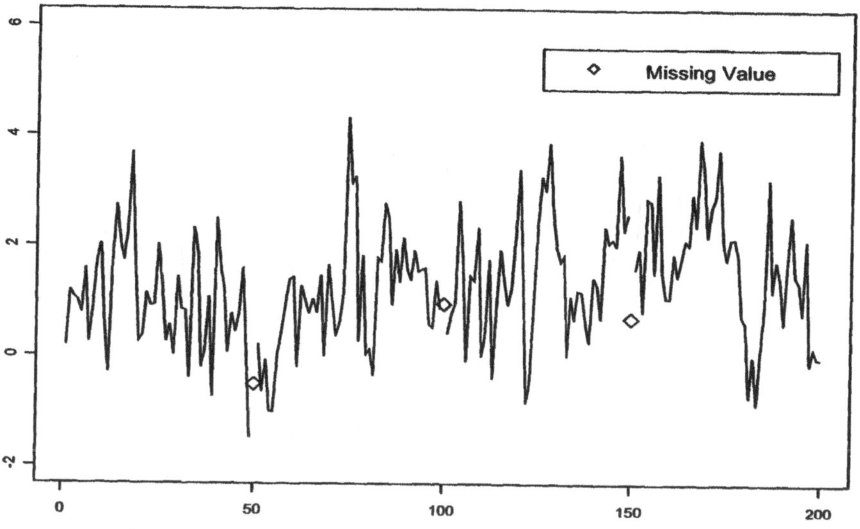

Figure 11.7 Simulated fractional noise process with $d = 0.4$, $\mu = 0$, $\sigma^2 = 1$, sample size $n = 200$, and three missing values located at $t = 50, 100, 150$.

11.6.2 A Simulated Example

To illustrate both the interpolation and the Bayesian imputation techniques, consider a simulated sample of a fractional noise process FN(d) with $d = 0.4$ and sample size $n = 200$. This series has three missing values located at $t = 50, 100, 150$ as shown in Figure 11.7.

The sample mean of the process is $\widehat{\mu} = 1.2084$ while the maximum-likelihood estimates for the parameters of this model are $\widehat{d} = 0.3921$, $\widehat{\sigma}_d = 0.0669$ ($t_d = 5.8540$), and $\widehat{\sigma} = 0.9691$.

The true values and the interpolations of the missing observations are reported in the second and third columns of Table 11.5, respectively. These best linear interpolators were calculated following formula (11.9).

Table 11.5 Missing Data Estimation

Missing Value	True Value	Interpolator	SD	Bayes Estimate	SD
y_{50}	-0.5107	-0.2308	0.9194	-0.2577	0.9084
y_{100}	0.9468	0.8658	0.9194	0.8720	0.9230
y_{150}	0.6976	2.0450	0.9194	2.0426	0.9249

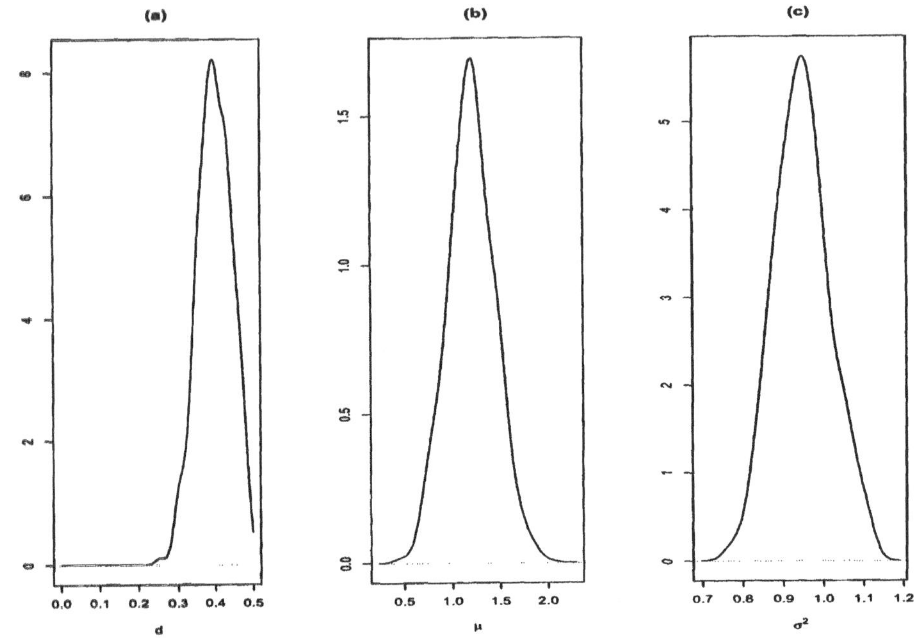

Figure 11.8 Posterior densities of the parameters of the fractional noise model. (a) d, (b) μ, and (c) σ^2.

Observe that while the interpolators of y_{50} and y_{100} seem to be relatively close to their true values, the interpolator of y_{150} seems to be comparatively far from its target. We shall come back to this issue after analyzing the results from the Bayesian approach.

On the other hand, the estimated standard deviations of these interpolators are similar (up to a four-digit rounding). Observe that in the optimal scenario where the full past and full future is available, the interpolation error standard deviation given by formula (11.10) is

$$\sigma_{\text{int}}(\widehat{d}) = 0.8934.$$

Thus, the interpolation efficiency achieved is 97% $(= 100 \times 0.8934/0.9194)$ in this case.

We now turn our attention to the estimates produced by the Bayesian methodology, which are also reported in Table 11.5. These estimates were obtained by applying the Metropolis-Hastings algorithm described in Chapter 8 and the missing data modifications discussed in Subsection 11.6.1.

Figure 11.8 shows kernel estimations of the posterior distributions of the parameters d, μ, and σ^2. Besides, the posterior means and posterior standard deviations of these parameters are reported in Table 11.6.

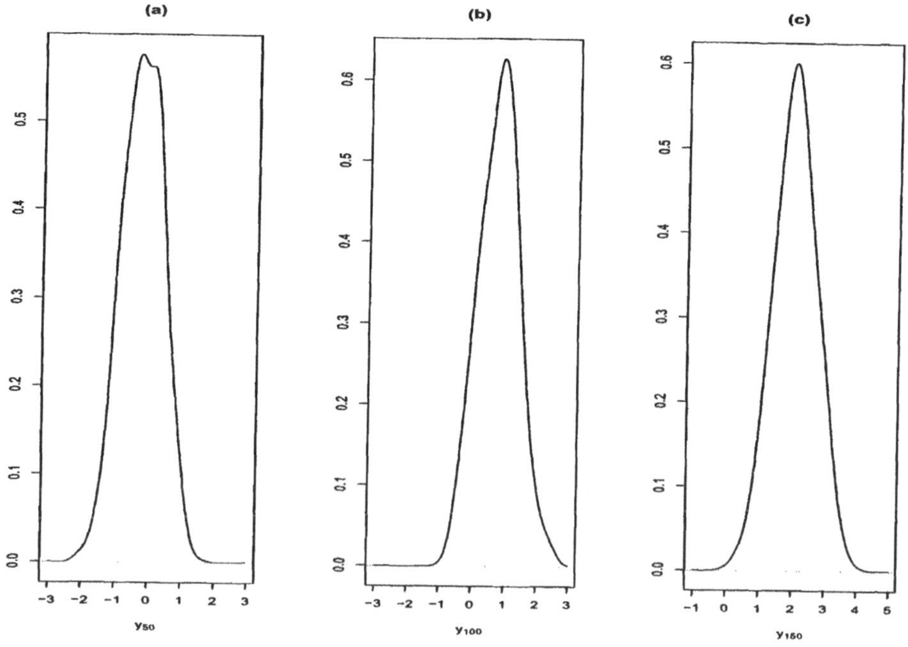

Figure 11.9 Predictive densities of the missing values. (a) y_{50}, (b) y_{100}, and (c) y_{150}.

Figure 11.9 exhibits the smoothed *predictive* densities of the missing values y_{50}, y_{100}, and y_{150}. The mean and sample standard deviation of these variables are shown in the fifth and sixth column of Table 11.5, respectively. The reported means and standard deviations from the Bayesian approach are very similar to those from the linear interpolation methodology. Furthermore, note that analogous to the linear interpolators, the Bayes estimates of y_{50} and y_{100} are close to their true values. However, the estimate of y_{150} is far off. In fact, the interpolator and the Bayes estimate of y_{150} are very similar. Why these two estimates seem to be so far from their target? An insight about what is happening is provided by Figure 11.10 and Figure 11.11. From these figures, observe that both estimates of y_{150} follow very closely the local behavior of the data.

Table 11.6 Bayesian Estimation: Posterior Mean and Posterior Standard Deviation of the Parameters

	d	μ	σ^2
Posterior Mean	0.3997	1.2040	0.9483
Posterior SD	0.0449	0.2389	0.0686

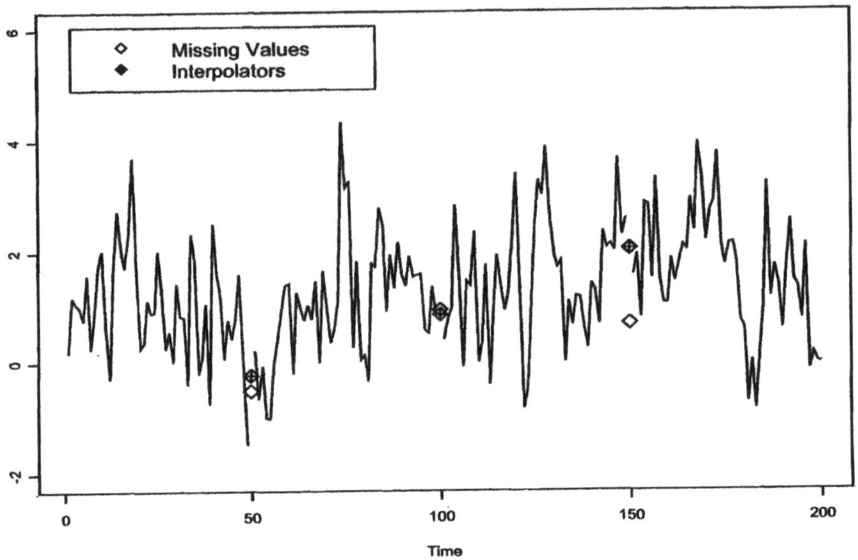

Figure 11.10 Interpolation of the missing values.

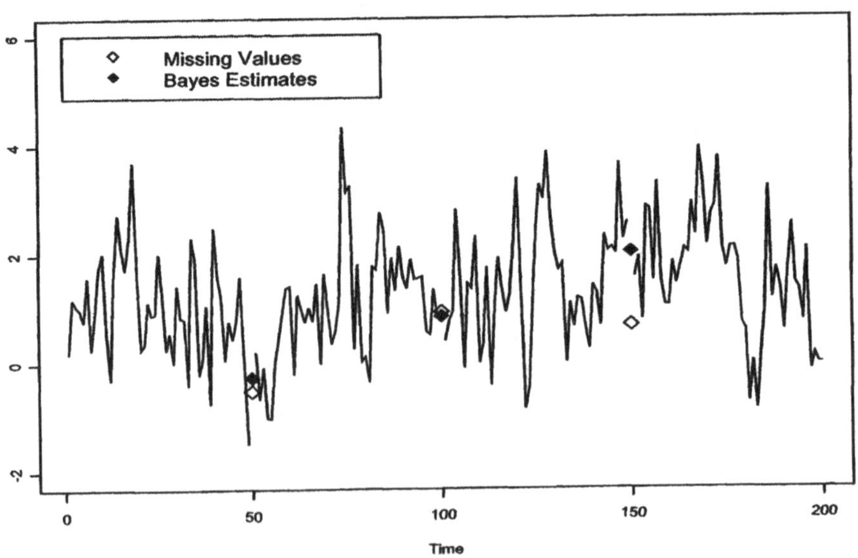

Figure 11.11 Bayes estimates of the missing values.

In fact, the local mean around $t = 150$ is

$$\bar{y}_{\rm loc} = \frac{1}{11} \sum_{t=145}^{155} y_t = 2.1313.$$

Therefore, as the interpolator and the Bayes estimate mimic the local behavior of the series around $t = 150$, they miss the true value $y_{150} = 0.6976$.

11.7 BIBLIOGRAPHIC NOTES

The time series literature on missing values is extensive. For instance, Jones (1980) develops a Kalman filter approach to deal with missing values in ARMA models. Ansley and Kohn (1983), Harvey and Pierse (1984), and Kohn and Ansley (1986) extend Jones' result to ARIMA processes. Further extensions of state space model techniques to ARFIMA models are considered by Palma and Chan (1997) and Palma (2000). Chapter 4 of Shumway and Stoffer (2000) discusses the Kalman filter modifications to account for missing data. An alternative approach for computing the maximum-likelihood estimates called the *expectation maximization (EM)* method to deal with data gaps is also discussed extensively in that book. Other methods for treating missing values are studied by Wilson, Tomsett, and Toumi (2003).

A good general reference to the problem of defining the likelihood function for incomplete data is the book by Little and Rubin (2002). Theorems 11.2, 11.3, 11.4, and 11.5 are proved in Palma and del Pino (1999). This paper also discusses the problem of missing values in the Nile River data.

Estimation and interpolation of time series with missing values have been studied by many authors; see ,for example, Cheng and Pourahmadi (1997), Bondon (2002), Damsleth (1980), Pourahmadi (1989), and Robinson (1985), among others. In addition, Yajima and Nishino (1999) address the problem of estimating the autocovariance function from an incomplete time series. A detailed study about interpolation of missing data appears in Chapter 8 of Pourahmadi (2001).

Bayesian methods for handling models with missing data are described, for instance, in Robert and Casella (2004, Chapter 9) and Little and Rubin (2002, Chapter 10). Other Bayesian techniques for dealing with time series models with missing values are addressed, for example, by Wong and Kohn (1996).

Problems

11.1 Let $\{y_t\}$ be a first-order autoregressive process such that

$$y_t = \phi y_{t-1} + \varepsilon_t,$$

where the white noise sequence $\{\varepsilon_t\}$ follows a standard normal distribution. Suppose that we have observed the values $\{y_1, y_2, y_3, y_5\}$ but y_4 is missing.

a) Calculate the joint density of y_1, y_2, y_3, y_4, y_5, $f(y_1, y_2, y_3, y_4, y_5)$.

b) Find z, the value that maximizes $f(y_1, y_2, y_3, y_4, y_5)$ with respect to y_4.

c) Show that z corresponds to the *smoother* of y_4, that is,

$$z = E[y_4 | y_1, y_2, y_3, y_5].$$

11.2 Suppose that $\{y_t : t \in \mathbb{Z}\}$ is a discrete-time stochastic process and we have observed the values $y_1, \ldots, y_{m-1}, y_{m+1}, \ldots, y_n$ but y_m has not been observed. Let f be the joint density of $y_1, \ldots, y_n$ and let g be the unimodal conditional density of y_m given the observed values. Assume that the mode and the mean of g coincide. Let z be the value that maximizes f with respect to y_{mis}. Show that under these circumstances,

$$z = E[y_m | y_1, \ldots, y_{m-1}, y_{m+1}, \ldots, y_n].$$

11.3 Consider an invertible process with autoregressive decomposition, $y_t = \varepsilon_t - \sum_{j=1}^{\infty} \pi_j y_{t-j}$ and $\sigma = 1$. Let $e_t(1) = y_t - \widehat{y}_t$, where $\widehat{y}_t = P_{t-1} y_t$, P_{t-1} is the projection operator onto $\mathcal{H}_{t-1} = \overline{\text{sp}}\{\text{observed } y_{t-1}, y_{t-2}, \ldots\}$. Define the one-step prediction error by $\sigma_t^2(1) = \text{Var}[e_t(1)]$. Suppose that y_t is the first missing value in the time series. Since up to time t there are no missing observations, $\widehat{y}_t = \sum_{j=1}^{\infty} \pi_j y_{t-j}$, so that $y_t - \widehat{y}_t = \varepsilon_t$ and $\sigma_t^2(1) = \text{Var}[y_t - \widehat{y}_t] = \text{Var}[\varepsilon_t] = 1$.

a) Show that since at time $t+1$, $\mathcal{H}_t = \mathcal{H}_{t-1}$ (no new information is available at time t) we have

$$\sigma_{t+1}^2(1) = \text{Var}[y_{t+1} - \widehat{y}_{t+1}] = \text{Var}[y_{t+1} - P_t y_{t+1}],$$

with

$$P_t y_{t+1} = -\pi_1 P_t y_t - \pi_2 y_{t-1} - \pi_3 y_{t-2} - \cdots$$

and

$$y_{t+1} - P_t y_{t+1} = \varepsilon_{t+1} - \pi_1 (y_t - P_t y_t) = \varepsilon_{t+1} - \pi_1 (y_t - P_{t-1} y_t).$$

b) Verify that

$$\begin{aligned} \sigma_{t+1}^2(1) &= \pi_1^2 \text{Var}[y_t - P_{t-1} y_t] + \text{Var}[\varepsilon_{t+1}] \\ &= \pi_1^2 \sigma_t^2(1) + 1 = \pi_1^2 + 1 \geq \sigma_t^2(1). \end{aligned}$$

11.4 Implement computationally the modified Kalman filter equations of Subsection 11.2.4 for an fractional noise process.

11.5 Show that for $k \geq 1$,

$$\sigma_{t+k}^2(1) - 1 \geq \left\{ \frac{\pi_k}{\pi_{k+1}} \right\}^2 [\sigma_{t+k+1}^2(1) - 1].$$

11.6 Verify that for a fractional noise process, the one-step MSPE after a missing value is strictly decreasing. That is, for all $k \geq 1$,

$$\sigma_{t+k+1}^2(1) < \sigma_{t+k}^2(1).$$

11.7 Show that the one-step MSPE of an ARMA process with unit variance innovations converges to 1 at a exponential rate, that is,

$$|\sigma^2_{t+k}(1) - 1| \leq ca^{-k},$$

as $k \to \infty$ where $|a| < 1$.

11.8 Verify that the one-step MSPE of a fractional ARIMA process with unit variance innovations converges to 1 at a hyperbolic rate, that is,

$$|\sigma^2_{t+k}(1) - 1| \leq ck^{-\alpha},$$

as $k \to \infty$ with $c > 0$ and $\alpha = 2 - 2d > 0$.

11.9 Let $\widetilde{y}_t$ be the best linear interpolator of y_t based on $\overline{\text{sp}}\{y_j, j \neq t\}$ and let $\widetilde{\sigma}^2$ be its error variance

$$\widetilde{\sigma}^2 = \text{Var}(y_t - \widetilde{y}_t).$$

Let x_t be the standardized two-side innovation defined by

$$x_t = \frac{y_t - \widetilde{y}_t}{\widetilde{\sigma}^2}.$$

Let $\alpha(z) = 1 + \sum_{j=1}^{\infty} \alpha_j z^j + \sum_{j=1}^{\infty} \alpha_j z^{-j}$ where α_j are the coefficients appearing in the best linear interpolator (11.6) and $z \in \mathbb{C}$ with $|z| = 1$.

a) Verify that x_t may be formally expressed as

$$x_t = \frac{1}{\widetilde{\sigma}^2} \alpha(B) \psi(B) \varepsilon_t,$$

where $\psi(B)$ is the Wold decomposition of the process (11.5).

b) Prove that

$$\alpha(z) = c\pi(z)\pi(z^{-1}),$$

where

$$c = \frac{1}{\sum_{j=0}^{\infty} \pi_j^2},$$

π_j are the coefficients in the AR(∞) expansion (11.5) and $|z| = 1$.

c) Based on the previous results, show that

$$x_t = \frac{c}{\widetilde{\sigma}^2} \pi(B^{-1}) \varepsilon_t.$$

d) Let f be the spectral density of y_t. Verify that the spectral density of the process x_t is given by

$$f_x(\lambda) = \frac{1}{(2\pi)^2 f(\lambda)}.$$

11.10 Let $\rho(k)$ be the autocorrelation of order k of a second-order stationary process. Verify that the coefficients α_j of the best linear interpolator of y_0 based on $\overline{\text{sp}}\{y_t, t \neq 0\}$ satisfy the formula

$$\sum_{j=-\infty}^{\infty} \alpha_j \rho(k-j) = 0, \tag{11.12}$$

for any $k \in \mathbb{Z}$, $k \neq 0$.

11.11 Let $\{y_t : t \in \mathbb{Z}\}$ be the AR(1) process described by

$$y_t = \phi y_{t-1} + \varepsilon_t,$$

where $\{\varepsilon_t\}$ is a white noise sequence with variance σ^2.

a) Show that for this AR(1) model, the best linear interpolator of y_0 with full past $\{y_t, t < 0\}$ and full future $\{y_t, t > 0\}$ is given by

$$\widetilde{y}_0 = \frac{\phi(y_1 + y_{-1})}{1+\phi^2}.$$

b) Prove that the interpolation error variance is

$$\sigma^2_{\text{int}} = \frac{\sigma^2}{1+\phi^2}.$$

c) Verify formula (11.12) in this case, with

$$\alpha_1 = \alpha_{-1} = -\frac{\phi}{1+\phi^2}.$$

11.12 Consider the MA(1) process $y_t = \varepsilon_t - \theta\varepsilon_{t-1}$ where $\{\varepsilon_t\}$ is a white noise process with variance σ^2 and $|\theta| < 1$.

a) Show that the coefficients α_j of the best linear interpolator are given by

$$\alpha_j = \theta^j,$$

for $j \geq 0$ and then the best linear interpolator of y_0 based on $\overline{\text{sp}}\{y_t, t \neq 0\}$ is

$$\widetilde{y}_0 = -\sum_{j=1}^{\infty} \theta^j (y_j + y_{-j}).$$

b) Verify that in this case, formula (11.12) may be written as

$$\alpha_j - \left(\frac{1+\theta^2}{\theta}\right)\alpha_{j-1} + \alpha_{j-2} = 0,$$

for $k \in \mathbb{Z}$, $k \neq 0$.

c) Show that interpolation error variance is given by $\sigma_{\text{int}}^2 = \sigma^2(1 - \theta^2)$.

11.13 Prove that for a fractional noise process FN(d), the coefficients α_j of the best linear interpolator satisfy

$$\alpha_j \sim \frac{\Gamma(1+d)}{\Gamma(-d)} j^{-1-2d},$$

as $j \to \infty$.

11.14 Calculate the coefficients α_j of the best linear interpolator for the ARFIMA$(1, d, 0)$ discussed in Example 11.5. Recall that

$$\pi_j = \begin{cases} 1, & j = 0, \\ \eta_j + \phi\eta_{j-1}, & j \geq 1, \end{cases}$$

and that

$$\sum_{i=0}^{\infty} \eta_i \eta_{i+j} = \frac{\gamma_0(j)}{\sigma^2},$$

where $\gamma_0(j)$ is the autocovariance function of a fractional noise.

CHAPTER 12

SEASONALITY

In many practical applications researchers have found time series exhibiting both long-range dependence and cyclical behavior. For instance, this phenomenon occurs in revenue series, inflation rates, monetary aggregates, gross national product series, shipping data, and monthly flows of the Nile River; see Section 12.6 for specific references. Consequently, several statistical methodologies have been proposed to model this type of data including the Gegenbauer autoregressive moving-average processes (GARMA), seasonal autoregressive fractionally integrated moving-average (SARFIMA) models, k-factor GARMA processes, and flexible seasonal fractionally integrated processes (flexible ARFISMA), among others. In this chapter we review some of these statistical methodologies. A general long-memory seasonal process is described in Section 12.1. This section also discusses some large sample properties of the MLE and Whittle estimators such as consistency, central limit theorem, and efficiency. Calculation of the asymptotic variance of maximum-likelihood and quasi-maximum-likelihood parameter estimates is addressed in Section 12.2. The finite sample performance of these estimators is studied in Section 12.4 by means of Monte Carlo simulations while Section 12.5 is devoted to the analysis of a real-life data illustration of these estimation methodologies. Further reading on this topic are suggested in Section 12.6 and several problems are listed at the end of this chapter.

Long-Memory Time Series. By Wilfredo Palma

12.1 A LONG-MEMORY SEASONAL MODEL

A general class of Gaussian seasonal long-memory processes may be specified by the spectral density

$$f(\lambda) = g(\lambda)|\lambda|^{-\alpha} \prod_{i=1}^{r} \prod_{j=1}^{m_i} |\lambda - \lambda_{ij}|^{-\alpha_i}, \tag{12.1}$$

where $\lambda \in (-\pi, \pi]$, $0 \leq \alpha, \alpha_i < 1$, $i = 1, ..., r$, $g(\lambda)$ is a symmetric, strictly positive, continuous, bounded function and $\lambda_{ij} \neq 0$ are poles for $j = 1, ..., m_i$, $i = 1, ..., r$. To ensure the symmetry of f, we assume that for any $i = 1, ..., r$, $j = 1, ..., m_i$, there is one and only one $1 \leq j' \leq m_i$ such that $\lambda_{ij} = -\lambda_{ij'}$.

As shown in the following examples, the spectral densities of many widely used models such as the seasonal ARFIMA process and the k-factor GARMA process satisfy (12.1). In Example 12.1 we review the seasonal ARFIMA model while in Example 12.2 we examine the so-called k-factor GARMA process.

EXAMPLE 12.1

Consider a seasonal ARFIMA model with multiple periods $s_1, \ldots, s_r$:

$$\phi(B) \prod_{i=1}^{r} \Phi_i(B^{s_i}) y_t = \theta(B) \prod_{i=1}^{r} \left[\Theta_i(B^{s_i})(1 - B^{s_i})^{-d_{s_i}}\right] (1 - B)^{-d} \varepsilon_t, \tag{12.2}$$

where $\phi(B)$, $\Phi_i(B^{s_i})$, $\theta(B)$, $\Theta_i(B^{s_i})$ are autoregressive and moving-average polynomials, for $i = 1, \ldots, r$.

The spectral density of the model described by (12.2) is given by

$$f_{s_1,\ldots,s_r}(\lambda) = \frac{\sigma^2}{2\pi} \frac{|\theta(e^{i\lambda})|^2}{|\phi(e^{i\lambda})|^2} |1 - e^{i\lambda}|^{-2d} \prod_{i=1}^{r} \frac{|\Theta_i(e^{i\lambda s_i})|^2 |1 - e^{i\lambda s_i}|^{-2d_{s_i}}}{|\Phi_i(e^{i\lambda s_i})|^2}.$$

Observe that this spectral density may be written as

$$f_{s_1,\ldots,s_r}(\lambda) = H(\lambda)|\lambda|^{-2d-2d_{s_1}-\cdots-2d_{s_r}} \prod_{i=1}^{r} \prod_{j=1}^{s_i} |\lambda - \lambda_{ij}|^{-2d_{s_i}},$$

which is a special case of (12.1) where

$$\begin{aligned} H(\lambda) &= \frac{\sigma^2}{2\pi} \frac{|\theta(e^{i\lambda})|^2}{|\phi(e^{i\lambda})|^2} \prod_{i=1}^{r} \frac{|\Theta_i(e^{i\lambda s_i})|^2}{|\Phi_i(e^{i\lambda s_i})|^2} \\ &\times \frac{|\lambda|^{2d+2d_{s_1}+\cdots+2d_{s_r}} \prod_{i=1}^{r} \prod_{j=1}^{s_i} |\lambda - \lambda_{ij}|^{2d_{s_i}}}{|1 - e^{i\lambda}|^{2d} \prod_{i=1}^{r} |1 - e^{i\lambda s_i}|^{2d_{s_i}}}, \end{aligned}$$

and $\lambda_{ij} = 2\pi j/s_i$ for $i = 1, ..., r$, $j = 1, ..., [s_i/2]$, $\lambda_{ij} = 2\pi([s_i/2] - j)/s_i$ for $i = 1, ..., r$, $j = [s_i/2] + 1, \ldots, s_i$, $\alpha = 2d + 2d_{s_1} + \cdots + 2d_{s_r}$, and $\alpha_i = 2d_{s_i}$.

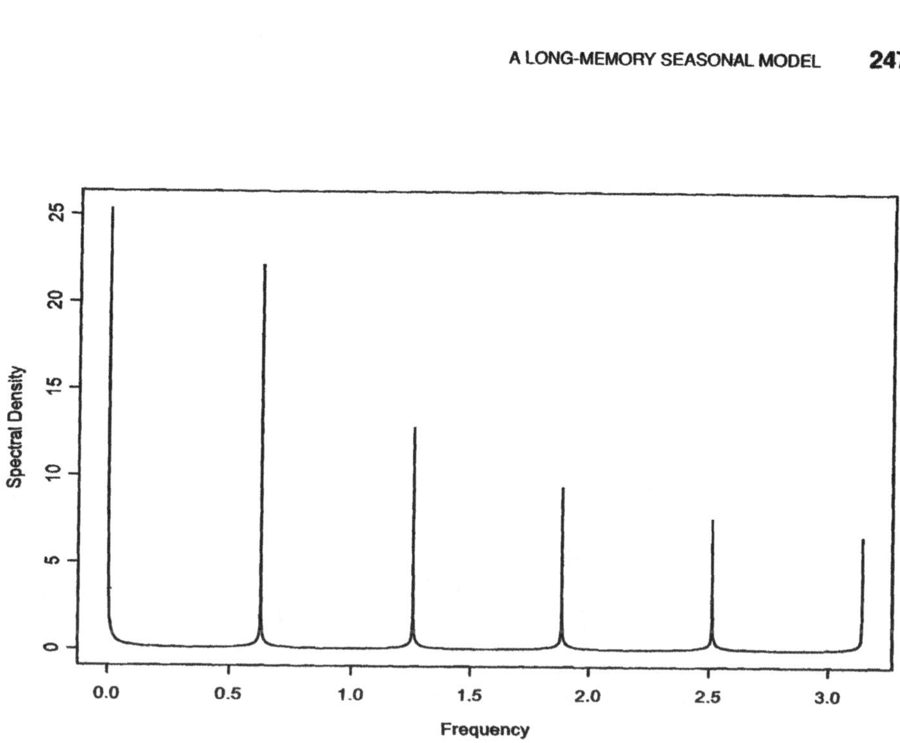

Figure 12.1 Spectral density of a SARFIMA$(0, d, 0) \times (0, d_s, 0)_s$ process with $d = 0.1$, $d_s = 0.3$, and $s = 10$.

From Figure 12.1 and Figure 12.2 we may visualize the shape of the spectral density of a SARFIMA model for two sets of parameters. Figure 12.1 displays the spectral density of a SARFIMA$(0, d, 0) \times (0, d_s, 0)_s$ process with $d = 0.1$, $d_s = 0.3$, $s = 10$, and $\sigma^2 = 1$. On the other hand, Figure 12.2 shows the spectral density of a SARFIMA$(0, d, 0) \times (1, d_s, 1)_s$ process with $d = 0.1$, $d_s = 0.3$, $\Phi = -0.8$, $\Theta = 0.1$, $s = 10$, and $\sigma^2 = 1$. As expected, these plots have poles at the frequencies $\lambda = 2\pi j/s$, $j = 0, 1, \ldots, 5$.

EXAMPLE 12.2

The spectral density of a k-factor GARMA process [see Woodward, Cheng, and Gray (1998)] is given by

$$f(\lambda) = c|\theta(e^{i\lambda})|^2 |\phi(e^{i\lambda})|^{-2} \prod_{j=1}^{k} |\cos\lambda - u_j|^{-d_j}, \tag{12.3}$$

where $c > 0$ is a constant, u_j are distinct values, $d_j \in (0, \frac{1}{2})$ when $|u_j| = 1$, and $d_j \in (0, 1)$ when $|u_j| \neq 1$.

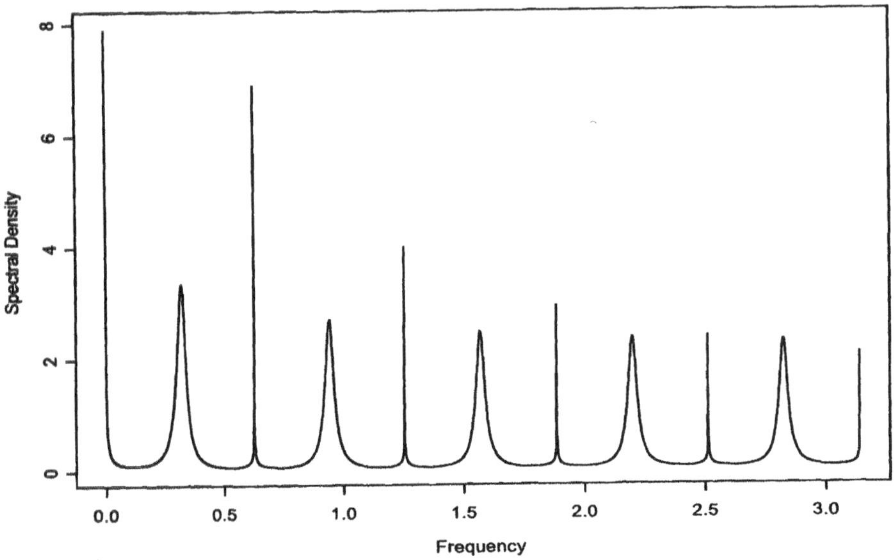

Figure 12.2 Spectral density of a SARFIMA$(0, d, 0) \times (1, d_s, 1)_s$ process with $d = 0.1$, $d_s = 0.3$, $\Phi = -0.8$, $\Theta = 0.1$ and $s = 10$.

For $|u_j| \leq 1$, we may write $u_j = \cos \lambda_j$ and this spectral density may be written in terms of (12.1) as follows:

$$f(\lambda) = H(\lambda) \prod_{j=1}^{k} |\lambda - \lambda_j|^{-d_j} |\lambda + \lambda_j|^{-d_j},$$

where

$$H(\lambda) = c|\theta(e^{i\lambda})|^2 |\phi(e^{i\lambda})|^{-2} \prod_{j=1}^{k} \left| \frac{\cos \lambda - \cos \lambda_j}{\lambda^2 - \lambda_j^2} \right|^{-d_j}$$

is a strictly positive, symmetric, continuous function with

$$\lim_{\lambda \to \pm\lambda_\ell} H(\lambda) = c \frac{|\theta(e^{i\lambda_\ell})|^2}{|\phi(e^{i\lambda_\ell})|^2} \left| \frac{\sin \lambda_\ell}{2\lambda_\ell} \right|^{-d} \prod_{j \neq \ell}^{k} \left| \frac{\cos \lambda_\ell - \cos \lambda_j}{\lambda_\ell^2 - \lambda_j^2} \right|^{-d_j},$$

for $\lambda_\ell \neq 0$ and for $\lambda_\ell = 0$

$$\lim_{\lambda \to 0} H(\lambda) = 2^{d_\ell} c|\theta(1)|^2 |\phi(1)|^{-2} \prod_{j \neq l}^{k} \left| \frac{1 - \cos \lambda_j}{\lambda_j^2} \right|^{-d_j}.$$

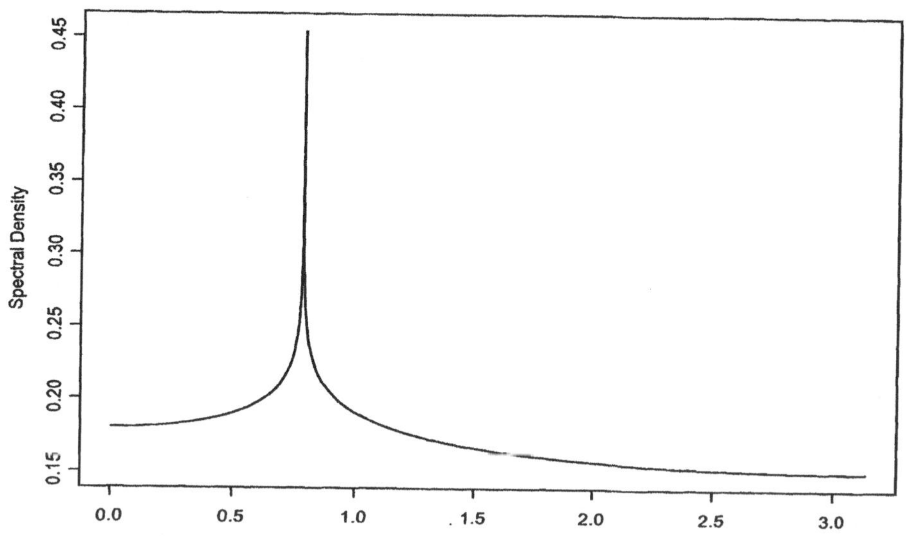

Figure 12.3 Spectral density of a k-factor GARMA process with $k = 1$, $\lambda_1 = \pi/4$, and $d_1 = 0.1$.

Observe that all these limits are finite and $H(\lambda)$ is a bounded function.

Figure 12.3 depicts the spectral density of a k-factor GARMA process with $k = 1$, $\lambda_1 = \pi/4$, and $d_1 = 0.1$. Notice that there is only one pole located at frequency $\pi/4$.

When the singularities λ_{ij} are known, the exact maximum-likelihood estimators of Gaussian time series models with spectral density (12.1) have the following large-sample properties; see Palma and Chan (2005).

Theorem 12.1. *Let $\widehat{\theta}_n$ be the exact MLE for a process satisfying* (12.1) *and θ_0 the true parameter. Then, under some regularity conditions we have*

(*a*) *Consistency: $\widehat{\theta}_n \to \theta_0$ in probability as $n \to \infty$.*

(*b*) *Central Limit Theorem: $\sqrt{n}(\widehat{\theta}_n - \theta_0) \to N(0, \Gamma^{-1}(\theta_0))$, as $n \to \infty$, where $\Gamma(\theta) = (\Gamma_{ij}(\theta))$ with*

$$\Gamma_{ij}(\theta) = \frac{1}{4\pi} \int_{-\pi}^{\pi} \left[\frac{\partial \log f_\theta(\lambda)}{\partial \theta_i} \right] \left[\frac{\partial \log f_\theta(\lambda)}{\partial \theta_j} \right] d\lambda, \tag{12.4}$$

and $f_\theta(\lambda)$ is the spectral density (12.1).

(*c*) *Efficiency: $\widehat{\theta}_n$ is an efficient estimator of θ_0.*

When the location of the pole is unknown, we still can obtain similar large-sample properties for quasi-maximum-likelihood parameter estimates. Consider, for example, the class of long-memory seasonal models defined by the spectral density

$$f(\lambda) = \frac{\sigma^2}{2\pi} \left| 4 \sin \frac{\lambda + \omega}{2} \sin \frac{\lambda - \omega}{2} \right|^{-\alpha} |h(\lambda, \tau)|^2, \tag{12.5}$$

where $\theta = (\alpha, \tau)$ are the parameters related to the long-memory and the short-memory components, and ω denotes the unknown location of the pole.

Define the function

$$S(\theta, \omega) = \frac{1}{\widetilde{n}} \sum_{j=0}^{\widetilde{n}} \frac{I(\lambda_j)}{k(\lambda_j, \theta, \omega)}, \tag{12.6}$$

where $\widetilde{n} = [n/2]$, $I(\lambda_j)$ is the periodogram given by (3.27) evaluated at the Fourier frequency $\lambda_j = 2\pi j/n$ and

$$k(\lambda, , \theta, \omega) = \frac{2\pi}{\sigma^2} f(\lambda).$$

Consider the estimators $\widehat{\theta}_n$ and $\widehat{\omega}_n$ which minimize $S(\theta, \omega)$:

$$\begin{pmatrix} \widehat{\theta}_n \\ \widehat{\omega}_n \end{pmatrix} = \text{argmin}_{\Theta \times Q} S(\theta, \lambda_q),$$

where $Q = \{q : q = 0, 1, \dots, \widetilde{n}\}$. Notice that ω belongs to a discrete set of frequencies $\lambda_0, \dots, \lambda_{\widetilde{n}}$.

Some asymptotic results about these estimators are summarized in the following theorem due to Giraitis, Hidalgo, and Robinson (2001).

Theorem 12.2. *Let θ_0 and ω_0 be the true parameters of the model described by the spectral density* (12.5). *Then*

(a) $\widehat{\theta}_n - \theta_0 = O_p(n^{-1/2})$,

(b) $\widehat{\omega}_n - \omega_0 = O_p(n^{-1})$,

(c) $\sqrt{n}(\widehat{\theta}_n - \theta_0) \to N(0, \Gamma^{-1})$, *in distribution as* $n \to \infty$, *where* Γ *is given by* (12.4) *evaluated with the spectral density* (12.5).

12.2 CALCULATION OF THE ASYMPTOTIC VARIANCE

Analytic expressions for the integral in (12.4) are difficult to obtain for an arbitrary period s. For a SARFIMA$(0, d, 0) \times (0, d_s, 0)_s$ model, the matrix $\Gamma(\theta)$ may be written as

$$\Gamma(\theta) = \begin{bmatrix} \frac{\pi^2}{6} & c(s) \\ c(s) & \frac{\pi^2}{6} \end{bmatrix}, \tag{12.7}$$

with $c(s) = (1/\pi)\int_{-\pi}^{\pi}\{\log|2\sin(\lambda/2)|\}\{\log|2\sin[s(\lambda/2)]|\}d\lambda$. An interesting feature of the asymptotic variance-covariance matrix of the parameter estimates (12.7) is that for a SARFIMA$(0,d,0)\times(0,d_s,0)_s$ process, the exact maximum-likelihood estimators $\widehat{d}$ and $\widehat{d}_s$ have the same variance.

An explicit expression for this integral can be given for $s=2$. In this case,

$$\Gamma(\boldsymbol{\theta}) = \frac{\pi^2}{12}\begin{bmatrix} 2 & 1 \\ 1 & 2 \end{bmatrix}.$$

For other values of s, the integral may be evaluated numerically. For instance, Figure 12.4 shows the evolution of $\mathrm{Var}(\widehat{d})$ as a function of the period s [see panel (a)] and the evolution of $\mathrm{Cov}(\widehat{d},\widehat{d}_s)$ as s increases [see panel (b)]. Both curves are based on the numerical evaluation of equation (12.7) and then inverting this matrix to obtain the asymptotic variance-covariance matrix of the parameters.

Observe that $\mathrm{Var}(\widehat{d}_s)$, equivalently $\mathrm{Var}(\widehat{d})$, starts at a value of $8/\pi^2$ and decreases to $6/\pi^2$ as $s\to\infty$. That is, for a very large period s, the asymptotic variance of $\widehat{d}_s$ is the same as the variance of $\widehat{d}$ from an ARFIMA$(0,d,0)$ model.

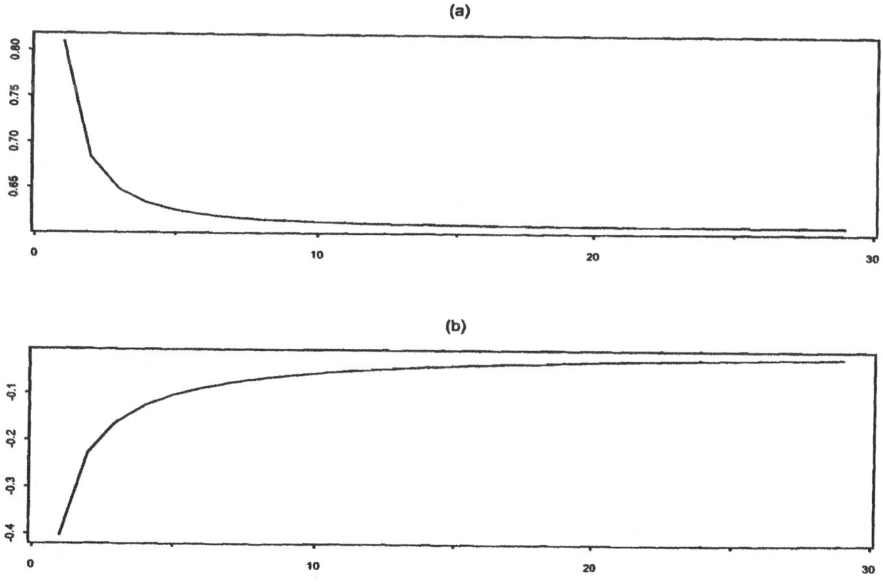

Figure 12.4 (a) Values of $\mathrm{Var}(\widehat{d}_s)$ as a function of the period s and (b) values of $\mathrm{Cov}(\widehat{d},\widehat{d}_s)$ as a function of s.

12.3 AUTOCOVARIANCE FUNCTION

Finding explicit formulae for the ACF of a general seasonal model is rather difficult. However, we can obtain an asymptotic expression as the lag increases. In particular, the following theorem is an immediate consequence of a result Lemma 1 of Oppenheim, Ould Haye, and Viano (2000), see also Leipus and Viano (2000, Lemma 8).

Theorem 12.3. *Assume that in the spectral density* (12.1) g *is a* $C^\infty([-\pi,\pi])$ *function and* $\alpha_1 > \alpha_2 \geq \cdots \geq \alpha_r$. *Let* $c_0, c_1, \ldots, c_{m_1}$ *be constants. Then, for large lag* h *the autocovariance function* $\gamma(h)$ *satisfies*

(a) If $\alpha > \alpha_1$, *then*

$$\gamma(h) = |h|^{\alpha-1}[c_0 + o(1)].$$

(b) If $\alpha \leq \alpha_1$, *then*

$$\gamma(h) = |h|^{\alpha_1-1}\left[c_0 \mathbf{1}_{\{\alpha=\alpha_1\}} + \sum_{j=1}^{m_1} c_j \cos(h\lambda_{1j}) + o(1)\right].$$

Notice that the large lag behavior of the ACF depends on the maximum value of the exponents $\alpha, \alpha_1, \ldots, \alpha_r$. For the SARFIMA process with $0 < d, d_{s_1}, \ldots, d_{s_r} < \frac{1}{2}$ and $d + d_{s_1} + \cdots + d_{s_r} < \frac{1}{2}$ the maximum exponent is always reached at zero frequency since $\alpha = d + d_{s_1} + \cdots + d_{s_r}$. Therefore in that case for large lag h the ACF behaves like

$$\gamma(h) = |h|^{2d+2d_{s_1}+\cdots+2d_{s_r}-1}[c_0 + o(1)].$$

EXAMPLE 12.3

As an illustration of the shape of the autocovariance function of a seasonal long-memory process consider the SARFIMA$(0, d, 0)\times(0, d_s, 0)_s$ process described by the discrete-time equation

$$y_t = (1 - B^s)^{-d_s}(1 - B)^{-d}\varepsilon_t,$$

where $\{\varepsilon_t\}$ is a zero mean and unit variance white noise.

In what follows, we plot the theoretical autocovariance function for three particular cases of this model from lag $h = 1$ to lag $h = 500$. The values of the ACF were calculated following the splitting method described in Subsection 4.1.3.

Figure 12.5 displays the theoretical ACF of a SARFIMA$(0, d, 0)\times(0, d_s, 0)_s$ process with parameters $d = 0.1$, $d_s = 0.3$, and $s = 12$.

Using the same method, Figure 12.6 shows the theoretical ACF of a SARFIMA $(0, d, 0)\times(0, d_s, 0)_s$ process with parameters $d = 0.3$, $d_s = 0.15$, and $s = 12$. Finally, Figure 12.7 exhibits the theoretical ACF of a SARFIMA$(0, d, 0)\times$ $(0, d_s, 0)_s$ process with parameters $d = 0.05$, $d_s = 0.44$, and $s = 24$.

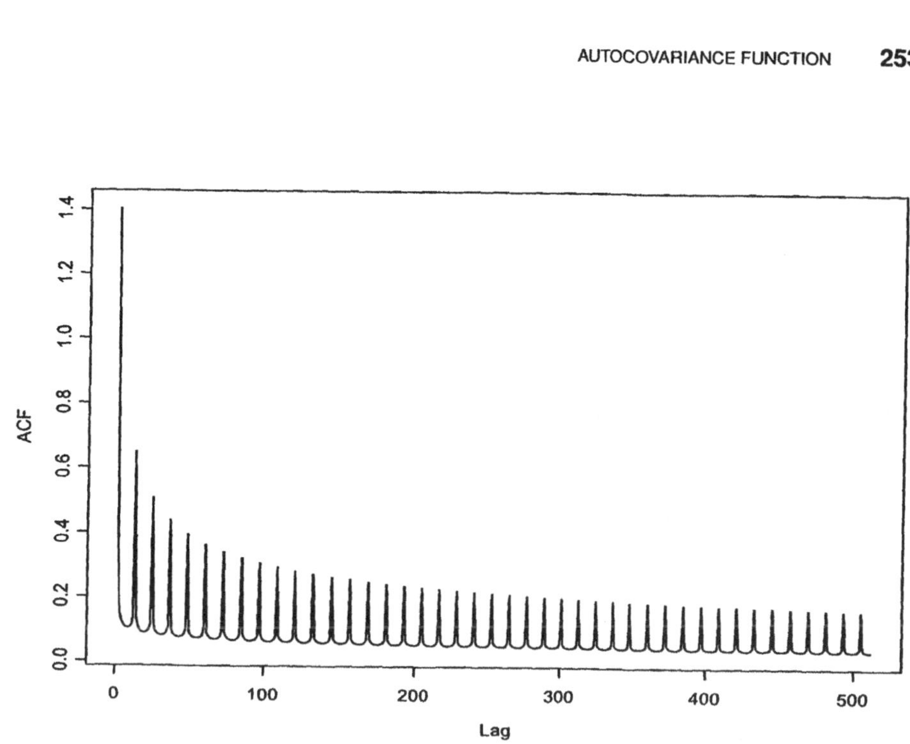

Figure 12.5 Autocovariance function of a SARFIMA$(0, d, 0) \times (0, d_s, 0)_s$ process with $d = 0.1$, $d_s = 0.3$, and $s = 12$.

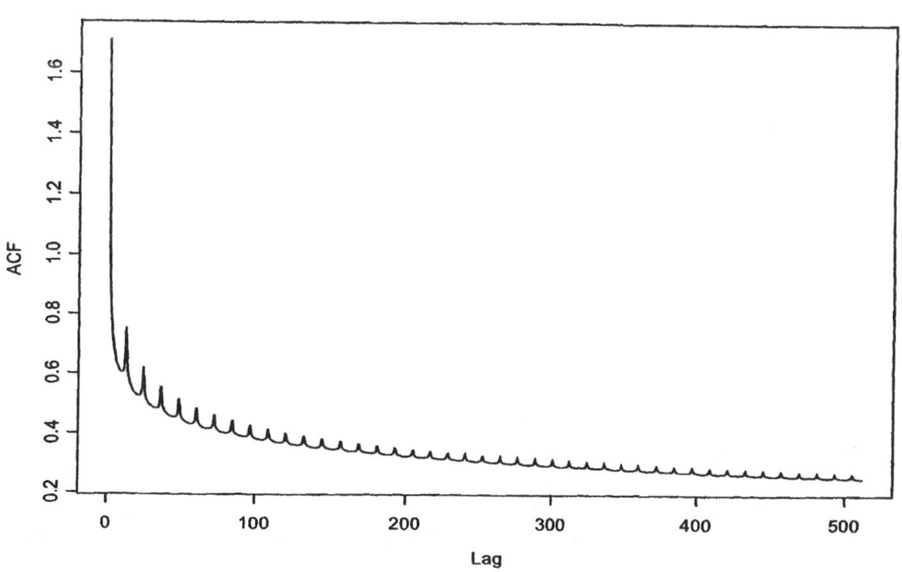

Figure 12.6 Autocovariance function of a SARFIMA$(0, d, 0) \times (0, d_s, 0)_s$ process with $d = 0.3$, $d_s = 0.15$, and $s = 12$.

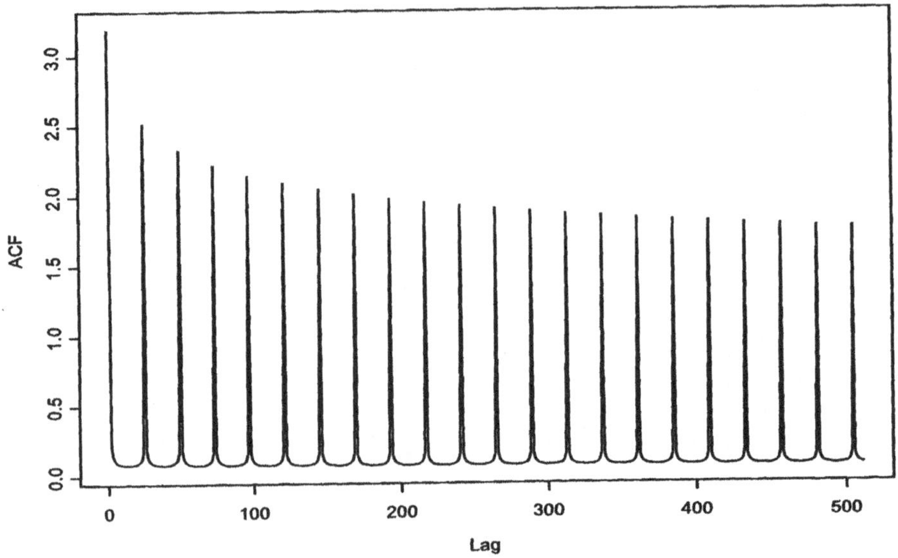

Figure 12.7 Autocovariance function of a SARFIMA$(0, d, 0) \times (0, d_s, 0)_s$ process with $d = 0.05$, $d_s = 0.44$, and $s = 24$.

12.4 MONTE CARLO STUDIES

In order to assess the finite sample performance of the ML estimates in the context of long-memory seasonal series, we show a number of Monte Carlo simulations for the class of SARFIMA models described by the difference equation

$$y_t - \mu = (1 - B^s)^{-d_s}(1 - B)^{-d}\varepsilon_t, \tag{12.8}$$

where μ is the mean of the series, and $\{\varepsilon_t\}$ are independent and identically distributed normal random variables with zero mean and unit variance.

Table 12.1 to Table 12.4 report the results from the Monte Carlo simulations for the SARFIMA$(0, d, 0) \times (0, d_s, 0)_s$ process (12.8) with mean $\mu = 0$ assumed to be either *known* or *unknown* depending on the experiment, for different values of d, d_s, sample size n, seasonal period s. The white noise variance is $\sigma^2 = 1$ in all the simulations.

The finite sample performance of the MLE is compared to the Whittle estimate and the Kalman filter approach with truncation $m = 80$.

This simulation setup is for finite sample comparison purposes only, since the asymptotic results provided by Theorem 12.1 or Theorem 12.2 do not necessarily apply to the Kalman method.

The results are based on 1000 repetitions, with seasonal series generated using the Durbin-Levinson algorithm with zero mean and unit variance Gaussian noise.

Table 12.1 SARFIMA Simulations: Sample Size $n = 256$ and Seasonal Period $s = 6$

			Known Mean					
			Exact		Whittle		Kalman	
d	d_s		$\widehat{d}$	$\widehat{d}_s$	$\widehat{d}$	$\widehat{d}_s$	$\widehat{d}$	$\widehat{d}_s$
0.1	0.3	Mean	0.0945	0.2928	0.0590	0.2842	0.0974	0.3080
		S.D	0.0020	0.0018	0.0024	0.0039	0.0024	0.0026
0.2	0.2	Mean	0.1924	0.1901	0.1574	0.1602	0.2098	0.1909
		S.D	0.0023	0.0023	0.0034	0.0036	0.0022	0.0028
0.3	0.1	Mean	0.2924	0.0948	0.2591	0.0610	0.3046	0.1003
		S.D	0.0022	0.0022	0.0035	0.0024	0.0024	0.0023
			Unknown Mean					
			Exact		Whittle		Kalman	
			$\widehat{d}$	$\widehat{d}_s$	$\widehat{d}$	$\widehat{d}_s$	$\widehat{d}$	$\widehat{d}_s$
0.1	0.3	Mean	0.0806	0.2842	0.0590	0.2842	0.0749	0.2987
		S.D	0.0020	0.0020	0.0024	0.0039	0.0020	0.0030
0.2	0.2	Mean	0.1768	0.1812	0.1574	0.1601	0.1851	0.1799
		S.D	0.0027	0.0024	0.0034	0.0036	0.0018	0.0029
0.3	0.1	Mean	0.2755	0.0863	0.2591	0.0610	0.2800	0.0867
		S.D	0.0025	0.0022	0.0035	0.0024	0.0027	0.0022

Table 12.2 SARFIMA Simulations: Sample Size $n = 256$ and Seasonal Period $s = 10$

			Known Mean					
			Exact		Whittle		Kalman	
d	d_s		$\widehat{d}$	$\widehat{d}_s$	$\widehat{d}$	$\widehat{d}_s$	$\widehat{d}$	$\widehat{d}_s$
0.1	0.3	Mean	0.0955	0.2912	0.0599	0.2886	0.0991	0.3175
		S.D	0.0022	0.0016	0.0025	0.0032	0.0032	0.0022
0.2	0.2	Mean	0.1975	0.1916	0.1583	0.1640	0.2005	0.1963
		S.D	0.0022	0.0020	0.0034	0.0032	0.0027	0.0026
0.3	0.1	Mean	0.2947	0.0953	0.2601	0.0621	0.2978	0.1014
		S.D	0.0022	0.0022	0.0037	0.0023	0.0026	0.0023
			Unknown Mean					
			Exact		Whittle		Kalman	
			$\widehat{d}$	$\widehat{d}_s$	$\widehat{d}$	$\widehat{d}_s$	$\widehat{d}$	$\widehat{d}_s$
0.1	0.3	Mean	0.0806	0.2840	0.0599	0.2886	0.0725	0.3110
		S.D	0.0023	0.0017	0.0025	0.0032	0.0025	0.0025
0.2	0.2	Mean	0.1814	0.1837	0.1583	0.1640	0.1811	0.1894
		S.D	0.0025	0.0021	0.0034	0.0032	0.0030	0.0026
0.3	0.1	Mean	0.2781	0.0871	0.2601	0.0621	0.2698	0.0897
		S.D	0.0027	0.0022	0.0037	0.0023	0.0028	0.0022

Table 12.3 SARFIMA Simulations: Sample Size $n = 512$ and Seasonal Period $s = 6$

			Known Mean					
			Exact		Whittle		Kalman	
d	d_s		$\widehat{d}$	$\widehat{d}_s$	$\widehat{d}$	$\widehat{d}_s$	$\widehat{d}$	$\widehat{d}_s$
0.1	0.3	Mean	0.0995	0.2951	0.0803	0.2942	0.1057	0.3118
		S.D	0.0012	0.0010	0.0014	0.0016	0.0012	0.0013
0.2	0.2	Mean	0.1966	0.1977	0.1795	0.1839	0.1952	0.2021
		S.D	0.0013	0.0011	0.0017	0.0015	0.0014	0.0013
0.3	0.1	Mean	0.2962	0.0980	0.2811	0.0792	0.3060	0.0964
		S.D	0.0011	0.0011	0.0014	0.0013	0.0014	0.0012
			Unknown Mean					
			Exact		Whittle		Kalman	
			$\widehat{d}$	$\widehat{d}_s$	$\widehat{d}$	$\widehat{d}_s$	$\widehat{d}$	$\widehat{d}_s$
0.1	0.3	Mean	0.0919	0.2900	0.0803	0.2942	0.0870	0.3045
		S.D	0.0012	0.0010	0.0014	0.0016	0.0011	0.0013
0.2	0.2	Mean	0.1880	0.1923	0.1795	0.1839	0.1765	0.1943
		S.D	0.0014	0.0012	0.0017	0.0015	0.0011	0.0013
0.3	0.1	Mean	0.2878	0.0932	0.2811	0.0792	0.2849	0.0864
		S.D	0.0012	0.0012	0.0014	0.0013	0.0014	0.0012

Table 12.4 SARFIMA Simulations: Sample Size $n = 512$ and Seasonal Period $s = 10$

			Known Mean					
			Exact		Whittle		Kalman	
d	d_s		$\widehat{d}$	$\widehat{d}_s$	$\widehat{d}$	$\widehat{d}_s$	$\widehat{d}$	$\widehat{d}_s$
0.1	0.3	Mean	0.0979	0.2959	0.0768	0.3006	0.0995	0.3134
		S.D	0.0012	0.0009	0.0014	0.0016	0.0016	0.0014
0.2	0.2	Mean	0.1994	0.1957	0.1813	0.1832	0.2028	0.2007
		S.D	0.0012	0.0011	0.0016	0.0016	0.0014	0.0014
0.3	0.1	Mean	0.2963	0.0968	0.2801	0.0783	0.3070	0.0948
		S.D	0.0011	0.0012	0.0015	0.0014	0.0015	0.0013
			Unknown Mean					
			Exact		Whittle		Kalman	
			$\widehat{d}$	$\widehat{d}_s$	$\widehat{d}$	$\widehat{d}_s$	$\widehat{d}$	$\widehat{d}_s$
0.1	0.3	Mean	0.0896	0.2913	0.0768	0.3006	0.0799	0.3074
		S.D	0.0012	0.0009	0.0014	0.0016	0.0014	0.0014
0.2	0.2	Mean	0.1908	0.1908	0.1813	0.1832	0.1809	0.1931
		S.D	0.0013	0.0012	0.0016	0.0016	0.0015	0.0014
0.3	0.1	Mean	0.2876	0.0921	0.2801	0.0783	0.2890	0.0862
		S.D	0.0012	0.0012	0.0015	0.0014	0.0013	0.0012

The autocovariance function was computed by the convolution method of Subsection 4.1.3.

In order to explore the effect of the estimation of the mean we have considered two situations: *known mean* where the process is assumed to have zero mean and *unknown mean* where the expected value of the process is estimated by the sample mean and then centered before the computations.

The exact MLE method has been implemented computationally by means of the Durbin-Levinson algorithm discussed in Chapter 4 with autocovariance calculated by the approach given in Subsection 4.1.3.

The Whittle method has been implemented by minimizing the following expression; see Section 4.4 and equation (12.6):

$$S(\theta) = \frac{2}{n} \sum_{k=1}^{[n/2]} \frac{I(\lambda_k)}{f_\theta(\lambda_k)},$$

where $\theta = (d, d_s)$ and $\lambda_k = 2\pi k/n$, with periodogram given by

$$I(\lambda_k) = \frac{1}{2\pi n} \left| \sum_{t=1}^{n} y_t e^{\imath t \lambda_k} \right|^2,$$

see definition (3.27), and spectral density

$$f_\theta(\lambda) = \frac{1}{2\pi} \left|1 - e^{\imath\lambda}\right|^{-2d} \left|1 - e^{\imath\lambda s}\right|^{-2d_s}.$$

The approximate Kalman filter ML estimates are based on a finite state space representation of the truncated MA(∞) expansion described in Section 4.3.

All these experiments were conducted with a Pentium IV (2.8-GHz) machine using a FORTRAN program in a Windows XP platform. Table 12.5 shows the *central processing unit* (CPU) average times in milliseconds for one evaluation of the respective method. The average times are computed over 2000 calls from the optimization routine. For all sample sizes studied, the fastest method is the Whittle.

From Table 12.1 to Table 12.4, it seems that for the *known mean* case the exact MLE and the Kalman methods display little bias for both sample sizes. On the other

Table 12.5 CPU Average Times in Milliseconds for Computing $\widehat{d}$ and $\widehat{d}_s$ with $d = 0.2$, $d_s = 0.2$, and $s = 6$

Sample Size	Exact	Whittle	Kalman
256	3.0	0.1	10.5
512	7.0	0.2	11.1
1024	17.9	0.5	12.0

Table 12.6 Asymptotic Standard Deviation of $\widehat{d}$ and $\widehat{d}_s$

n	$s = 6$	$s = 10$
256	0.0503	0.0503
512	0.0356	0.0356

hand, the Whittle method presents a noticeable downward bias for both estimators $\widehat{d}$ and $\widehat{d}_s$.

The sample standard deviations of the estimates are close to their theoretical counterparts, reported in Table 12.6, for the three methods considered. However, the exact MLE seems to have slightly lower sample standard deviations than the other methods, for both long-memory parameters and both sample sizes. The theoretical values of the standard deviations of the estimated parameters given in Table 12.6 are based on formula (12.7).

In the *unknown mean* case, all the estimates seem to display a downward bias, which is stronger for the Whittle method. However, the bias displayed by this estimate is similar to the *known mean* case, since the Whittle algorithm is not affected by the estimation of the mean. Similarly to the previous case, the estimated standard deviations are comparable to the theoretical values, and the exact MLE displays slightly lower sample standard deviations than the other methods for most long-memory parameters and sample size combinations.

12.5 ILLUSTRATION

In this section we apply the maximum-likelihood estimation to the analysis of a time series consisting of *hyper text transfer protocol* (HTTP) requests to a World Wide Web server at the University of Saskatchewan; see Palma and Chan (2005) for details about these data. It has been reported that communication network traffic may exhibit long-memory behavior; see, for example, Beran (1994b) and Willinger, Taqqu, Leland, and Wilson (1995) . The data analyzed here consist of the logarithm of the number of requests within one-hour periods.

The Internet traffic series is shown in Figure 12.8 while its sample autocorrelation function is displayed in Figure 12.9. Observe that the sample ACF decays slowly and exhibits a 24-hour periodicity. To account for these features we fit a SARFIMA model to this time series.

Table 12.7 reports the maximum-likelihood parameter estimates and the t-tests for the SARFIMA$(1,0,1) \times (0,d_s,0)_s$ with $s = 24$ process:

$$(1 - \phi B)(y_t - \mu) = (1 - \theta B)(1 - B^s)^{-d_s}\varepsilon_t,$$

where ε_t is a white noise sequence with variance σ^2.

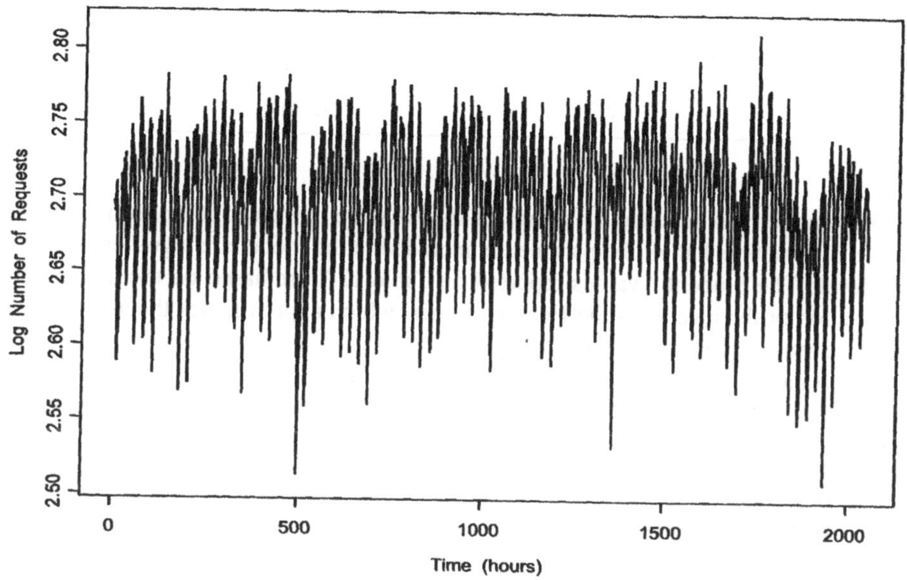

Figure 12.8 Log HTTP requests data.

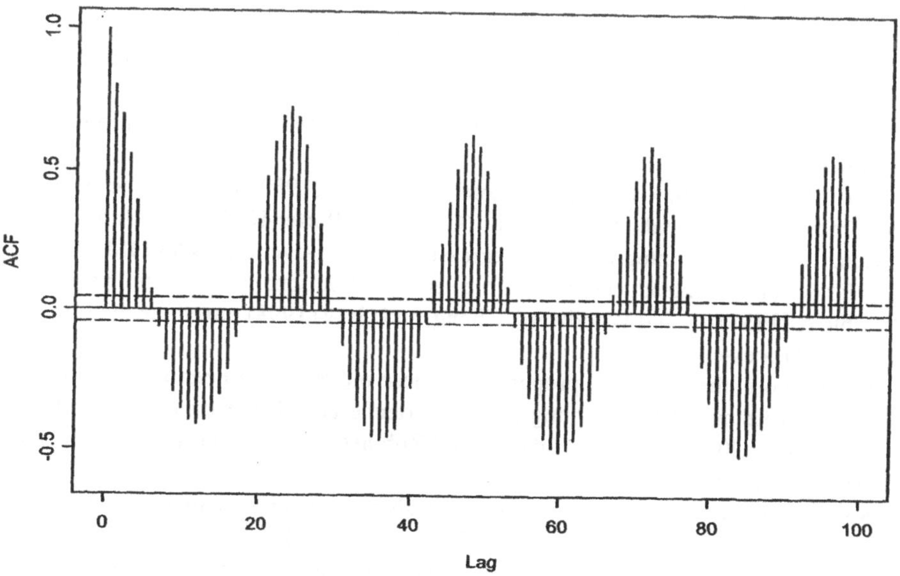

Figure 12.9 Log HTTP requests data: Sample autocorrelation function.

Table 12.7 Log Internet Traffic Data: Maximum-Likelihood Estimation of the SARFIMA$(1, 0, 1) \times (0, d_s, 0)_s$ Model

Parameter	d_s	ϕ	θ
Estimate	0.4456	0.8534	0.3246
Student-t	2.2558	7.5566	2.8623

This model was selected by means of the Akaike's information criterion. From Table 12.7, notice that all the parameters included in the model are significant at the 5% level. The Student-t values reported on Table 12.7 are based on the numerical calculation of the inverse of the Hessian matrix, which approximates the asymptotic variance-variance matrix of the parameters.

The standard deviation of the Internet traffic series is 0.6522 while the residual standard deviation is 0.2060. Thus, the fitted seasonal long-memory model explains roughly two thirds of the total standard deviation of the data.

12.6 BIBLIOGRAPHIC NOTES

Long-range-dependent data with seasonal behavior have been reported fields as diverse as economics, physics, and hydrology. For example, inflation rates are studied by Hassler and Wolters (1995), revenue series are analyzed by Ray (1993a), monetary aggregates are considered by Porter-Hudak (1990), quarterly gross national product and shipping data are discussed by Ooms (1995), and monthly flows of the Nile River are studied by Montanari, Rosso, and Taqqu (2000).

Many statistical methodologies have been proposed to model this seasonal long-range-dependent data. For example, Abrahams and Dempster (1979) extend the fractional Gaussian noise process [see Mandelbrot and Van Ness (1968)] to include seasonal components.

On the other hand, Gray, Zhang, and Woodward (1989) propose the generalized fractional or Gegenbauer processes (GARMA), Porter-Hudak (1990) discusses seasonal fractionally integrated autoregressive moving-average (SARFIMA) models, Hassler (1994) introduces the flexible seasonal fractionally integrated processes (flexible ARFISMA), and Woodward, Cheng, and Gray (1998) introduce the k-factor GARMA processes.

Furthermore, the statistical properties of these models have been investigated by Giraitis and Leipus (1995), Chung (1996), Arteche and Robinson (2000), Velasco and Robinson (2000), Giraitis, Hidalgo, and Robinson (2001), and Palma and Chan (2005), among others.

Some large sample properties of the eigenvalues of covariance matrices from seasonal long-memory models are discussed in Palma and Bondon (2003) while finite sample performances of a number of estimation techniques for fractional seasonal models are studied in the papers by Reisen, Rodrigues, and Palma (2006a,b).

Problems

12.1 For which values of α, $\alpha_1, \ldots, \alpha_r$ does the model described by the spectral density (12.1) exhibit long memory? Be specific about the definition of long memory that you are using.

12.2 Let s be a positive integer and $d \in (0, \frac{1}{2})$. Calculate the coefficients ψ_j in the expansion

$$(1 - B^s)^{-d} = \sum_{j=1}^{\infty} \psi_j B^j.$$

12.3 Let s be a positive integer and $d < 0$.

a) Prove that the expansion

$$(1 - B^s)^d = \sum_{j=1}^{\infty} \pi_j B^j,$$

is absolutely convergent.

b) Let z be a random variable such that $E(z^2) < \infty$. Show that

$$(1 - B^s)^d z = 0.$$

c) Consider the equation

$$(1 - B^s)^d y_t = \varepsilon_t, \tag{12.9}$$

where $\{\varepsilon_t\}$ is a white noise sequence with finite variance. Prove that there is a stationary solution $\{y_t\}$ of equation (12.9).

d) Show that the process $\{x_t\}$ defined as $x_t = y_t + z$ is also a stationary solution of equation (12.9).

12.4 Consider the SARFIMA$(0,0,0) \times (0, d_s, 0)_s$ process

$$y_t = (1 - B^s)^{-d_s} \varepsilon_t,$$

where ε_t is a white noise sequence with variance $\sigma = 1$. Let $\gamma_s(k)$ be the ACF of the process y_t and $\gamma(k)$ be the ACF of a fractional noise FN(d) process with unit noise variance. Verify that

$$\gamma_s(k) = \gamma(sk).$$

12.5 Consider the SARFIMA$(2, d, 2) \times (0, d_s, 0)_s$ process

$$(1 - \phi_1 B - \phi_2 B^2) y_t = (1 - \theta_1 B - \theta_2 B^2)(1 - B^s)^{-d_s} \varepsilon_t,$$

where ε_t is a white noise sequence with variance σ and

$$\boldsymbol{\theta} = (\phi_1, \phi_2, d, \theta_1, \theta_2, d_s, \sigma) = (0.2, 0.5, 0.3, -0.2, 0.5, 0.1, 1.3).$$

a) Is this process stationary?
b) Is this process invertible?

12.6 Calculate numerically the variance-covariance matrix $\Gamma(\theta)$ given in (12.7) for a SARFIMA$(0, d, 0) \times (0, d_s, 0)_s$ process with $\boldsymbol{\theta} = (d, d_s) = (0.1, 0.2)$ and $s = 12$.

12.7 Simulate a sample of 1000 observations from a Gaussian SARFIMA$(1, d, 1) \times (0, d_s, 0)_s$ process with parameters

$$\boldsymbol{\theta} = (\phi_1, \theta_1, d, d_s, \sigma) = (0.6, 0.3, 0.2, 0.2, 1).$$

12.8 Implement computationally the splitting method for calculating the theoretical ACF of a SARFIMA$(0, d, 0) \times (0, d_s, 0)_s$ process.

12.9 Write a state space system for the SARFIMA process described in the previous problem.

12.10 Calculate the MLE for the SARFIMA process in Problem 12.7 by means of the state space systems and the Whittle method.

12.11 Suppose that T is the variance-covariance matrix of a stationary Gaussian process $\{y_1, y_2, \ldots, y_n\}$ with spectral density satisfying equation (12.1).

a) Let $|T|$ be the Euclidean norm of T, that is,

$$|T| = [\mathrm{tr}(TT^*)]^{1/2}.$$

Show that

$$|T|^2 = \sum_{i=1}^{n}\sum_{j=1}^{n}\gamma^2(i-j) = n\sum_{j=0}^{n-1}\left(1-\frac{j}{n}\right)\gamma^2(j).$$

b) Verify that an application of Theorem 12.3 yields

$$\gamma(j) \leq K j^{\tilde{\alpha}-1},$$

for $j \geq 1$, where $\tilde{\alpha} = \max\{\alpha, \alpha_1, \ldots, \alpha_r\}$.

c) Show that

$$|T|^2 \leq n\sum_{j=0}^{n-1}\gamma^2(j) \leq Kn\sum_{j=1}^{n} j^{2\tilde{\alpha}-2}.$$

d) By applying Lemma 3.3, verify that

$$|T|^2 \leq Kn^{2\tilde{\alpha}},$$

and then

$$|T| \leq Kn.$$

12.12 Consider the GARMA process described by the spectral density

$$f(\lambda) = H(\lambda)|\lambda - \lambda_1|^{-d_1}|\lambda + \lambda_1|^{-d_1},$$

where $H(\lambda)$ is a $C^\infty([-\pi, \pi])$ function. Verify that the ACF of this process satisfies the asymptotic expression

$$\gamma(h) = |h|^{2d_1 - 1}[c_1 \cos(h\lambda_1) + o(1)],$$

as $|h| \to \infty$.

REFERENCES

M. Abrahams and A. Dempster. (1979). Research on seasonal analysis. progress report on the asa/census project on seasonal adjustment. Technical report. Department of Statistics, Harvard University, Boston, MA.

P. Abry, P. Flandrin, M. S. Taqqu, and D. Veitch. (2003). Self-similarity and long-range dependence through the wavelet lens. In P. Doukhan, G. Oppenheim, and M. S. Taqqu, editors, *Theory and Applications of Long-Range Dependence*. Birkhäuser, Boston, MA, pp. 527–556.

R. K. Adenstedt. (1974). On large-sample estimation for the mean of a stationary random sequence. *Annals of Statistics* 2, 1095–1107.

G. S. Ammar. (1998). Classical foundations of algorithms for solving positive definite Toeplitz equations. *Calcolo. A Quarterly on Numerical Analysis and Theory of Computation* 33, 99–113.

B. D. O. Anderson and J. B. Moore. (1979). *Optimal Filtering*. Prentice-Hall, New York.

C. F. Ansley and R. Kohn. (1983). Exact likelihood of vector autoregressive-moving average process with missing or aggregated data. *Biometrika* 70, 275–278.

M. Aoki. (1990). *State Space Modeling of Time Series*. Springer, Berlin.

J. Arteche and P. M. Robinson. (2000). Semiparametric inference in seasonal and cyclical long memory processes. *Journal of Time Series Analysis* 21, 1–25.

R. T. Baillie. (1996). Long memory processes and fractional integration in econometrics. *Journal of Econometrics* 73, 5–59.

R. T. Baillie, T. Bollerslev, and H. O. Mikkelsen. (1996). Fractionally integrated generalized autoregressive conditional heteroskedasticity. *Journal of Econometrics* 74, 3–30.

A. A. Barker. (1965). Monte Carlo calculations of the radial distribution functions for a proton-electron plasma. *Australian Journal of Physics* 18, 119–133.

O. E. Barndorff-Nielsen and N. Shephard. (2001). Modelling by Lévy processes for financial econometrics. In *Lévy Processes*. Birkhäuser, Boston, MA, pp. 283–318.

G. K. Basak, N. H. Chan, and W. Palma. (2001). The approximation of long-memory processes by an ARMA model. *Journal of Forecasting* 20, 367–389.

W. Bell and S. Hillmer. (1991). Initializing the Kalman filter for nonstationary time series models. *Journal of Time Series Analysis* 12, 283–300.

A. F. Bennett. (1992). *Inverse Methods in Physical Oceanography*. Cambridge Monographs on Mechanics. Cambridge University Press, Cambridge.

J. Beran. (1994a). On a class of M-estimators for Gaussian long-memory models. *Biometrika* 81, 755–766.

J. Beran. (1994b). *Statistics for Long-Memory Processes*, Vol. 61, *Monographs on Statistics and Applied Probability*. Chapman and Hall, New York.

J. Beran and N. Terrin. (1994). Estimation of the long-memory parameter, based on a multivariate central limit theorem. *Journal of Time Series Analysis* 15, 269–278.

S. Bertelli and M. Caporin. (2002). A note on calculating autocovariances of long-memory processes. *Journal of Time Series Analysis* 23, 503–508.

R. J. Bhansali and P. S. Kokoszka. (2003). Prediction of long-memory time series. In P. Doukhan, G. Oppenheim, and M. S. Taqqu, editors, *Theory and Applications of Long-Range Dependence*. Birkhäuser, Boston, MA, pp. 355–367.

N. H. Bingham, C. M. Goldie, and J. L. Teugels. (1987). *Regular Variation*, Vol. 27, *Encyclopedia of Mathematics and Its Applications*. Cambridge University Press, Cambridge.

P. Bloomfield. (1985). On series representations for linear predictors. 13, 226–233.

R. Bojanić and J. Karamata. (1963). On slowly varying functions and asymptotic relations. Technical Report 432. Math. Research Center, Madison, WI.

T. Bollerslev. (1986). Generalized autoregressive conditional heteroskedasticity. *Journal of Econometrics* 31, 307–327.

T. Bollerslev. (1988). On the correlation structure for the generalized autoregressive conditional heteroskedastic process. *Journal of Time Series Analysis* 9, 121–131.

T. Bollerslev and H. O. Mikkelsen. (1996). Modeling and pricing long memory in stock market volatility. *Journal of Econometrics* 73, 151–184.

P. Bondon. (2002). Prediction with incomplete past of a stationary process. *Stochastic Processes and Their Applications* 98, 67–76.

P. Bondon and W. Palma. (2006). Prediction of strongly dependent time series. *Working Paper*, Supelec, Paris.

P. Bondon and W. Palma. (2007). A class of antipersitent processes. *Journal of Time Series Analysis,* in press.

G. E. P. Box, G. M. Jenkins, and G. C. Reinsel. (1994). *Time Series Analysis*. Prentice Hall, Englewood Cliffs, NJ.

G. E. P. Box and G. C. Tiao. (1992). *Bayesian Inference in Statistical Analysis*. Wiley, New York.

F. J. Breidt, N. Crato, and P. de Lima. (1998). The detection and estimation of long memory in stochastic volatility. *Journal of Econometrics* 83, 325–348.

A. E. Brockwell. (2004). A class of generalized long-memory time series models. Technical Report 813. Department of Statistics, Carnegie Mellon University, Pittsburgh.

P. J. Brockwell and R. A. Davis. (1991). *Time Series: Theory and Methods*. Springer, New York.

N. L. Carothers. (2000). *Real Analysis*. Cambridge University Press, Cambridge.

N. H. Chan. (2002). *Time Series. Applications to Finance*. Wiley Series in Probability and Statistics. Wiley, New York.

N. H. Chan, J. B. Kadane, R. N. Miller, and W. Palma. (1996). Estimation of tropical sea level anomaly by and improved Kalman filter. *Journal of Physical Oceanography* 26, 1286–1303.

N. H. Chan and W. Palma. (1998). State space modeling of long-memory processes. *Annals of Statistics* 26, 719–740.

N. H. Chan and W. Palma. (2006). Estimation of long-memory time series models: A survey of different likelihood-based methods. In T. B. Fomby and D. Terrell, editors, *Econometric Analysis of Financial and Economic Time Series, Part B*, Vol. 20, *Adv. Econometrics*. Elsevier, Amsterdam, pp. 89–121.

N. H. Chan and G. Petris. (2000). Recent developments in heteroskedastic time series. In W. S. Chan, W. K. Li, and H. Tong, editors, *Statistics and Finance: An Interface*. Imperial College Press, London, pp. 169–184.

R. Cheng and M. Pourahmadi. (1997). Prediction with incomplete past and interpolation of missing values. *Statistics & Probability Letters* 33, 341–346.

K. Choy and M. Taniguchi. (2001). Stochastic regression model with dependent disturbances. *Journal of Time Series Analysis* 22, 175–196.

C. F. Chung. (1996). A generalized fractionally integrated autoregressive moving-average process. *Journal of Time Series Analysis* 17, 111–140.

J. B. Conway. (1990). *A Course in Functional Analysis*, Vol. 96, *Graduate Texts in Mathematics*. Springer, New York.

D. R. Cox. (1984). Long-range dependence: A review. In H. A. David and H. T. David, editors, *Statistics: An Appraisal*. Iowa State University Press, Ames, IA, pp. 55–74.

R. F. Curtain and H. Zwart. (1995). *An Introduction to Infinite-Dimensional Linear Systems Theory*, Vol. 21, *Texts in Applied Mathematics*. Springer, New York.

R. Dahlhaus. (1989). Efficient parameter estimation for self-similar processes. *Annals of Statistics* 17, 1749–1766.

R. Dahlhaus. (1995). Efficient location and regression estimation for long range dependent regression models. *Annals of Statistics* 23, 1029–1047.

R. Dahlhaus. (2006). Correction note: Efficient parameter estimation for self-similar processes. *Annals of Statistics* 34, 1045–1047.

E. Damsleth. (1980). Interpolating missing values in a time series. *Scandinavian Journal of Statistics. Theory and Applications* 7, 33–39.

A. Demos. (2002). Moments and dynamic structure of a time-varying parameter stochastic volatility in mean model. *Econometrics Journal* 5, 345–357.

R. S. Deo and C. M. Hurvich. (2003). Estimation of long memory in volatility. In P. Doukhan, G. Oppenheim, and M. S. Taqqu, editors, *Theory and Applications of Long-Range Dependence*. Birkhäuser, Boston, MA, pp. 313–324.

I. Dittmann and C. W. J. Granger. (2002). Properties of nonlinear transformations of fractionally integrated processes. *Journal of Econometrics* 110, 113–133.

J. A. Doornik and M. Ooms. (2003). Computational aspects of maximum likelihood estimation of autoregressive fractionally integrated moving average models. *Computational Statistics & Data Analysis* 42, 333–348.

J. Durbin. (1960). The fitting of time series models. *International Statistical Review* 28, 233–244.

J. Durbin and S. J. Koopman. (2001). *Time Series Analysis by State Space Methods*, Vol. 24, *Oxford Statistical Science Series*. Oxford University Press, Oxford.

R. F. Engle. (1982). Autoregressive conditional heteroscedasticity with estimates of the variance of United Kingdom inflation. *Econometrica* 50, 987–1007.

W. Feller. (1971). *An Introduction to Probability Theory and Its Applications. Vol. II.* Second edition. Wiley, New York.

P. Flandrin. (1999). *Time-Frequency/Time-Scale Analysis*, Vol. 10, *Wavelet Analysis and Its Applications*. Academic, San Diego, CA.

R. Fox and M. S. Taqqu. (1986). Large-sample properties of parameter estimates for strongly dependent stationary Gaussian time series. *Annals of Statistics* 14, 517–532.

R. Fox and M. S. Taqqu. (1987). Central limit theorems for quadratic forms in random variables having long-range dependence. *Probability Theory and Related Fields* 74, 213–240.

A. Gelman and D. B. Rubin. (1992). Inference from iterative simulation using multiple sequences. *Statistical Science* 7, 457–511.

J. Geweke. (1989). Bayesian inference in econometric models using Monte Carlo integration. *Econometrica* 57, 1317–1339.

J. Geweke and S. Porter-Hudak. (1983). The estimation and application of long memory time series models. *Journal of Time Series Analysis* 4, 221–238.

E. Ghysels, A. C. Harvey, and E. Renault. (1996). Stochastic volatility. In *Statistical Methods in Finance*, Vol. 14, *Handbook of Statistics*. North-Holland, Amsterdam, pp. 119–191.

L. Giraitis, J. Hidalgo, and P. M. Robinson. (2001). Gaussian estimation of parametric spectral density with unknown pole. *Annals of Statistics* 29, 987–1023.

L. Giraitis, P. Kokoszka, R. Leipus, and G. Teyssière. (2003). Rescaled variance and related tests for long memory in volatility and levels. *Journal of Econometrics* 112, 265–294.

L. Giraitis and R. Leipus. (1995). A generalized fractionally differencing approach in long-memory modeling. *Matematikos ir Informatikos Institutas* 35, 65–81.

L. Giraitis and D. Surgailis. (1990). A central limit theorem for quadratic forms in strongly dependent linear variables and its application to asymptotical normality of Whittle's estimate. *Probability Theory and Related Fields* 86, 87–104.

G. H. Givens and J. A. Hoeting. (2005). *Computational Statistics.* Wiley Series in Probability and Statistics. Wiley, Hoboken, NJ.

I. S. Gradshteyn and I. M. Ryzhik. (2000). *Table of Integrals, Series, and Products.* Academic, San Diego, CA.

C. W. J. Granger and R. Joyeux. (1980). An introduction to long-memory time series models and fractional differencing. *Journal of Time Series Analysis* 1, 15–29.

C. W. L. Granger and A. P. Andersen. (1978). *An Introduction to Bilinear Time Series Models.* Vandenhoek and Ruprecht, Gottingen.

H. L. Gray, N. F. Zhang, and W. A. Woodward. (1989). On generalized fractional processes. *Journal of Time Series Analysis* 10, 233–257.

U. Grenander. (1954). On the estimation of regression coefficients in the case of an autocorrelated disturbance. *Annals of Mathematical Statistics* 25, 252–272.

U. Grenander and M. Rosenblatt. (1957). *Statistical Analysis of Stationary Time Series.* Wiley, New York.

P. Hall. (1997). Defining and measuring long-range dependence. In *Nonlinear Dynamics and Time Series (Montreal, PQ, 1995)*, Vol. 11, *Fields Inst. Commun.* Amer. Math. Soc., Providence, RI, pp. 153–160.

E. J. Hannan. (1970). *Multiple Time Series.* Wiley, New York.

E. J. Hannan and M. Deistler. (1988). *The Statistical Theory of Linear Systems.* Wiley, New York.

A. C. Harvey. (1989). *Forecasting Structural Time Series and the Kalman Filter.* Cambridge University Press, Cambridge.

A. C. Harvey and R. G. Pierse. (1984). Estimating missing observations in economic time series. *Journal of the American Statistical Association* 79, 125–131.

A. C. Harvey, E. Ruiz, and N. Shephard. (1994). Multivariate stochastic variance models. *Review of Economic Studies* 61, 247–265.

U. Hassler. (1994). (Mis)specification of long memory in seasonal time series. *Journal of Time Series Analysis* 15, 19–30.

U. Hassler and J. Wolters. (1995). Long memory in inflation rates: International evidence. *Journal of Business and Economic Statistics* 13, 37–45.

J. Hasslett and A. E. Raftery. (1989). Space-time modelling with long-memory dependence: Assessing Ireland's wind power resource. *Journal of Applied Statistics* 38, 1–50.

W. K. Hastings. (1970). Monte Carlo sampling methods using Markov chains and their applications. *Biometrika* 57, 97–109.

C. He and T. Teräsvirta. (1999a). Properties of moments of a family of GARCH processes. *Journal of Econometrics* 92, 173–192.

C. He and T. Teräsvirta. (1999b). Properties of the autocorrelation function of squared observations for second-order GARCH processes under two sets of parameter constraints. *Journal of Time Series Analysis* 20, 23–30.

C. He, T. Teräsvirta, and H. Malmsten. (2002). Moment structure of a family of first-order exponential GARCH models. *Econometric Theory* 18, 868–885.

M. Henry. (2001). Averaged periodogram spectral estimation with long-memory conditional heteroscedasticity. *Journal of Time Series Analysis* 22, 431–459.

K. W. Hipel and A. I. McLeod. (1994). *Time Series Modelling of Water Resources and Environmental Systems*. Elsevier, Amsterdam.

J. R. M. Hosking. (1981). Fractional differencing. *Biometrika* 68, 165–176.

J. R. M. Hosking. (1996). Asymptotic distributions of the sample mean, autocovariances, and autocorrelations of long-memory time series. *Journal of Econometrics* 73, 261–284.

H. E. Hurst. (1951). Long-term storage capacity of reservoirs. *Transactions of the American Society of Civil Engineers* 116, 770–779.

C. M. Hurvich, R. Deo, and J. Brodsky. (1998). The mean squared error of Geweke and Porter-Hudak's estimator of the memory parameter of a long-memory time series. *Journal of Time Series Analysis* 19, 19–46.

I. A. Ibragimov and Y. A. Rozanov. (1978). *Gaussian Random Processes*, Vol. 9, *Applications of Mathematics*. Springer, New York.

P. Iglesias, H. Jorquera, and W. Palma. (2006). Data analysis using regression models with missing observations and long-memory: an application study. *Computational Statistics & Data Analysis* 50, 2028–2043.

A. Inoue. (1997). Regularly varying correlation functions and KMO-Langevin equations. *Hokkaido Mathematical Journal* 26, 457–482.

A. Inoue. (2002). Asymptotic behavior for partial autocorrelation functions of fractional ARIMA processes. *Annals of Applied Probability* 12, 1471–1491.

R. H. Jones. (1980). Maximum likelihood fitting of ARMA models to time series with missing observations. *Technometrics* 22, 389–395.

R. E. Kalman. (1961). A new approach to linear filtering and prediction problems. *Transactions of the American Society of Mechanical Engineers* 83D, 35–45.

R. E. Kalman and R. S. Bucy. (1961). New results in linear filtering and prediction theory. *Transactions of the American Society of Mechanical Engineers* 83, 95–108.

J. Karamata. (1930). Sur un mode de croissance régulière des fonctions. *Mathematica (Cluj)* 4, 38–53.

M. Karanasos. (1999). The second moment and the autocovariance function of the squared errors of the GARCH model. *Journal of Econometrics* 90, 63–76.

M. Karanasos. (2001). Prediction in ARMA models with GARCH in mean effects. *Journal of Time Series Analysis* 22, 555–576.

M. Karanasos and J. Kim. (2003). Moments of the ARMA-EGARCH model. *Econometrics Journal* 6, 146–166.

T. Kawata. (1972). *Fourier Analysis in Probability Theory*. Academic, New York.

T. Kobayashi and D. L. Simon. (2003). Application of a bank of Kalman filters for aircraft engine fault diagnostics. Technical Report E-14088. National Aeronautics and Space Administration, Washington, DC.

R. Kohn and C. F. Ansley. (1986). Estimation, prediction, and interpolation for ARIMA models with missing data. *Journal of the American Statistical Association* 81, 751–761.

P. S. Kokoszka and M. S. Taqqu. (1995). Fractional ARIMA with stable innovations. *Stochastic Processes and Their Applications* 60, 19–47.

H. Künsch. (1986). Discrimination between monotonic trends and long-range dependence. *Journal of Applied Probability* 23, 1025–1030.

H. Künsch, J. Beran, and F. Hampel. (1993). Contrasts under long-range correlations. *Annals of Statistics* 21, 943–964.

R. Leipus and M. C. Viano. (2000). Modelling long-memory time series with finite or infinite variance: a general approach. *Journal of Time Series Analysis* 21, 61–74.

N. Levinson. (1947). The Wiener RMS (root mean square) error criterion in filter design and prediction. *Journal of Mathematical Physics* 25, 261–278.

W. K. Li. (2004). *Diagnostic Checks in Time Series*. Chapman & Hall/CRC, New York.

W. K. Li and T. K. Mak. (1994). On the squared residual autocorrelations in non-linear time series with conditional heteroskedasticity. *Journal of Time Series Analysis* 15, 627–636.

W. K. Li and A. I. McLeod. (1986). Fractional time series modelling. *Biometrika* 73, 217–221.

S. Ling and W. K. Li. (1997a). Diagnostic checking of nonlinear multivariate time series with multivariate arch errors. *Journal of Time Series Analysis* 18, 447–464.

S. Ling and W. K. Li. (1997b). On fractionally integrated autoregressive moving-average time series models with conditional heteroscedasticity. *Journal of the American Statistical Association* 92, 1184–1194.

R. J. A. Little and D. B. Rubin. (2002). *Statistical Analysis with Missing Data.* Wiley Series in Probability and Statistics. Wiley, Hoboken, NJ.

I. N. Lobato and N. E. Savin. (1998). Real and spurious long-memory properties of stock-market data. *Journal of Business & Economic Statistics* 16, 261–283.

B. B. Mandelbrot. (1975). Limit theorems on the self-normalized range for weakly and strongly dependent processes. *Z. Wahrscheinlichkeitstheorie und Verw. Gebiete* 31, 271–285.

B. B. Mandelbrot. (1976). Corrigendum: Limit theorems on the self-normalized range for weakly and strongly dependent processes [Z. Wahrscheinlichkeitstheorie Verw. Gebiete **31** (1974/75), 271–285]. *Z. Wahrscheinlichkeitstheorie und Verw. Gebiete* 33, 220.

B. B. Mandelbrot and J. W. Van Ness. (1968). Fractional Brownian motions, fractional noises and applications. *SIAM Review* 10, 422–437.

R. N. Mantegna and H. E. Stanley. (2000). *An Introduction to Econophysics. Correlations and Complexity in Finance.* Cambridge University Press, Cambridge.

A. Maravall. (1983). An application of nonlinear time series forecasting. *Journal of Business & Economic Statistics* 1, 66–74.

A. I. McLeod and W. K. Li. (1983). Diagnostic checking ARMA time series models using squared-residual autocorrelations. *Journal of Time Series Analysis* 4, 269–273.

N. Metropolis, A. W. Rosenbluth, A. H. Teller, and E. Teller. (1953). Equations of state calculations by fast computing machines. *Journal of Chemical Physics* 21, 1087–1092.

S. P. Meyn and R. L. Tweedie. (1993). *Markov Chains and Stochastic Stability.* Communications and Control Engineering Series. Springer, London.

R. Möhring. (1990). Parameter estimation in gaussian intermediate-memory time series. Technical Report 90-1. Institut fur Mathematische Stochastik, University of Hamburg, Hamburg.

A. Montanari, R. Rosso, and M. S. Taqqu. (2000). A seasonal fractional ARIMA model applied to Nile River monthly flows at Aswan. *Water Resources Research* 36, 1249–1259.

E. Moulines and P. Soulier. (2003). Semiparametric spectral estimation for fractional processes. In *Theory and applications of long-range dependence.* Birkhäuser Boston, Boston, MA, pp. 251–301.

J. C. Naylor and A. F. M. Smith. (1982). Applications of a method fGr the efficient computation of posterior distributions. *Journal of the Royal Statistical Society. Series C. Applied Statistics* 31, 214–225.

D. B. Nelson. (1991). Conditional heteroskedasticity in asset returns: a new approach. *Econometrica* 59, 347–370.

M. Ooms. (1995). Flexible seasonal long memory and economic time series. Technical Report EI-9515/A. Econometric Institute, Erasmus University, Rotterdam.

G. Oppenheim, M. Ould Haye, and M.-C. Viano. (2000). Long memory with seasonal effects. *Statistical Inference for Stochastic Processes* 3, 53–68.

J. S. Pai and N. Ravishanker. (1998). Bayesian analysis of autoregressive fractionally integrated moving-average processes. *Journal of Time Series Analysis* 19, 99–112.

W. Palma. (2000). Missing values in ARFIMA models. In W. S. Chan, W. K. Li, and H. Tong, editors, *Statistics and Finance: An Interface*. Imperial College Press, London, pp. 141–152.

W. Palma and P. Bondon. (2003). On the eigenstructure of generalized fractional processes. *Statistics & Probability Letters* 65, 93–101.

W. Palma and N. H. Chan. (1997). Estimation and forecasting of long-memory processes with missing values. *Journal of Forecasting* 16, 395–410.

W. Palma and N. H. Chan. (2005). Efficient estimation of seasonal long-range-dependent processes. *Journal of Time Series Analysis* 26, 863–892.

W. Palma and G. del Pino. (1999). Statistical analysis of incomplete long-range dependent data. *Biometrika* 86, 965–972.

W. Palma and M. Zevallos. (2004). Analysis of the correlation structure of square time series. *Journal of Time Series Analysis* 25, 529–550.

C.-K. Peng, S. V. Buldyrev, S. Havlin, M. Simons, H. E. Stanley, and A. L. Goldberger. (1994). Mosaic organization of DNA nucleotides. *Physical Review E* 49, 1685–1689.

D. B. Percival and A. T. Walden. (2006). *Wavelet Methods for Time Series Analysis*, Vol. 4, *Cambridge Series in Statistical and Probabilistic Mathematics*. Cambridge University Press, Cambridge.

S. Porter-Hudak. (1990). An application of the seasonal fractionally differenced model to the monetary aggregates. *Journal of the American Statistical Association, Applic. Case Studies* 85, 338–344.

M. Pourahmadi. (1989). Estimation and interpolation of missing values of a stationary time series. *Journal of Time Series Analysis* 10, 149–169.

M. Pourahmadi. (2001). *Foundations of Time Series Analysis and Prediction Theory*. Wiley, New York.

S. J. Press. (2003). *Subjective and Objective Bayesian Statistics*. Wiley Series in Probability and Statistics. Wiley, Hoboken, NJ.

W. H. Press, S. A. Teukolsky, W. T. Vetterling, and B. P. Flannery. (1992). *Numerical Recipes in FORTRAN*. Cambridge University Press, Cambridge.

G. Rangarajan and M. Ding, editors. (2003). *Procesess with Long-Range Correlations*. Springer, Berlin.

N. Ravishanker and B. K. Ray. (1997). Bayesian analysis of vector ARFIMA processes. *Australian Journal of Statistics* 39, 295–311.

B. K. Ray. (1993a). Long-range forecasting of IBM product revenues using a seasonal fractionally differenced ARMA model. *International Journal of Forecasting* 9, 255–269.

B. K. Ray. (1993b). Modeling long-memory processes for optimal long-range prediction. *Journal of Time Series Analysis* 14, 511–525.

B. K. Ray and R. S. Tsay. (2002). Bayesian methods for change-point detection in long-range dependent processes. *Journal of Time Series Analysis* 23, 687–705.

V. A. Reisen, A. L. Rodrigues, and W. Palma. (2006a). Estimating seasonal long-memory processes: a Monte Carlo study. *Journal of Statistical Computation and Simulation* 76, 305–316.

V. A. Reisen, A. L. Rodrigues, and W. Palma. (2006b). Estimation of seasonal fractionally integrated processes. *Computational Statistics & Data Analysis* 50, 568–582.

B. Ripley. (1987). *Stochastic Simulation*. Wiley, New York.

C. P. Robert. (2001). *The Bayesian Choice*. Springer Texts in Statistics. Springer, New York.

C. P. Robert and G. Casella. (2004). *Monte Carlo Statistical Methods*. Springer Texts in Statistics. Springer, New York.

P. M. Robinson. (1985). Testing for serial correlation in regression with missing observations. *Journal of the Royal Statistical Society, Series B* 47, 429–437.

P. M. Robinson. (1991). Testing for strong serial correlation and dynamic conditional heteroskedasticity in multiple regression. *Journal of Econometrics* 47, 67–84.

P. M. Robinson. (1995a). Gaussian semiparametric estimation of long range dependence. *Annals of Statistics* 23, 1630–1661.

P. M. Robinson. (1995b). Log-periodogram regression of time series with long range dependence. *Annals of Statistics* 23, 1048–1072.

P. M. Robinson. (2001). The memory of stochastic volatility models. *Journal of Econometrics* 101, 195–218.

P. M. Robinson and M. Henry. (1999). Long and short memory conditional heteroskedasticity in estimating the memory parameter of levels. *Econometric Theory* 15, 299–336.

P. M. Robinson and F. J. Hidalgo. (1997). Time series regression with long-range dependence. *Annals of Statistics* 25, 77–104.

M. Rosenblatt. (1961). Independence and dependence. In *Proc. 4th Berkeley Sympos. Math. Statist. and Prob.* Univ. California Press, Berkeley, CA, pp. 431–443.

Y. A. Rozanov. (1967). *Stationary Random Processes*. Holden-Day, San Francisco.

W. Rudin. (1976). *Principles of Mathematical Analysis*. McGraw-Hill, New York.

M. Schlather. (2006). *Simulation and Analysis of Random Fields*. The RandomFields Package, Contributed R Package.

N. Shephard. (1996). Statistical aspects of ARCH and stochastic volatility. In D. R. Cox, D. B. Hinkley, and O. E. Barndorff-Nielsen, editors, *Time Series Models: In Econometrics, Finance and Other Fields*. Chapman Hall, London, .

R. H. Shumway and D. S. Stoffer. (2000). *Time Series Analysis and Its Applications*. Springer, New York.

P. Sibbertsen. (2001). S-estimation in the linear regression model with long-memory error terms under trend. *Journal of Time Series Analysis* 22, 353–363.

F. Sowell. (1992). Maximum likelihood estimation of stationary univariate fractionally integrated time series models. *Journal of Econometrics* 53, 165–188.

L. T. Stewart. (1979). Multiparameter univariate Bayesian analysis. *Journal of the American Statistical Association* 74, 684–693.

W. F. Stout. (1974). *Almost Sure Convergence*. Academic, New York–London.

M. Taniguchi and Y. Kakizawa. (2000). *Asymptotic Theory of Statistical Inference for Time Series*. Springer Series in Statistics. Springer, New York.

M. S. Taqqu. (2003). Fractional Brownian motion and long-range dependence. In P. Doukhan, G. Oppenheim, and M. S. Taqqu, editors, *Theory and Applications of Long-Range Dependence*. Birkhäuser, Boston, MA, pp. 5–38.

M. S. Taqqu, V. Teverovsky, and W. Willinger. (1995). Estimators for long-range dependence: an empirical study. *Fractals* 3, 785–788.

S. J. Taylor. (1986). *Modelling Financial Time Series*. Wiley, New York.

N. Terrin and C. M. Hurvich. (1994). An asymptotic Wiener-Itô representation for the low frequency ordinates of the periodogram of a long memory time series. *Stochastic Processes and Their Applications* 54, 297–307.

G. Teyssière and A. Kirman, editors. (2007). *Long Memory in Economics*. Springer, Berlin.

G. C. Tiao and R. S. Tsay. (1994). Some advances in non linear and adaptive modelling in time series. *Journal of Forecasting* 13, 109–131.

L. Tierney. (1994). Markov chains for exploring posterior distributions. *Annals of Statistics* 22, 1701–1762.

L. Tierney and J. B. Kadane. (1986). Accurate approximations for posterior moments and marginal densities. *Journal of the American Statistical Association* 81, 82–86.

L. Tierney, R. E. Kass, and J. B. Kadane. (1989). Fully exponential Laplace approximations to expectations and variances of nonpositive functions. *Journal of the American Statistical Association* 84, 710–716.

O. Toussoun. (1925). Mémoire sur l'Histoire du Nil. *Mémoires de l'Institut d'Egypte* 18, 366–404.

D. Veitch and P. Abry. (1999). A wavelet-based joint estimator of the parameters of long-range dependence. *Institute of Electrical and Electronics Engineers. Transactions on Information Theory* 45, 878–897.

C. Velasco and P. M. Robinson. (2000). Whittle pseudo-maximum likelihood estimation for nonstationary time series. *Journal of the American Statistical Association* 95, 1229–1243.

P. Whittle. (1951). *Hypothesis Testing in Time Series Analysis*. Hafner, New York.

W. Willinger, M. S. Taqqu, W. E. Leland, and D. V. Wilson. (1995). Self-similarity in high-speed packet traffic: Analysis and modeling of Ethernet traffic measurements. *Statistical Science* 53, 165–188.

P. S. Wilson, A. C. Tomsett, and R. Toumi. (2003). Long-memory analysis of time series with missing values. *Physical Review E* 68, 017103 (1)–(4).

H. Wold. (1938). *A Study in the Analysis of Stationary Time Series*. Almqvist and Wiksell, Uppsala.

C. M. Wong and R. Kohn. (1996). A Bayesian approach to estimating and forecasting additive nonparametric autoregressive models. *Journal of Time Series Analysis* 17, 203–220.

W. A. Woodward, Q. C. Cheng, and H. L. Gray. (1998). A k-factor GARMA long-memory model. *Journal of Time Series Analysis* 19, 485–504.

Y. Yajima. (1985). On estimation of long-memory time series models. *Australian Journal of Statistics* 27, 303–320.

Y. Yajima. (1988). On estimation of a regression model with long-memory stationary errors. *Annals of Statistics* 16, 791–807.

Y. Yajima. (1989). A central limit theorem of Fourier transforms of strongly dependent stationary processes. *Journal of Time Series Analysis* 10, 375–383.

Y. Yajima. (1991). Asymptotic properties of the LSE in a regression model with long-memory stationary errors. *Annals of Statistics* 19, 158–177.

Y. Yajima and H. Nishino. (1999). Estimation of the autocorrelation function of a stationary time series with missing observations. *Sankhyā. Indian Journal of Statistics. Series A* 61, 189–207.

E. Zivot and J. Wang. (2003). *Modeling Financial Time Series with S-Plus®*. Springer, New York.

A. Zygmund. (1959). *Trigonometric Series. Vol. I.* Cambridge University Press, Cambridge.

TOPIC INDEX

AUTHOR INDEX

Printed and bound by CPI Group (UK) Ltd, Croydon, CR0 4YY
16/09/2026
14963463-0001